CRITICAL PERSPECTIVES ON SUBURBAN INFRASTRUCTURES

Contemporary International Cases

Edited by Pierre Filion and Nina M. Pulver

Most new urban growth takes place in the suburbs; consequently, infrastructures in these areas are in a constant state of playing catch-up, often leading to repeated service crises. However, the push to address the tensions stemming from this rapid growth also positions suburbs as a major source of urban innovation. Taking a critical social science perspective to identify political, economic, social, and environmental issues related to suburban infrastructures, this collection highlights the similarities and differences between infrastructure conditions in the suburbs of the global north and global south.

Adopting an international approach grounded in case studies from three continents, this book discusses infrastructure issues within various suburban and societal contexts: low-density, infrastructure-rich, global north suburban areas; rapidly developing Chinese suburbs; and deeply socially stratified suburbs in poor global south countries. While acknowledging the stark differences between these regions, this collection points to the features common to all suburban areas and underscores the similarities in the infrastructure issues confronting them.

(Global Suburbanisms)

PIERRE FILION is a professor in the School of Planning at the University of Waterloo.

NINA M. PULVER is a PhD candidate in the School of Planning at the University of Waterloo.

GLOBAL SUBURBANISMS

Series Editor: Roger Keil, York University

Urbanization is at the core of the global economy today. Yet, crucially, suburbanization now dominates twenty-first century urban development. This book series is the first to systematically take stock of worldwide developments in suburbanization and suburbanisms today. Drawing on methodological and analytical approaches from political economy, urban political ecology, and social and cultural geography, the series seeks to situate the complex processes of suburbanization as they pose challenges to policymakers, planners, and academics alike.

For a list of the books published in this series, see page 409.

EDITED BY PIERRE FILION
AND NINA M. PULVER

Critical Perspectives on Suburban Infrastructures

Contemporary International Cases

UNIVERSITY OF TORONTO PRESS
Toronto Buffalo London

Toronto Buffalo London
utorontopress.com
Printed in the U.S.A.

ISBN 978-1-4875-0495-3 (cloth) ISBN 978-1-4875-2361-9 (paper)

♾ Printed on acid-free paper with vegetable-based inks.

Global Suburbanisms

Library and Archives Canada Cataloguing in Publication

Title: Critical perspectives on suburban infrastructures : contemporary international cases / edited by Pierre Filion and Nina M. Pulver.
Names: Filion, Pierre, 1952– editor. | Pulver, Nina M., 1978– editor.
Series: Global suburbanisms.
Description: Series statement: Global suburbanisms | Includes bibliographical references and index.
Identifiers: Canadiana 20190049405 | ISBN 9781487523619 (softcover) | ISBN 9781487504953 (hardcover)
Subjects: LCSH: Suburbs – Developed countries – Case studies. | LCSH: Suburbs – Developing countries – Case studies. | LCSH: Infrastructure (Economics) – Developed countries – Case studies. | LCSH: Infrastructure (Economics) – Developing countries – Case studies. | LCGFT: Case studies.
Classification: LCC HT351 .C75 2019 | DDC 307.74—dc23

University of Toronto Press acknowledges the financial assistance to its publishing program of the Canada Council for the Arts and the Ontario Arts Council, an agency of the Government of Ontario.

Contents

Illustrations, Figures, and Tables

Illustrations

Figures

Tables

Acknowledgments

The editors acknowledge financial support from the Social Science and Humanities Research Council Major Collaborative Research Initiative Grant 412-2010-1003, entitled "Global Suburbanisms: Governance, Land, and Infrastructure in the 21st Century." A number of chapters in this book present findings of research funded by this grant. The editors are also grateful for the participation of many contributors to this book in the Global Suburban Infrastructure workshop, which was held at the University of Waterloo. Most of the chapters originated from this workshop, which was funded by a SSHRC Connection grant, number 611-2014-0225, and the Faculty of the Environment of the University of Waterloo.

CRITICAL PERSPECTIVES ON SUBURBAN INFRASTRUCTURES

Contemporary International Cases

1 Introduction: The Scope and Scales of Suburban Infrastructure

PIERRE FILION, ROGER KEIL, AND NINA M. PULVER

Introduction: A Book about Suburban Infrastructures

Infrastructures order urban and suburban form and provide conditions essential to life and to the functioning of society. The conditions that lead to infrastructure implementation range from financial, technical, and organizational requirements to interests with a stake in certain infrastructural outcomes, as well as planning and decision-making processes. The term "infrastructure" suggests the combination of a supportive role ("infra") and the buttressing of techno-societal arrangements ("structure"). The "structure" part of the term can also be interpreted as connoting the networked and interconnected nature of the components of most infrastructures. The purpose of infrastructures is not to pursue their own ends, but rather to create conditions that promote the achievement of wider societal objectives. Infrastructures are enablers. They influence behaviour patterns and make it possible for urban areas to exist and operate as economic entities. In this sense, infrastructures can be seen not only as integrators, bringing together conditions essential to the functioning of society, but also as factors of partition that advantage certain interests at the expense of others. Inevitably, infrastructures act as driving forces of political and social dynamics in that certain geographical areas and constituencies enjoy resource benefits at the expense of other population groups and regions.

In this introductory chapter, we consider both upstream and downstream aspects, that is, aspects that determine the form of infrastructures and are their consequences, as well as infrastructures themselves. Metaphorically, we look at infrastructures as if they were a tree: the decision-making that supports the design and maintenance of infrastructures represents the roots; the actual infrastructures – either physically tangible or consisting of organizations delivering services – that make it possible for society to function are the trunk; and the reverberation of infrastructures on society are the leaves and branches. In this volume, infrastructures are understood broadly to include physical artefacts as well as services and immaterial systems such as the law and finance. At the core of its understanding of infrastructures is the interaction between infrastructures and society. System builders and operators work in tandem across large technical systems on a constant renewal and reconstruction of technologies and organizational structures that shape society as much as they are shaped by it (Hughes, 1987). This volume seeks to explore the multiple dimensions of infrastructures from a *suburban* perspective. Why concentrate on suburbs? There are three primary rationales for focusing attention on this particular form of settlement.

First, the edge of the urbanized perimeter is where most new development takes place, and, accordingly, there is a constant need for new infrastructures within these suburban areas – a requirement that can pose significant challenges. Therefore, perhaps more than any other form of human settlement, suburbs can suffer from a lag effect regarding infrastructures – infrastructure development is constantly under pressure to keep pace with expanding development, which can lead to service crises. Suburbs are therefore ideally suited to the illustration of infrastructure tensions and their consequences. Second, it is often in suburbs where most urban region residents live and economic activity takes place. Disparities in economic opportunity or social well-being that are created and reinforced by infrastructures are frequently more pronounced within and between suburban areas than in other parts of metropolitan regions. Third, suburbs are highly contrasted environments, especially when considered from an international perspective. Their social, economic, and functional diversity thus lends itself to the consideration of a broad range of circumstances that give rise to infrastructure-related tensions.

This book offers an international approach to suburban infrastructures. Its chapters present cases from a number of countries and regions. Three chapters introduce cases from the United States, two from Canada, and two more from the United Kingdom. And cases from Sweden, Eastern Europe, India, China, and Vietnam are each the object of a chapter. While not, strictly speaking, global in its reach – there are no chapters on Latin America, Africa, and Oceania – the volume nevertheless offers sufficient international material to highlight stark contrasts in the nature of suburbs across the world, both in their need for infrastructures and in the capacity of administrations to provide these infrastructures. Furthermore, the international dimension of the book shows how infrastructures can adapt to a diversity of suburban conditions witnessed worldwide.

The collection of cases here fits within a growing critical reflection on infrastructures from a social science perspective. Social science–informed considerations of infrastructures have broadened the scope of a field of study that had traditionally been the domain of engineering, planning, and economics. Mainstream infrastructure studies centre on the technical and economic dimensions of infrastructures, as well as on their impact on economic development. While it would be misleading to describe explorations of the relation between infrastructures and different aspects of society as an exclusively recent phenomenon – Harold Innis (1962 [1956]) was already investigating the relationship between national infrastructure networks and government structuring in the first half of the twentieth century – there is of late rapidly growing concern about the adaptation of infrastructures to new economic circumstances and financial systems, and especially to increasing concern for the environment and infrastructural resilience as effects of climate change become more apparent (Brown, 2014; Coutard & Rutherford, 2015; MacCleary, Peterson, & Stern, 2012; Tomalty, 2014). There is also rising apprehension of the social inequity dimension of infrastructures, which are portrayed as advantaging certain constituencies while marginalizing others (Bakker, 2011; Graham & Marvin, 2001). To understand the source of social inequality engendered by infrastructures, researchers have turned to the governance mechanisms undergirding infrastructure-related decisions (Parkin, 1994; Torrance, 2009).

For this collection, contributors were invited to write a chapter reflecting their own work on the suburban infrastructure theme and the conceptual frameworks that guide their research. The advantage of this approach is that it allows contributors to advance their own perspectives and rely on the empirical material that is most familiar to them, hence the variety of empirical cases and conceptual perspectives on suburban infrastructures present in the book. It is thus suited to the demonstration of the existence of numerous research trajectories within this field of study. However, while the content of the book addresses many aspects of suburban infrastructures seen from social, political, financial, economic, and environmental perspectives, it cannot cover all dimensions of this object of study. As with most edited books, the specificity of the research material introduced by the contributors is to some extent at the expense of a systematic treatment of the different facets of its object of study. The purpose of this introductory chapter is not to offer a unifying common framework for the volume, as that would violate the multiple conceptual trajectories adopted in its chapters, but rather to provide a sense of the broad range of forms the suburban infrastructure phenomenon takes and an awareness of their far-reaching consequences.

This introductory chapter focuses on the integration and fragmentation impacts of infrastructures; it is about people who are affected by infrastructure-induced integration and fragmentation. The dual nature of infrastructures manifests itself in the inequity of their impacts. Here, the tone is set for the entire volume by taking a critical perspective on suburban infrastructures, on what they do and do not achieve. It thus follows in the footsteps of previous social science critical reflection on infrastructures in general and on suburban infrastructures (see, for example, Addie & Keil, 2015; Bloch, Papachristodoulou, & Brown, 2013; Filion, 2013a; 2013b; Star, 1999; Young, Burke Wood, & Keil, 2011; Young & Keil, 2014).

Infrastructures are presented in most planning documents as fulfilling an integrative role (notwithstanding a growing interest in some of these documents for "bypass" infrastructures targeted at specific categories of users, such as express rail connections to airports, which turn their back on populations living along their corridors). Integration is key to the definition of the societal roles that are attributed to infrastructures. From a geographical perspective, they are indeed given a connective mandate involving the provision of services to different areas and facilitating interactions between these places. The integrative mission of infrastructures thus consists in the interconnection of districts within an ensemble, as, for example, suburban neighbourhoods within the metropolitan whole. Plans typically frame the objective of transportation infrastructures in these terms. For example, in the words of a Toronto region plan, "the transportation system will be fully integrated. It will be easy to make a decision on how to get somewhere or ship something thanks to seamless integration" (Metrolinx, 2008: 18). From this idealized perspective, infrastructures also aim to provide individuals and groups with conditions for a healthy and productive existence – in other words, to ensure their full integration with and participation in society. It is easy to see how, for example, physical infrastructures like transportation, water, and sewerage, as well as social infrastructures including education and health services, aspire to conform to this integrative role. The integrative dimension of infrastructures is also depicted by their networking features, causing them to expand their territorial reach and connect with other infrastructure systems across multiple regions (Easterling, 2010; 2014).

At the same time, a profound gap exists between the idealized official discourse on infrastructures, which portrays them as essentially integrative, and their far more nuanced consequences. Just as infrastructures are agents of integration, they are also sources of fragmentation. Infrastructures favour certain geographical areas and segments of the population at the expense of others, marginalizing social groups by making it difficult for them to participate in society. It is obviously tempting for decision-makers to manipulate this Janus-faced characteristic of infrastructures in attempts to advance their own interests and their views of society.

Vantage Points on Infrastructures

To capture the wide diversity of forms suburban infrastructures can take as well as their numerous and far-reaching ramifications, the next paragraphs look at infrastructures from three vantage points: from above, from the ground level, and from underneath (the underground infrastructures). Immaterial infrastructures, such as legal codes and financial systems, are also considered.

A View from the Air

To gain a sense of the wide diversity of suburban infrastructures, their interconnection, and their structural effects on land use, as well as of the extreme unevenness of their distribution, we take a view from the air, as from an airplane. It is thus natural to begin our aerial tour of infrastructures with airports. This departure point assists in situating our suburban focus, as airports are themselves typically located in suburban areas while being a critical infrastructural element that a significant percentage of metropolitan region residents rely upon, irrespective of where they live in the region. Metropolitan-wide infrastructures, such as airports, are usually found within the suburban realm because their development coincided with the period when suburban growth was taking place (virtually all major airports were created at different times since the Second World War) and their need for abundant space can only be accommodated within this realm. Our aerial view of metropolitan regions illustrates the connection between major infrastructures, such as airports, and urban development. Across the world, airports have served as growth poles within suburban environments by concentrating jobs and services, thus pulling urban development in their direction (Addie, 2014; Freestone & Baker, 2011).

Transportation networks servicing airports demonstrate the extent to which infrastructures can consume suburban space. In rich countries, airports are indeed connected to expressways, arterials, and, in many instances, rail transit. (And, less visibly, these transportation infrastructures are matched by vital water, sewage, communication, and electricity connections.) Overall, the airport and attendant service facilities contribute to the important proportion of space devoted to infrastructures in suburbs. Other large infrastructure consumers of suburban space include railway marshalling yards, water reservoirs, sewage treatment plants, expressway interchanges, and landfill sites.

The aerial perspective illustrates the major role transportation infrastructures play in organizing suburban space. For example, in North America the superblock delineated by arterials and expressways is the organizing principle of the suburb. It serves to create suburban morphologies that separate functions and income groups (Filion, 2018). The view from the air also displays how expressway, arterial, and rail networks operate at a metropolitan scale and beyond (Keil & Young, 2008). As the aerial perspective zooms out, it shows how these infrastructures connect the region to its continent and, via airports and harbour facilities, the rest of the world. Similarly, green infrastructures manifest themselves in suburban regions. Peri-urban space is characterized by more abundant green space than found in urban cores, including private residential yards, parks, and greenbelts. Such green areas are not only major contributors to metropolitan area recreation opportunities, food production, carbon uptake, flood protection, and mitigation of air and water pollution, among many other services that supplement or augment the services of more conventional "grey" infrastructures, but they also serve to structure urban and suburban form (Amati & Taylor, 2010; Benedict & McMahon, 2006; Yigitcanlar, 2010). Greenbelts or "emerald necklaces" that encircle cities like Toronto, Ottawa, and Portland, Oregon, were delineated to protect open green space and limit sprawling urban development. Despite this intent, suburban development has "leapfrogged" over greenbelts and extends into areas ever more distant from city cores, a result of suburban expansion that is clearly evident from the air.

The aerial outlook reveals the different types of suburbs across the world, which are distinguished by their density and predominant built form and land-use pattern. The main distinction in terms of suburban infrastructures seen from the air is between rich global north and poor global south countries. Of course, not all global north suburban areas exhibit rich-country infrastructure networks, just as not all global south suburbs can be characterized as infrastructure poor. In the examples that follow, we intentionally choose examples from relatively rich global north and relatively poor global south countries to accentuate differences in suburban infrastructure networks. A typical global south suburban area tends to be criss-crossed by boulevards rather than expressways, which are more expensive to build and consume more space. This pattern is but one example of the limited investment capacity that results in fewer and less expensive infrastructures than those found in the global north. Some global south metropolitan regions, such as Bogota, exhibit innovative ways of providing transportation infrastructures at a lower cost. For example, the aerial perspective reveals heavily used bus rapid transit networks, a cost-effective mass transportation solution that takes advantage of existing boulevards (Hidalgo & Grafiteaux, 2008). A less positive manifestation of global north and global south suburban infrastructure differences is the dearth of circular open-air tanks indicating wastewater treatment plants in global south suburbs, a feature that is common in global north suburban landscapes. The view from above therefore points to a much-reduced capacity in the global south to afford technological fixes to deal with the environmental consequences of urbanization.

Another distinction between global north and global south suburban infrastructures is a more uneven distribution of infrastructures and sharper differences in network configurations in the global south (Coutard & Rutherford, 2015). For example, in global south suburbs, the amount of green space varies considerably from district to district, and at night there are vast darkened sectors within these metropolitan regions, indicating the presence of improvised settlements (also referred to as informal or self-built settlements) where the availability of electricity is limited. At night from the air, disparity in access to electricity illustrates the stark inequalities in global south suburban infrastructure distribution. Certain districts seem to contain infrastructures on a par with those found in rich-nation suburbs. In both settings, there are arterials, schools, and hospitals along with plentiful green space and electricity. But, in the case of private security infrastructures, rich global south suburban communities are often better supplied than global north suburbs where there is less crime. Meanwhile, it is difficult to discern from above infrastructure networks within the large improvised global south suburban settlements. In many such places, made up of a jumble of tightly clustered low-rise buildings, narrow asymmetrical dirt roads constitute the only evidence of physical infrastructure visible from the air (Boo, 2012).

A View from the Ground

Ground-level perspectives present an entirely different outlook on infrastructures. From the ground, we can concentrate on how people interact with infrastructures and how they are affected by them. We can observe how infrastructures condition behaviour. The rhythm of suburbs and cities is set by infrastructures. People organize their time according to the possibilities afforded by infrastructures: for example, the time it takes to go from one location to another. The effect of infrastructures on how people use the city has been highlighted by time budget studies (Chapin, 1974; Farber, Neutens, Miller, & Li, 2013).

From the ground level, it is also possible to detect social consequences of unequal access to infrastructures. In suburban areas, people who can rely on the automobile typically are able to engage in many more activities than those who are limited to public transit. There is therefore a life opportunity dimension to different modes of transportation. While public transportation patrons wait at the station or move slowly on circuitous bus routes, people using the car can often reach several destinations or access locations that are not served by public transit routes. Inequity can have much more dramatic effects when infrastructures essential to health and life, such as clean water, and/or those critical for enabling participation in society, such as education, are absent. Areas of deprivation, especially improvised developments in global south suburbs, typically exhibit all of these infrastructure impediments. They are poorly connected to the adjacent urban area, lack basic water and sanitary infrastructures, and their supply of basic public sector services, such as schools, is deficient. Sometimes, to add insult to injury, the only land available for such settlements is in areas that are less desirable for urban development, causing them to endure the negative externalities of infrastructures their residents cannot access, such as the noise and fumes of expressways and airports.

In poor suburban settlements with insufficient political might and purchasing power, deprived of infrastructures because they have been overlooked by governments and large companies involved in the provision of private infrastructures, people themselves become the infrastructures. Such areas develop mechanisms, as imperfect as they may be, to provide necessities of life and basic community needs. Low-paid labour substitutes itself for services that are dispensed in other circumstances by physical and formal social infrastructures (Roy, 2011a; Simone, 2004). Situations where people become the infrastructure can, for example, consist of individuals operating latrines and selling feces as fertiliser, bringing potable water to homes, and dispensing traditional medicine (Desai, McFarlane, & Graham, 2015; Gandy, 2006; McFarlane, Desai, & Graham, 2014; Simone, 2010; Varley, 2013; Wilson, Vellis, & Cheeseman, 2006).

Ground-level observation, by focusing on the behaviour of people in urban and suburban areas, further draws attention to the complexity of infrastructures that intermingle with the remainder of the built environment. It is indeed difficult to distinguish what is infrastructure and what is not within these environments. When does space cease to be a link and instead become a destination? We are, in the first case, in the presence of an infrastructure, the purpose of which is to provide accessibility to destination spaces. There is no problem distinguishing infrastructure space from destination space when vehicles are involved, because people use their vehicles to reach destinations. Things are murkier, however, when dealing with pedestrians. One way of distinguishing the two types of space is to observe movements of people in urban and suburban areas – the direction they take, their gestures, their rhythm. Spaces where people walk straight at a relatively high speed and where interactions are limited can be considered as linking spaces, thus playing the role of infrastructures. By contrast, destination spaces can be identified by people looking around, stopping, and interacting. In this approach, it is the behaviour of people that defines what is and what is not a space of infrastructure (Takaki, 2012). Such a perspective on the city draws attention to how connection space links more animated parts of the urban texture, thereby composed of points of varying intensity.

A View from Below

The third vantage point on suburban infrastructures is taken from below, concentrating on underground infrastructures. This category of infrastructures corresponds to a common conceptualization of infrastructures, perceived as invisible conduits providing services that are often taken for granted and unlikely to raise political enthusiasm. Because they are not seen, buried infrastructures (pipes, electrical systems, and communication networks) are not generally a major source of rewards for politicians, but disruptions to underground systems can carry heavy political costs, which points to the vital role these infrastructures play (Latour & Hermant, 1998).

Despite their invisibility from the ground or the air, the presence and absence of underground infrastructures are a hallmark of infrastructure inequality. Virtually all rich-country urban and suburban areas are provided with underground infrastructure networks. In poorer parts of the world, these networks are a luxury available only in formally developed parts of urban areas, which often account for a minority of their total population. The absence of underground water and sewerage pipes are a foremost factor of deficient public health conditions in improvised settlements.

Intangible Aspects

Beyond these three vantage points, there are aspects of suburban infrastructures that remain invisible. These are the immaterial dimensions of infrastructures, which include decision-making processes, laws and regulations, financial arrangements, institutional structures, and community-level organizations. Such dimensions are essential conditions that underlay the existence of tangible infrastructures. The enabling role of intangible infrastructures qualifies them as infrastructures in their own right; they indeed sustain the functioning of all facets of society. This intangible dimension of infrastructures, along with the three vantage points described in this section, exposes the broad range of phenomena that fall under the infrastructure label. It also illustrates how infrastructures intersect with all aspects of society.

The Suburban Dimension

Types of Suburbs

By exploring infrastructures in suburban settings from an international perspective, this volume considers the relation between infrastructures and multiple suburban forms (McFarlane & Rutherford, 2008; Stanilov & Scheer, 2004). Different types of infrastructures support different urban and suburban forms, and, conversely, different urban and suburban forms determine the types of infrastructures that can operate within a given area. Suburban morphologies can frequently be distinguished by their density and land-use patterning. There is the North American model characterized by its low density and rigid functional and social segregation. The North American model is the product of a post–Second World War generalization of automobile use supported by massive public sector investment in expressways, at the expense of other forms of transportation, as well as a planning system narrowly focused on zoning (Filion, Bunting, & Warriner, 1999; Hirt, 2014). It has proven difficult to wean this suburban model from its dependence on the automobile. Distances are generally too long for walking and cycling, and low density, coupled with a dispersal of origins and destinations – the result of strict land-use specialization – is unfavourable to public transit. North American suburban development has been criticized for its high infrastructure expense, largely a consequence of the need to supply sufficient roads and expressways to cater to a near universal dependence on the automobile, and the high cost of providing these and other horizontal infrastructures in a low-density environment. The North American model has been exported, but elsewhere it tends to be hybridized with other suburban forms. Also in North America, Canada has taken a different development path than much of the United States, with density variations in older and newer suburbs and a trajectory of post-suburbanization such as the creation of high-density and multifunctional suburban centres, that has led to a set of infrastructure issues unique to that country's suburban constellation (Charmes & Keil, 2015; Tomalty & Mallach, 2015; Young & Keil, 2014).

In Western Europe, suburbs post much higher densities due to the presence of high-rise housing concentrated in Le Corbusier–inspired *grands ensembles* and to low-rise configurations that are more compact than in North America (Dufaux & Fourcaut, 2004). Land uses also tend to be less rigidly separated in Western European suburbs. These suburbs are therefore more walking, cycling, and public transit conducive than their North American counterparts. They are typically well connected to metro and commuter rail networks. Also, the per capita cost of infrastructures is less in these dense suburbs than it is in low-density configurations. Outside Western Europe, there are two major variations on the theme of the planned (in opposition to improvised) high-density suburb. First, there are suburbs developed since the Second World War in the former Soviet Union and its satellite countries (Gutnov et al., 1971).

They consist of large districts taking the form of high-rise residential buildings aligned along wide boulevards and separated by generous amounts of open space. Land uses are highly specialized in this type of suburb, notwithstanding retail facilities built since 1989 on the open space that originally separated residential buildings from major arterials. The second variation on the high-density suburb theme is taking shape in the booming metropolises of China, where the suburban presence of high-density *grands ensembles* is generalized (Yu & Ng, 2007). In Chinese suburbs, different residential developments, each with its own configuration and density level, are situated side by side. In both variants, suburbs are generally well integrated into metropolitan infrastructure networks, most notably the public transportation system, to the extent that the term "suburban" is almost non-applicable (see chapter 13). Presently, Chinese metropolises are actively engaged in the extension of their metro systems to "new town" suburban developments.

It is in the global south that suburban morphologies are most diverse. Within the suburban realm of a global south metropolis there is usually a mixture of low-density residential sectors evocative of the North American model, high-density areas either in *grands ensembles* or more piecemeal configurations, and peripheral extensions of central-city patterns, along with the crowded improvised settlements lacking basic infrastructures and services. Perhaps more than in any other suburban configuration, these suburban areas are highly fragmented from a social, functional, and accessibility perspective.

There are wide national differences in the imagery attached to suburbs. For example, North American suburbs have traditionally been depicted as leafy havens for the middle class. In contrast, the disparaging declarations of Nicolas Sarkozy, minister of the interior during the fall 2005 Paris suburb riots, echo the perception of French suburbs as accretions of poor immigrants in the grip of violent youths (Moran, 2011). Today, these same suburbs are perceived as a breeding ground for terrorist cells.

Even within a same national or continental context, as in the case of North America, the concept of suburb defies easy definition. Attempts at defining suburbs have concentrated on location, dominant modes of transportation, lifestyle and culture, and physical appearance (Forsyth, 2012). In a Canadian context, Gordon and Janzen (2013) have formulated a model using density and transportation modal shares to identify different types of suburbs. According to their understanding of suburbs, 80 per cent of the population of Canadian metropolitan regions and 66 per cent of the national population reside in suburban-type areas.

Shared Features of Suburbs

Despite the numerous ways of defining suburbs and the many worldwide suburban configurations, it is possible to identify attributes common to all suburbs. Richard Harris (2010: 27–9) has proposed three defining criteria: peripheral location, residential density between that of the city and the country, and relative newness. Moreover, while suburbs worldwide present a range of morphologies and dynamics, they tend to differ from those of their central city. Related to the newness criterion is the tendency for most urban regional growth to take place in the suburbs, a consequence of which is a lag in most suburban areas between infrastructure development and demographic and economic activity growth. Another shared feature of suburbs is their complex relation with infrastructures due to their growth, peripheral location, and social and economic diversity. It is the purpose of this volume to chart and make sense of this complexity. In chapter 2, Jean-Paul Addie proposes a framework, classifying infrastructures as *in*, *of*, and *for* the suburb, which conceptualizes different relations between suburbs and infrastructures.

Infrastructures from a Suburban Perspective

The suburban context is especially well suited to an examination of infrastructures. As regards infrastructures, suburbs are zones of tension. Suburban areas are characterized by rapid change, mostly growth but also, occasionally, decline and retrofitting, and so infrastructures relating to different suburban circumstances are constantly in flux or playing catch-up. In many instances, this fluctuation is due to a lack of government financial capacity in poor countries, which face exploding suburban population growth. In rich countries, part of the problem with adapting infrastructures to societal needs is because infrastructure investments are largely dictated by macroeconomic objectives. Infrastructure investment has become a victim of increasingly frequent episodes of spending austerity meant to contain public debt, inflation, or tax levels. In contrast, infrastructure expenditure plays an important part in expansionary macroeconomic policies, which depend on public sector spending to stimulate the economy. Presently, China is taking such a macroeconomic approach to generate employment and expand the domestic consumer market. In shrinking cities and suburbs, there are challenges of service delivery through networked infrastructures originally designed for a booming, expanding urban environment (Bernt & Naumann, 2006). In areas where suburban retrofits are underway, incompatibilities between technologies, users, and functions need to be addressed (Hodson & Marvin, 2015; Jessen & Roost, 2016). Suburbs are also places of innovation. New development formulas are mostly tried in suburbs where the urban phenomenon reinvents itself. Suburbs contain infrastructures needed for their own operation, for their connection to metropolitan regions and beyond, and for the requirements of their metropolitan regions in their entirety, including those concerning their incorporation into global networks.

Suburban Infrastructures in the Present Context

In this section, we explore four issues that relate to infrastructures in general, their suburban manifestations, and the present societal context. The first examines infrastructures as means of perpetuating political and economic power, and therefore as instruments of social inequality. The second focuses on infrastructure strategies in the present neoliberal climate and their consequences. The third involves clashes between the need for adjustment and innovation, and path dependences built into infrastructure systems. The final issue consists of semiotics associated with infrastructures and their influence on investment choices.

Power, Inequality, and Suburban Infrastructures

Because of their durability and role in setting behavioural norms and organizing society, infrastructures are a major player in power struggles. They make it possible to extend and perpetuate power relations. Control over the development and administration of infrastructures contributes to the possibility for dominant interests to shape society in a way that reflects their values and rewards them financially, thus further enhancing their power (Swyngedouw, 2004). Infrastructures indeed have the potential to set up path dependencies with deep impacts on society, and consequently perpetuate prevailing power relations (Parkin, 1994). Infrastructures influence the location, form, and nature of urban and suburban development as well as day-to-day behaviour (Leigh & Blakely, 2013; Suchman, Blomberg, Orr, & Trigg, 1999; Zimmerman, 2001). It can then be assumed that those who control infrastructure-related decisions dictate the future of society. The public sector and, increasingly, with the rise of neoliberalism and the attendant depletion of public coffers, private corporations are involved in the shaping and operation of infrastructures (Bakker, 2010; Clarke Annez, 2006; Haughton & McManus, 2012; Pethe & Ghodke, 2002; Siemiatycki, 2013; Torrance, 2009; Wellman & Spiller, 2012). The influence of power struggles on decisions pertaining to infrastructures, however, tends to be obscured by the technocratic discourse justifying these decisions. It pictures infrastructure-related decisions as the emanation of rational, evidence-based planning processes (see, for example, Guignier & Madanat, 1999; Karvonen & Brand, 2009; Ranmert, 1997; Rubio & Fogué, 2013; Wesselink, Buchanan, Georgiadou, & Turnhout, 2013). Alongside authority over law-making, the police and the military, institution crafting, ideology, and public sector spending, infrastructures are key means by which it is possible to influence the trajectory of society.

Throughout history, infrastructures have played a critical role in nation building. Over the last centuries, the deployment of nationwide infrastructures such as railways, expressways, communication networks, or education systems has taken place in ways that reflect the intermingling of dominant political and economic forces. The same goes for urban areas. Infrastructures interconnect different parts of metropolitan regions. Of course, private corporations have managed to take advantage of unified national and metropolitan markets, which ease access to the labour force, foster inter-firm linkages, and make it possible to reach large consumer pools. Transportation, education, and communication infrastructures have been instrumental in forging national values, lifestyles, and identities. Similar, but much attenuated, effects also happen at the scale of urban regions, which explains differences in metropolitan cultures (Reese & Rosenfeld, 2012). Interconnections that have happened at national and urban area levels increasingly take place at the international scale (Brenner, 1999).

We observe here the integrative dimension of infrastructures. It involves the interconnection of space into national or metropolitan units, and increasingly international networks, and the allocation of conditions making it possible for individuals to participate in society, such as basic requirements for health as well as education and transportation. However, the role of infrastructures in fostering integration takes different forms, each favouring certain interests at the expense of others. Thus, decision-making around infrastructure development has serious and far-reaching implications.

There are vast differences, related to its financial and economic capacity, in the extent to which a jurisdiction can achieve integration through infrastructures (Bjorvatn, 2000). These variances relate to the economic and political circumstances of the polity under consideration – but it is not a simple matter of rich global north and poor global south (Varley, 2013). Vast gross national product differences among global south countries translate into varying infrastructure development possibilities and capacities. Differences also occur in the priority countries give to infrastructure provision, and in the subsequent repercussions for social equity resulting from those decisions.

While central to idealized visions of infrastructures, their integrative dimension accounts for only part of their reality, for infrastructures also cause fragmentation. Fragmentation effects can be unintended consequences of infrastructures or deliberate outcomes of infrastructure-related decision-making. By their very nature, infrastructures are factors of economic and social inequality (Schlosberg, 2007; Soja, 2010). Normally, when one sector is advantaged by the presence of an infrastructure, another must suffer because of its absence. The mapping of property values reveals increments in sectors of high accessibility close to transportation infrastructures (Chau & Ng, 1998). Historically, there has been dogged competition between urban regions for the location of railway terminals and expressways (Banister & Berechman, 2000; Logan & Molotch, 1987; Wylie, 1996). And competition still prevails for infrastructures such as high-speed train stations, hub airports, and container ports. There is a notable exception to this tendency, however. Infrastructures that are sources of negative externalities (locally undesirable land uses [LULUs]) depress values of close-by properties. This is the case, for example, of landfills and incinerators, and forms of transportation that generate noise and air pollution. In all these circumstances, infrastructures breed fragmentation by creating constituencies that benefit from, and others that are hindered by, their presence or absence (Graham, 2002). In extreme cases, tunnel effects result in exposure to the negative externalities of infrastructures without access to their benefits, a situation commonly encountered in poor areas.

Spatial proximity is not the only factor of uneven accessibility to infrastructures. Fees for access can splinter the public into those who can and those who cannot afford to pay for certain infrastructural services (Graham & Marvin, 2001). Such social fracturing has become more common with the deteriorating financial circumstances of the public sector, which had traditionally been the foremost purveyor of infrastructures. The tendency has indeed been for public sector agencies to reduce funding for infrastructures and/or partially or fully privatize them (Graham, 2000). These two reactions have led to rising fees, making it difficult, in extreme circumstances, for some individuals to maintain access to the necessities of life (Bakker, 2011; Gandy, 2004; Kooy & Bakker, 2008; Monstadt & Schramm, 2013). The principle of universal access that guided the pricing of infrastructures in rich countries has made way to higher access fees in the wake of public sector financial crises and additional reliance on the private sector. Thus, a widening gap exists between those who can and those who cannot afford access to infrastructures. This gap has been observed in Detroit, where multiple households who could not pay their bill had their water cut off (Guillen, 2015). The unfortunate decision to do so stemmed from an administrative reorganization of Detroit water services in the wake of the bankruptcy of the city.

Infrastructure-related fragmentation is not only the consequence of conditions allowing or denying access to infrastructures; it also stems from differences in infrastructure standards. The quality of public transit varies between suburbs, as does that of schools and health services (see, for example, Hertel, Keil, & Collens, 2015; Keil & Young, 2014). In these circumstances, social splintering stems from life conditions and life chances associated with different grades of infrastructures.

Possibilities afforded by infrastructures result in the setting of social norms regarding expected behaviour and performance (Hechter & Opp, 2001). Infrastructures therefore become factors of marginalization for individuals who cannot access them and meet the societal norms that ensue from their existence. For example, in an automobile-oriented suburb it is difficult for an individual to function without the use of a car, as activities are scattered at long distances from each other. Likewise, generalized reliance on the internet makes it hard to survive socially and economically without going online. Social contacts, dealing with public sector services, many consumer transactions, and the search and application for jobs, increasingly take place online. The film *I, Daniel Blake*, which won the Palme d'Or at Cannes in 2016, describes difficulties in dealing with the UK welfare system encountered by someone who does not have access to the internet to fill in and submit forms. The same goes for expected personal hygiene standards in a context where nearly everyone can bathe or shower. Norms of social and economic inclusion become factors of exclusion for those who cannot avail themselves of the required infrastructures (such as the homeless for whom washing regularly is a problem), a further source of infrastructure-induced social fragmentation.

Infrastructure deprivation is highly visible in the suburban landscape. It is especially noticeable in pockets of extreme poverty found within the suburban realm of rich countries. These areas are consistently characterized by the low quality of their infrastructures, such as lack of public transit limiting accessibility and poor school standards impeding integration into the employment market. The geography of these deprived suburban districts tends to coincide with municipal boundaries or sectors with distinct urban forms (Carpio, Irazábal, & Pulido, 2011). Among notorious examples of suburban deprivation is the US municipality of Ferguson (a suburb of St. Louis, Missouri), where an unarmed eighteen-year-old African American man, Michael Brown, was fatally shot by police in August 2014. The killing sparked a long period of unrest in Ferguson. Suburban poverty is also associated with France's *grands ensembles*, where low-rent housing is concentrated. Destitute suburbs in the United States, and to a lesser extent in France, are inhabited by racial minorities (Dikec, 2007; 2011; Kimmelman, 2015), thus reinforcing racial inequalities.

We have noted the extent of suburban infrastructure inequality in the global south. Such a situation is caused by several factors. First, infrastructure decision-makers frequently have political and economic affiliations with classes residing in wealthy suburbs. In these circumstances, the needs of the constituencies that most matter to decision-makers are met, while those of poorer constituencies are largely overlooked. In addition, when infrastructures operate on a cost recovery or profit-making rather than a social equity basis, only wealthy suburban areas can afford costly infrastructures. Infrastructures are also seen as tools meant to bring societies into the modern age. But limited funding availability does not permit modernist standards to be reached everywhere, such that only select areas benefit and others are neglected. Infrastructures are also intended as instruments of economic development, targeting firms along with the individuals managing those enterprises, so that infrastructures are concentrated in areas containing offices, industries, or tourist attractions, and in upper- and middle-class residential districts. All these factors explain the paucity of state- or corporate-sponsored infrastructures in global south suburban improvised settlements and reinforce the deep infrastructure divide witnessed in global south suburbs.

Infrastructures are not only *causes* of social fragmentation – they are also *responses* to fragmentation. Decisions regarding infrastructures can be seen as driven by patterns of social fragmentation, which explains why, once in place, infrastructures often provoke a reinforcement of these patterns. An extreme illustration is the setting up of defensive infrastructures – closed circuit television, walls, fences, and the use of distance are infrastructures put in place to set income groups apart from each other (Jacobs & Lees, 2013). These infrastructures are a response to crime associated with intense social polarization and, by their very existence, contribute to further separate social classes.

While control over infrastructure-related decisions extends and perpetuates power through the integration and fragmentation effects of infrastructures, it is important not to underplay the unpredictability inherent in infrastructure outcomes. There is a measure of uncertainty in the use of infrastructure strategies to achieve societal objectives; two reasons account for this situation. First, we have noted the numerous categories of infrastructures and the different ways of developing and managing them. Infrastructures operate at different scales and are managed and operated by various types of organizations, from large public sector agencies and private corporations to informal arrangements as found in poor global south suburban settlements. Such diversity makes it difficult for one entity to control the infrastructure landscape. Second, it is difficult to foresee long-term consequences of infrastructures, which raises questions about the reliability of infrastructures as instruments to attain long-term societal objectives.

Infrastructures are often used in unanticipated ways (Hodson, Marvin, Robinson, & Swilling, 2012; Poitras, 2011). As Marshall McLuhan has demonstrated, this situation is especially the case when infrastructures involve new technologies (McLuhan, 1994 [1964]; McLuhan & Fiore, 1967). For example, in the 1950s and 1960s when the use of the automobile became generalized in North American cities and governments engaged in the construction of massive urban and suburban expressway networks, it was widely assumed by planners that expressway hubs would foster concentrations of activities and maintain or even accentuate the centralized structure of metropolitan regions. As it turned out, contrary to this vision, pretty much everywhere cars and expressways proved to be factors of urban decentralization, inimical to high-density multifunctional clusters of activities, especially downtowns. In this sense, the transition to near universal car use over these decades was a powerful factor of suburbanization, determining the form and dynamics of the North American suburb and, to a lesser extent, of suburbs elsewhere in the world (Marshall, 2000).

Equally mistaken were early predictions regarding the urban consequences of a general reliance on information technology. Early in the information technology revolution, it was assumed that the technology would cause a spatial explosion of urban areas, as work, shopping, and education journeys would become less necessary (Gaspar & Glaeser, 1998). In these circumstances, it would be possible for residents to choose their residential location primarily, if not exclusively, on the basis of local amenities, often defined in terms of plentiful space. While certain aspects of urban life have been depleted by the information technology revolution – most obviously the disappearance of book and music stores and video outlets – overall impacts on urban dynamics have been surprisingly subdued. Employment standards that require physical presence in the workplace, the desire to see and feel merchandise, and the gregarious instincts of the human race explain the persistence of prevailing urban journey patterns. A telling illustration of the blending of information technology with the lasting appeal of human proximity is the presence of people working on their computers in cafés rather than at home. In light of these examples, one can understand why it is difficult to foresee the reverberations on urban dynamics of a next large technological innovation on the horizon that is likely to impact urban and suburban dynamics: electric self-driven vehicles.

Neoliberalism and Suburban Infrastructures

While the previous subsection has considered the power dimension of infrastructures and the social inequity that ensues, we now explore the form these universal features of infrastructures take within the current neoliberal context. The concept of neoliberalism, which reappears throughout this volume, refers to a political economy that has been replacing, since the late 1970s, the Keynesian welfare state originally set up in response to the Great Depression (Jessop, 1996; 2002). Governments took a key redistributive role within the Keynesian welfare state, meant to stimulate consumption and stave off recessions caused by excessive productive capacity. The state was also active in the provision of infrastructures required to support consumption and production. The foremost example of an infrastructure strategy fulfilling these roles was the federally funded Interstate Highways system in the United States, launched in 1956 (Lewis, 1999). Furthermore, as an important part of public sector expenditure, infrastructures were prominent components of counter-cyclical macroeconomic strategies.

Such an active involvement of the state in the economy clashed with the neoliberal belief in the superior economic and societal outcomes of laissez-faire economics over state intervention. Neoliberalism also advocates free market–inspired solutions to different social problems. The neoliberal agenda therefore calls for a deregulation of the economy, free trade, and tax cuts. Central to the neoliberal credo is a shrivelling of the state to ensure that it forsakes its steering of the economy and efforts at organizing society (Harvey, 2005; Kotz, 2015). This neoliberal perspective was framed and publicized by F.A. Hayek and Milton Friedman, among others, and became the political program of Reagan and Thatcher, which influenced subsequent administrations across the world, irrespective of their political stripes (Mirowski & Plehwe, 2009). The neoliberal turn and associated globalization have been associated with many late twentieth and early twenty-first century economic trajectories: the crisis-prone nature of the economy that culminated with the 2008 Great Recession; a "financialization" of the economy; income polarization that precipitated a withering of the middle class; and a slowing down of economic growth, especially in rich countries (Firzli, 2016).

Neoliberalism and its consequences have had different impacts on infrastructures. Perhaps most obvious are infrastructure deficits resulting from reduced state financial capacity due to tax cuts and competition from other pressing sectors of expenditure. In global north countries, the sequels of the financial crisis of the state are compounded by an increase in the cost of infrastructure development that outpaces the cost of living. Such a phenomenon is due to a disproportional increase in productivity in some economic sectors and the lowered cost of outsourced products and services. The consumer price index thus reflects the decline in the cost of these products and services. However, at the same time, it is also impacted by the rise in the value of domestically produced goods and services – especially those that are resistant to important productivity gains. Such goods and services include healthcare, education, and other public sector services, along with infrastructure construction and delivery. The problem for public sector agencies is that, at a given taxation rate, increases in fiscal entries largely correspond to the evolution of the consumer price index. Meanwhile, the cost of infrastructures increases at a faster pace.

There is a glaring infrastructure dimension to the roll-back and roll-out phases of neoliberalism. The roll-back phase consists of the dismantling of Keynesian welfare state arrangements, largely through public sector budget cuts. The roll-out phase involves the replacement of these arrangements by neoliberal-inspired mechanisms (Peck & Tickell, 2002). Transposed to the world of infrastructures, the first phase translates into the deferral or cancelling of new infrastructure projects and reductions in the maintenance and service levels of existing infrastructures due to tight finances and shifts in priorities. The roll-out phase, on the other hand, takes the form of public–private partnerships or fully privatized infrastructures. Privatization can target existing public sector infrastructures. Private interests can also be given the responsibility of developing and running new infrastructures on a profit-making basis. The growing involvement of private corporations in the provision and management of infrastructures can be interpreted as a reflection of the tendency for capitalism to colonize different aspects of the economy. In the neoliberal context, this tendency reflects the "financialization" of the economy, causing private capital to search for new spheres of investment. Privatization jeopardizes universal access principles driving many of the public sector infrastructure programs of the Keynesian welfare state era. The profit-making motive driving private sector infrastructure projects generally causes them to exclude those who cannot afford their access fees and to bypass poor areas. But the forfeiture of universality principles has also affected public sector infrastructures, as governments have increasingly come to rely on these infrastructures to raise revenue. Such is the case with toll express lanes directed at drivers capable of paying fees, which are often hefty, to avoid congestion on the free lanes – hence their nickname "Lexus lanes."

Privatization and the abandonment of universal access principles are creating two parallel worlds: one for the rich who can afford top-grade infrastructures and another for the masses who must make do with legacy public infrastructures pre-dating the neoliberal era. In this sense, infrastructures contribute to extend the neoliberal-induced polarization of society by differentiating conditions available to different social classes and further separating classes from each other.

Infrastructures and Path Dependencies

We have noted the inertia built into infrastructures when discussing their role in perpetuating power relations. Infrastructure inertia stems from path dependencies assuring the persistence of prevailing systems of infrastructures and their effects on urban and suburban areas, behaviour, and society. Such durability can be attributed to the vast amounts of capital sunk into these infrastructures, the emergence over time of a symbiosis with their environment, and the creation of habits around existing infrastructures. These circumstances impede significant infrastructure shifts, and thus potentially obstruct innovation and adaptation to changing circumstances. For example, current urban infrastructures are ill suited to the increased frequency of extreme climate events, which places residents and built structures at risk (Broto & Bulkeley, 2013; Chang, 2003; Coaffee, 2010; Dodson, 2014; Graham, 2010; 2011; Webb, 2007). Lack of infrastructure adaptability is equally felt in the case of human-induced disasters such as wars and economic crises (Filion, 2014). Present examples of impediments to change caused by infrastructure path dependences include the snail-paced introduction of electric automobiles, in large part due to complications in setting up electricity supply networks for these vehicles. Petrol oligopolies, which control gas stations, have not shown much enthusiasm for providing electrical recharging facilities.

The difficulty of changing the development trajectory of the North American suburb, as awareness rises concerning the environmental damage it causes, further illustrates the effects of infrastructure path dependencies. Despite the popularity of sustainable development and smart growth perspectives, most North American suburban areas under development remain automobile dependent. The main reason for this state of affairs is that the numerous interests involved in the production of this type of environment benefit from a perpetuation of the car-oriented dispersed formula. Politicians and planners can be confident that developers will buy into the type of community they propose and, accordingly, that the land they have zoned will not be left fallow. Engineering companies and developers are familiar with construction norms inherent in this formula and know how to manipulate the model to maximize profit. At the end of the line, consumers find a product they know and expect. It is difficult to break this chain because of resistance on the part of interests vested in it. Infrastructures and embedded development path dependencies intersect with political and economic power relations. A significant problem with the North American automobile-oriented suburban model is that it was devised for a large, expanding consumerist middle class. This model is not as well suited to the economically polarized post-industrial society shaped by the neoliberal agenda (Burchell, Downs, McCann, & Mukherji, 2005; Power, 2001). The 2008 Great Recession can indeed be interpreted partly as a consequence of households in precarious financial circumstances overextending their credit to gain access to the North American suburban lifestyle (Mian & Sufi, 2014).

As the evolution of infrastructures demonstrates, path dependencies may impede innovation but often fail to prevent it. Considerable benefits stemming from innovations can propel radical changes notwithstanding the inertia resulting from the presence of existing infrastructure systems. Despite their ample financial means and efforts at improving their services, large North American railway companies rapidly lost passengers to cars and airplanes, which were both benefiting in the 1950s from massive public investment. These competitors to trains were widely perceived as faster and more flexible alternatives. Railway companies then faced overcapacity, which in many cases contributed to their bankruptcy. We may be on the cusp of another infrastructure revolution, which could result in massive stranded investments. What if photovoltaic cells and electricity storage devices gain in efficiency and affordability to the extent that the public rapidly switches to these devices to power homes and automobiles? What would then happen to the extant large electrical generation facilities and power transmission networks? Would they become confined to the supply of energy to industries and very high-density urban environments? What of the large sums invested in electrical generation and transmission?

Just as the existence of path dependencies can inhibit the adoption of infrastructure innovations, the absence of such dependencies can have the opposite effect. Some places have skipped stages in the development of infrastructures, which has translated over the long run into lower infrastructure costs and equal, if not superior, services. For example, countries in Africa and Central and South America have omitted much of the railway stage of development and now rely instead on road transportation and air travel. Such situations are most clearly illustrated by communication technologies. In many places, wireless technologies have made it possible to jump the land-based telephone line stage, all the while providing better services at a much-reduced cost (Dourish & Bell, 2007). Such a situation does not apply to basic water and electricity distribution because of the inescapable reliance on pipes and wires (at least until now in the case of electricity), but it can be seen in the evolution of materials used, as for instance the switch from galvanized iron to copper piping, and construction techniques.

The Semiotics of Infrastructures

There is another, rarely acknowledged, side to infrastructures, a possible contributor to political cohesion and fragmentation: messages infrastructures transmit regarding the condition of society and one's position therein. As infrastructures are essential to the physical, social, and economic survival, as well as the well-being, of individuals, and interacting with infrastructures accounts for a substantial portion of their daily experience, it is to be expected that time spent using infrastructures will contribute to shape their view of society. The availability and condition of infrastructures can be seen as indicators of the overall state of society and of its treatment of different constituencies. Resulting perceptions can translate into the adoption of opposing political stands, hence contributing to the coalescence of different political constituencies. It has been argued that it is in part because of their heavier reliance on public sector infrastructures and services – notably public transit – that central-city residents are more likely than their suburban counterparts to support interventionist governments (Dunleavy, 1980; Walks, 2006). The experiential aspect of infrastructures thus has the potential of widening the divide between calls for government involvement in the steering of society and laissez-faire politics.

The fact that infrastructures can serve as barometers of the current condition of society, and mirror its uneven treatment of different constituencies, constitutes another reason for powerful interests to value infrastructures. Interest groups want to ensure that the picture projected by infrastructures matches the view of society they seek to broadcast (see, for example, Rivadulla & Bocarejo, 2014; Roy, 2011b). Hence, the large stylish airports of global south nations conform to the image of modernity and internationalism that elites want to project (Kaika & Swyngedouw, 2000). These elites are particularly sensitive to the perception of their country by international business people and tourists. Few infrastructure contrasts are as stark as the large airports meeting planetary standards located cheek by jowl with adjacent improvised settlements with near-total infrastructure deprivation, as seen in Mumbai, for example. While official documents promoting these nations on the world stage portray their large airports (or a modern metro, as in Delhi) as symbols of modernism, they conveniently ignore these nearby improvised settlements (Siemiatycki, 2006).

The Content of the Book

By taking a broad outlook on the understanding of infrastructures and suburbs, the introductory chapter has set the tone for the diversity of perspectives found in the chapters to follow. It has shown the multiple forms infrastructures can take, which range from the stereotypical hard infrastructures (such as roads, railways, pipes, and wires) to social services (education and health) and, finally, to intangible infrastructures such as legal and financial systems. It has also differentiated infrastructures based on their system of delivery: state, corporate, and community-based. The chapter has concentrated on the social equity dimension of integration- and fragmentation-related repercussions of infrastructures. It has associated integration with the rewards – political, social, or economic – which tend to be captured by constituencies with the capacity to influence infrastructure-related decision-making. In contrast, fragmentation effects are sources of marginalization by making it difficult to function in society for social categories that are bypassed by infrastructure strategies or victimized by the negative externalities of infrastructures. Just as infrastructures were defined in broad terms in the chapter, so was the suburban phenomenon. In the case of suburbs, the different definitions attempted to do justice to the different configurations found across the globe. To provide background information for the variety of perspectives on infrastructures found in the book's chapters, the introduction has attempted to be comprehensive in its approach to infrastructures and suburbs. At the same time, the introductory chapter has provided conceptual validation for the social justice concerns shared by all chapters and for their common social science approach to suburban infrastructures.

The book is divided into three sections that progress from conceptualizations of suburban infrastructures, to challenges faced by suburban infrastructures, to ways that suburban infrastructures are being reshaped for the future – both positively and negatively. In the first section, "Situating Suburban Infrastructures," the context is set for the multitude of ways that suburban infrastructure can be defined and analysed from political, economic, and social angles. In chapter 2, Jean-Paul Addie proposes a framework to enable analysis and comparison of the myriad forms and manifestations of suburbanism. Critically, he points out that suburban infrastructures are not always constructs limited to the geographical and functional confines of suburban areas – they belong to much broader networks with political, economic, and social dimensions that must be understood as complex systems with multiple intersecting relationships, characteristics, and tensions. Chapter 2 shows that suburban infrastructures are in some instances designed to benefit the suburbs, while in other cases they are meant to serve the entire metropolitan area and beyond, often with negative consequences for the suburban areas where they are located. Addie's proposed analytical framework anchors the section, and indeed the entire book, critically acknowledging the complexity of suburban infrastructure as well as the tensions between the social forces that influence its functioning.

Janice Morphet describes in chapter 3 how demographic shifts between city and suburb, coupled with policy transitions at the international and national levels, are fundamentally altering the way suburbs function within their metropolitan contexts. She suggests that traditional suburban administrative boundaries may no longer be appropriate for efficient infrastructure distribution and investment, and that rescaling according to economic spaces can help policymakers to better address the infrastructure shortcomings of suburbs. At the same time, she cautions that realigning administrative boundaries to mirror economic clusters should be seen as a transitory state, rather than a permanent shift to neoliberal geographic divisions. The final chapter of this section is by Alan Walks and analyses suburban financial infrastructures in North America. In chapter 4 he argues that the financial arrangements, which historically supported suburban development, are now draining the suburbs of economic value, thereby making their populations highly susceptible to economic fluctuations. This shift is in large part because assets fundamental to the suburban lifestyle – namely, detached homes and automobiles – have been easily financed through credit, allowing household debt to accumulate. Financial structures in the form of mortgage lending, borrowing on home equity, and automobile loan consolidation have allowed for consumptive lifestyles potentially leading to insurmountable household debt.

The second section, "Suburban Infrastructures in Crisis," teases apart the multitude of issues facing suburbs. Included chapters show how economic and political factions dictate the provision of suburban infrastructures. Neoliberal governance strategies increasingly influence suburban infrastructure provision, causing profitability rather than social equity motives to influence the distribution of infrastructures. In chapter 5, water infrastructure in Eastern Europe offers two municipal case studies to illustrate how global neoliberal shifts are affecting the management of infrastructures and are having repercussions on social equity. Frederick Peters uses the metaphor of a computer operating system that is gradually debugged in evolving "versions" to illustrate how different forms of public–private partnerships have evolved over time to address the management of water systems. Chapter 5 is probably the least focused on suburban, in opposition to urban, infrastructures in this volume. It does, however, provide the sharpest expression of the influence of neoliberalism on infrastructure development and management, a theme that is echoed by several other chapters. In chapter 6, Shubhra Gururani paints the picture of a large city, Gurgaon, just outside of Delhi, with no sewerage system. She explores how different individuals deal with this situation: the intense pride in having sanitary facilities in the home and the difficult arrangements the poor confront in order to cope with the absence of sewerage facilities. The chapter points to the severe environmental and health consequences of such situations in India. Gururani depicts how absent, insufficient, or faulty sewerage systems have become a major object of preoccupation across the country. However, this issue remains a problem without easy solutions due to an absence of institutional capacity to deal with the matter. The chapter opens a window on the deep-rooted politics of neglect in India.

In chapter 7, Igor Vojnovic, Zeenat Kotval-K, Jeanette Eckert, and Xiaomeng Li take a close look at the impacts of a specific economic policy agenda on infrastructure distribution. Using the state of Michigan as a case study, they point out how misguided policy, or lack of policy response altogether, can have deleterious effects on the maintenance of highway infrastructures and exacerbate socioeconomic disparities between income groups. The chapter interprets the described Michigan infrastructure climate as a possible prelude to privatization. Jonathan Rutherford explores in chapter 8 the pressures of population growth in an area of Sweden traditionally characterized by low density and transitory inhabitants. Rutherford casts light on the "politics of infrastructure" in an area undergoing pressure to connect to formalized, characteristically urban and suburban, infrastructure systems. He identifies the threats of such a transformation to the atmosphere of a valued rural setting, and the disproportional financial burden it would represent for long-term low-income residents. Chapter 9 deals with water and sewerage systems in suburban Hanoi. Sophie Schramm and Lucía Wright-Contreras depict the mosaic of water and sewerage systems servicing these suburbs: a small public system and a multiplicity of private systems of different scopes and effectiveness. As expected, wealthy areas benefit from better services than poor sectors. Yet, all inhabitants endure the severe water pollution caused by the fact that only a small proportion of effluents are treated. The most directly affected, however, are poorer residents living close to pools of stagnant polluted water. The chapter highlights the adverse consequences of an uncoordinated development of water and sewerage facilities.

The final section, "Reshaping Suburban Infrastructures," is comprised of cases that illustrate the relation between the form taken by suburban areas and infrastructures, and speculates on the future of suburban infrastructures – exposing both conditions that are currently reshaping existing suburbs as well as opportunities for making suburbs more sustainable. This section identifies the pressures that are changing the face of suburbs, which present significant challenges for the future of suburban infrastructure governance and management. At the same time, some of its chapters offer distinctively optimistic views on how suburbs can evolve and identify positive trajectories for development. In chapter 10, David Wachsmuth presents two US case studies that illustrate how new economic forces might radically change our understanding of suburbanism. He describes the blurring of the boundaries between the urban, suburban, and rural by contemporary transportation and information infrastructures. The rise of multi-city growth regions has changed how cities and suburbs function, causing suburban infrastructures that are distant from urban cores to become strategically central. This situation has historically been the case for infrastructural entities such as airports and energy plants, but greater information and transportation flows between different urban regions create new opportunities for economic development in underserved regions. In this case, Wachsmuth suggests that economic conditions may interact with suburban infrastructures in new ways that redefine local development. Sara Saboonian and Pierre Filion analyse in chapter 11 how green infrastructures can harmonize with built-environment intensification and new public transportation infrastructures to counteract the negative environmental impacts of the dispersed suburban model. Using case studies from the Toronto region, the authors investigate how green infrastructures and densification, seemingly at odds due to the important spatial requirements of green infrastructures, can be reconciled to improve the sustainability of suburban land-use configurations.

The focus of chapter 12, by Sara Macdonald and Lucy Lynch, is also on green infrastructures. The authors discuss the green infrastructure concept at a regional scale, highlighting how the Ontario Greater Golden Horseshoe Greenbelt, a geographically defined territory encircling the Toronto metropolitan area, interacts with hard and soft infrastructures inside and outside its boundary. Set aside as protected green space, this area is subject to extreme development pressures. In chapter 13, Xuefei Ren describes Chinese megaprojects driven by macroeconomic stimulation objectives and financed by questionable partnerships between local governments and private investors, with little oversight from higher-level governments. Displacement of numerous residents, high infrastructure user fees to fund and sustain infrastructure projects, and ballooning government debt all stem from unfettered overinvestment in suburban expansion. These conditions signal the need for Chinese cities to address social and financial challenges stemming from the suburban expansion this infrastructure bonanza causes. Rebecca Ince and Simon Marvin explore in chapter 14 the tension between local and national approaches to the financing of suburban infrastructures, specifically in the context of housing retrofits. Their analysis of top-down and bottom-up dynamics and the differing priorities exhibited by local and national governing bodies highlights how the success or failure of suburban infrastructure distribution depends heavily on political power relationships. The chapter also illustrates how infrastructure programs can be designed in ways that exclude certain social groups, in this case the ones that would have most benefited from housing retrofits. In chapter 15, Markus Moos and Jonathan Woodside concentrate on one manifestation of infrastructure inequality: public transportation infrastructure as a trigger for gentrification. The chapter points to increasing interest on the part of a large proportion of the urban population for lifestyles that involve walking and public transit use. It finds such proclivities to be especially pronounced among young adults. In these circumstances, residential areas along axes that provide good quality public transit (such as rail transit) are likely to be attractive to the wealthier segments of the population interested in walking and transit, at the detriment of lower income groups for whom access to public transit may be even more vital because of their absence of modal choice.

In chapter 16, the conclusion, the co-editors of the book, Pierre Filion and Nina Pulver, bring to light the main themes raised in the book. The conclusion shows how many chapters expressed concern for the social justice dimension of infrastructures – of existing and missing infrastructures. Another running theme is that of infrastructure deficits. Most chapters lament an insufficient presence of infrastructures and explore reasons for this situation. Of course, consequences in poor global south cities are far more severe than they are in global north cities. Two other unifying themes addressed in the conclusion are the transitions affecting suburbs, a factor accounting for mismatch between infrastructures and suburbs, and institutional deficiencies responsible for insufficient and uncoordinated responses to infrastructure needs.

REFERENCES

Addie, J.-P.D. (2014). Flying high (in the competitive sky): Conceptualizing the role of airports in global city-regions through "aero-regionalism." *Geoforum, 55*(1), 87–99. https://doi.org/10.1016/j.geoforum.2014.05.006

Addie, J.-P.D., & Keil, R. (2015). Real existing regionalism: The region between talk, territory and technology. *International Journal of Urban and Regional Research, 39*(2), 407–17. https://doi.org/10.1111/1468-2427.12179

Amati, M., & Taylor, L.E. (2010). From green belt to green infrastructure. *Planning Practice and Research, 25*(2), 143–55. https://doi.org/10.1080/02697451003740122

Bakker, K. (2010). *Privatizing water: Governance failure and the world's urban water crisis*. Ithaca, NY: Cornell University Press.

Bakker, K. (2011). Splintered urbanisms: Water, urban infrastructure and the modern social imaginary. In Matthew Gandy (Ed.), *Urban constellations* (pp. 62–4). Berlin: Jovis.

Banister, D., & Berechman, J. (2000). *Transport investment and economic development*. London: UCL Press.

Benedict, M.A., & McMahon, E.T. (2006). *Green infrastructure: Linking landscapes and communities*. Washington, DC: Island Press.

Bernt, M., & Naumann, M. (2006). Wenn der Hahn zu bleibt. Wasserversorgung in schrumpfenden Städten [When the faucet remains shut off. Water supply in shrinking cities]. In S. Frank and M. Gandy (Eds.), *Hydropolis: Wasser und die Stadt der Moderne* [Water and the modern city] (pp. 210–28). Frankfurt A.M.: Campus.

Bjorvatn, K. (2000). Urban infrastructure and industrialization. *Journal of Urban Economics, 48*(2), 205–18. https://doi.org/10.1006/juec.1999.2162

Bloch, R., Papachristodoulou, N., & Brown, D. (2013). Suburbs at risk. In R. Keil (Ed.), *Suburban constellations* (pp. 95–101). Berlin: Jovis.

Boo, K. (2012). *Behind the beautiful forevers: Life, death, and hope in a Mumbai undercity*. New York: Random House.

Brenner, N. (1999). Globalisation as reterritorialisation: The re-scaling of urban governance in the European Union. *Urban Studies, 36*(3), 431–51. https://doi.org/10.1080/0042098993466

Broto, V.C., & Bulkeley, H. (2013). Maintaining climate change experiments: Urban political ecology and the everyday reconfiguration of urban infrastructure. *International Journal of Urban and Regional Research, 37*(6), 1934–48. https://doi.org/10.1111/1468-2427.12050

Brown, A. (2014). *Next generation infrastructure*. Washington, DC: Island Press.

Burchell, R.W., Downs, A., McCann, B., & Mukherji, S. (2005). *Sprawl costs: Economic impacts of unchecked development*. Washington, DC: Island Press.

Carpio, G., Irazábal, C., & Pulido, L. (2011). Right to the suburb? Rethinking Lefebvre and immigrant activism. *Journal of Urban Affairs, 33*(2), 185–208. https://doi.org/10.1111/j.1467-9906.2010.00535.x

Chang, S.E. (2003). Evaluating disaster mitigations: Methodology for urban infrastructure systems. *Natural Hazard Review, 4*(4), 186–96. https://doi.org/10.1061/(asce)1527-6988(2003)4:4(186)

Chapin, F.S. (1974). *Human activity patterns in the city: Things people do in time and in space*. New York: John Wiley & Sons.

Charmes, E., & Keil, R. (2015). The politics of post-suburban densification in Canada and France. *International Journal of Urban and Regional Research, 39*(3), 581–602. https://doi.org/10.1111/1468-2427.12194

Chau, K.W., & Ng, F.F. (1998). The effects of improvement in public transportation capacity on residential price gradient in Hong Kong. *Journal of Property Valuation and Investment, 16*(4), 397–410. https://doi.org/10.1108/14635789810228204

Clarke Annez, P. (2006). *Urban infrastructure finance from private operators: What have we learned from recent experience?* (World Bank Research Paper 4045). Washington, DC: The World Bank.

Coaffee, J. (2010). Protecting vulnerable cities: The UK's resilience response to defending everyday urban infrastructure. *International Affairs, 86*(4), 939–54. https://doi.org/10.1111/j.1468-2346.2010.00921.x

Coutard, O., & Rutherford, J. (Eds). (2015). *Beyond the networked city: Infrastructure reconfigurations and urban change in the North and South*. London: Routledge.

Desai, R., McFarlane. C., & Graham, S. (2015). The politics of open defecation: Informality, body, and infrastructure in Mumbai. *Antipode, 47*(1), 98–120. https://doi.org/10.1111/anti.12117

Dikec, M. (2007). *Badlands of the republic: Space, politics and urban policy*. Oxford, UK: Blackwell.

Dikec, M. (2011). The politics of the *banlieue*. In M. Gandy (Ed.), *Urban constellations* (pp. 58–61). Berlin: Jovis.

Dodson, J.R. (2014). Suburbia under an energy transition: A socio-technical perspective. *Urban Studies, 51*(7), 1487–1505. https://doi.org/10.1177/0042098013500083

Dourish, P., & Bell, G. (2007). The infrastructure of experience and the experience of infrastructure: Meaning and structure in everyday encounters with space. *Environment and Planning B, 34*(3), 414–30. https://doi.org/10.1068/b32035t

Dufaux, F., & Fourcaut, A. (Eds.). (2004). *Le monde des grands ensembles*. Paris: Créaphis.

Dunleavy, P. (1980). *Urban political analysis: The politics of collective consumption*. London: Macmillan.

Easterling, K. (2010). Disposition and active form. In K. Scholl and S. Lloyd (Eds.), *Infrastructure as architecture: Designing composite networks* (pp. 96–9). Berlin: Jovis.

Easterling, K. (2014). *Extrastatecraft: The power of infrastructure space*. London: Verso.

Farber, S., Neutens, T., Miller, J., & Li, X. (2013). The social interaction potential of metropolitan regions: A time-geographic measurement approach using joint accessibility. *Annals of the Association of American Geographers, 103*(3), 483–504. https://doi.org/10.1080/00045608.2012.689238

Filion, P. (2013a). Automobiles, highways, and suburban dispersion. In R. Keil (Ed.), *Suburban constellations: Governance, land, and infrastructure in the 21st century* (pp. 79–84). Berlin: Jovis.

Filion, P. (2013b). The infrastructure is the message: Shaping the suburban morphology and lifestyle. In R. Keil (Ed.), *Suburban constellations: Governance, land, and infrastructure in the 21st century* (pp. 39–45). Berlin: Jovis.

Filion, P. (2014). Fading resilience? Creative destruction, neoliberalism and mounting risks. *Sapiens, 6*(1). http://sapiens.revues.org/1523

Filion, P. (2018). The morphology of dispersed suburbanism: The land use patterns of the dominant North American urban form. In R. Harris and U. Lehrer (Eds.), *The suburban land question: A global survey* (pp. 122–44). Toronto, ON: University of Toronto Press.

Filion, P., Bunting, T., & Warriner, K. (1999). The entrenchment of urban dispersion: Residential preferences and location patterns in the dispersed city. *Urban Studies, 36*(8), 1317–47. https://doi.org/10.1080/0042098993015

Firzli, M.N.J. (2016). The end of "globalization"? Economic policy in the post-neocon age. *Analyse financière, 60*, 8–10.

Forsyth, A. (2012). Defining suburbs. *Journal of Planning Literature, 27*(3), 270–81. https://doi.org/10.1177/0885412212448101

Freestone, R., & Baker, D. (2011). Spatial planning models of airport-driven urban development. *Journal of Planning Literature, 26*(3), 263–79. https://doi.org/10.1177/0885412211401341

Gandy, M. (2004). Rethinking urban metabolism: Water, space and the modern city. *City: Analysis of Urban Trends, Culture, Theory, Policy, Action, 8*(3), 363–79. https://doi.org/10.1080/1360481042000313509

Gandy, M. (2006). Planning, anti-planning and the infrastructure crisis facing metropolitan Lagos. *Urban Studies, 43*(2), 371–96. https://doi.org/10.1080/00420980500406751

Gaspar, J., & Glaeser, E.L. (1998). Information technology and the future of cities. *Journal of Urban Economics*, *43*(1), 136–56. https://doi.org/10.1006/juec.1996.2031

Gordon, D.L.A., & Janzen, M. (2013). Suburban nation? Estimating the size of Canada's suburban population. *Journal of Architectural and Planning Research*, *30*(3), 197–220.

Graham, S. (2000). Constructing premium network spaces: Reflections on infrastructure networks and contemporary urban development. *International Journal of Urban and Regional Research*, *24*(1), 183–200. https://doi.org/10.1111/1468-2427.00242

Graham, S. (2002). FlowCity: Networked mobilities and the contemporary metropolis. *Journal of Urban Technology*, *9*(1), 1–20. https://doi.org/10.1080/106307302317379800

Graham, S. (Ed.). (2010). *Disrupted cities: When infrastructure fails*. London: Routledge.

Graham, S. (2011). Disruptions. In M. Gandy (Ed.), *Urban constellations* (pp. 65–70). Berlin: Jovis.

Graham, S., & Marvin, S. (2001). *Splintering urbanism: Networked infrastructures, technological mobilities and the urban condition*. London: Routledge.

Guignier, F., & Madanat, S. (1999). Optimization of infrastructure systems maintenance and improvement policies. *Journal of Infrastructure Systems*, *5*(4), 124–34. https://doi.org/10.1061/(asce)1076-0342(1999)5:4(124)

Guillen, J. (2015, 24 May). Detroit water shutoffs to begin Tuesday. *Detroit Free Press*. https://www.freep.com/story/news/local/michigan/detroit/2015/05/24/detroit-water-shutoffs-poverty-unpaid-bills/27852135/

Gutnov, A., Baburov, A., Djumenton, G., Kharitonova, S., Lezava, I., & Sadovskij, S. (1971). *The ideal communist city*. New York: George Braziller.

Harris, R. (2010). Meaningful types in a world of suburbs. In M. Clapson & R. Hutchinson (Eds.), *Suburbanization in global society*. (Research in Urban Sociology, 10)(pp. 15–47). Bringley, UK: Emerald Group Publishing.

Harvey, D. (2005). *A brief history of neoliberalism*. Oxford: Oxford University Press.

Haughton, G., & McManus, P. (2012). Neoliberal experiments with urban infrastructure: The Cross City Tunnel, Sydney. *International Journal of Urban and Regional Research*, *36*(1), 90–105. https://doi.org/10.1111/j.1468-2427.2011.01019.x

Hechter, M., & Opp, K.-D. (2001). *Social norms*. New York: Russell Sage Foundation.

Hertel, S., Keil, R., & Collens, M. (2015). *Stitching tracks: Towards transit equity in the Greater Toronto and Hamilton Area*. (Report for Metrolinx). Toronto, ON: The City Institute at York University.

Hidalgo, D., & Grafiteaux, P. (2008). Bus rapid transit systems in Latin America and Asia: Results and difficulties in 11 cities. *Transportation Research Record*, *2072*(1), 77–88. https://doi.org/10.3141/2072-09

Hirt, S.A. (2014). *Zoned in the USA: The origins and implications of American land-use regulations*. Ithaca, NY: Cornell University Press.

Hodson, M., & Marvin, S. (Eds.). (2015). *Retrofitting cities: Priorities, governance and experimentation*. Abingdon, UK: Routledge.

Hodson, M., Marvin, S., Robinson, B., & Swilling, M. (2012). Reshaping urban infrastructure: Material flow analysis and transitions analysis in an urban context. *Journal of Industrial Ecology*, *16*(6), 789–800. https://doi.org/10.1111/j.1530-9290.2012.00559.x

Hughes, T.P. (1987). The evolution of large technological systems. In W.E. Bijker, T.P. Hughes, and T.J. Pinch (Eds.), *The social construction of technological systems* (pp. 50–82). Cambridge, MA: The MIT Press.

Innis, H.A. (1962 [1956]) *Essays on Canadian economic history*. Toronto, ON: University of Toronto Press.

Jacobs, J.M., & Lees, L. (2013). Defensible space on the move: Revisiting the urban geography of Alice Coleman. *International Journal of Urban and Regional Research*, *37*(5), 1559–83. https://doi.org/10.1111/1468-2427.12047

Jessen, J., & Roost, F. (Eds.). (2016). *Refitting suburbia*. Berlin: Jovis.

Jessop, B. (1996). Post-Fordism and the state. In B. Greve (Ed.), *Comparative welfare systems: The Scandinavian model in a period of change* (pp. 165–83). London: Palgrave Macmillan.

Jessop, B. (2002). Liberalism, neoliberalism, and urban governance: A state-theoretical perspective. *Antipode*, *34*(3), 452–72. https://doi.org/10.1111/1467-8330.00250

Kaika, M., & Swyngedouw, E. (2000). Fetishizing the modern city: The phantasmagoria of urban technical networks. *International Journal of Urban and Regional Research*, *24*(1), 120–38. https://doi.org/10.1111/1468-2427.00239

Karvonen, A., & Brand, R. (2009). Technical expertise, sustainability, and the politics of specialized knowledge. In G. Kütting & R. Lipschutz (Eds.), *Environmental governance: Power and knowledge in a local and global world* (pp. 38–59). Oxford: Routledge.

Keil, R., & Young, D. (2008). Transportation: The bottleneck of regional competitiveness in Toronto. *Environment and Planning C: Government and Policy*, *26*(4), 728–51. https://doi.org/10.1068/c68m

Keil, R., & Young, D. (2014). In-between mobility in Toronto's new (sub)urban neighbourhoods. In P. Watt & P. Smets (Eds.), *Neighbourhoods, mobilities and belonging in the city and suburb*. Basingstoke, UK: Palgrave Macmillan.

Kimmelman, M. (2015, 12 February). Paris aims to embrace its estranged suburbs. *The New York Times*. Retrieved from http://www.nytimes.com/2015/02/13/world/europe/paris-tries-to-embrace-suburbs-isolated-by-poverty-and-race.html?_r=1

Kooy, M., & Bakker, K. (2008). Splintered networks: The colonial and contemporary waters of Jakarta. *Geoforum, 39*(6), 1843–58. https://doi.org/10.1016/j.geoforum.2008.07.012

Kotz, D. (2015). *The rise and fall of neoliberal capitalism*. Cambridge, MA: Harvard University Press.

Latour, B., & Hermant, E. (1998). *Paris: Ville invisible*. Paris: La Découverte.

Leigh, N.G., & Blakely, J. (2013). *Planning local economic development: Theory and practice* (5th ed.). Thousand Oaks, CA: Sage.

Lewis, T. (1999). *Divided highways: Building the interstate highways, transforming American life*. New York: Penguin.

Logan, J.R., & Molotch, H.L. (1987). *Urban fortunes: The political economy of place*. Berkeley, CA: University of California Press.

MacCleary, R., Peterson, C., & Stern, J.D. (2012). *Shifting suburbs: Reinventing infrastructure for compact development*. Washington, DC: Urban Land Institute.

Marshall, A. (2000). *How cities work: Suburbs, sprawl, and the roads not taken*. Austin, TX: University of Texas Press.

McFarlane, C., Desai, R., & Graham, S. (2014). Informal urban sanitation: Everyday life, poverty, and comparison. *Annals of the Association of American Geographers, 104*(5), 989–1011. https://doi.org/10.1080/00045608.2014.923718

McFarlane, C., & Rutherford, J. (2008). Political infrastructures: Governing and experiencing the fabric of the city. *International Journal of Urban and Regional Research, 32*(2), 363–74. https://doi.org/10.1111/j.1468-2427.2008.00792.x

McLuhan, M. (1994 [1964]). *Understanding media: The extensions of man*. Cambridge, MA: MIT Press.

McLuhan, M., & Fiore, Q. (1967). *The medium is the message: An inventory of effects*. New York: Random House.

Metrolinx. (2008). *The big move: Transforming transportation in the Greater Toronto and Hamilton Area*. Toronto, ON: Author.

Mian, A., & Sufi, A. (2014). *House of debt: How they (and you) caused the Great Recession, and how we can prevent it from happening again*. Chicago, IL: University of Chicago Press.

Mirowski, P., & Plehwe, D. (2009). *The road from Mont-Pèlerin: The making of the neoliberal thought collective*. Cambridge, MA: Harvard University Press.

Monstadt, J., & Schramm, S. (2013). Beyond the networked city? Suburban constellations in water and sanitation systems. In R. Keil (Ed.), *Suburban constellations* (pp. 85–94). Berlin: Jovis.

Moran, M. (2011). Opposing exclusion: The political significance of the riots in French suburbs (2005–2007). *Modern and Contemporary France, 19*(3), 297–312. https://doi.org/10.1080/09639489.2011.588793

Parkin, J. (1994). A power model of urban infrastructure decision-making. *Geoforum, 25*(2), 203–11. https://doi.org/10.1016/0016-7185(94)90016-7

Peck, J, & Tickell, A. (2002). Neoliberalizing space. *Antipode, 34*(3), 380–404. https://doi.org/10.1111/1467-8330.00247

Pethe, A., & Ghodke, M. (2002). Funding urban infrastructure: From government to markets. *Economic and Political Weekly, 37*(25), 2467–70.

Poitras, C. (2011). A city on the move: The surprising consequences of highways. In S. Castonguay & M. Dagenais (Eds.), *Metropolitan natures: Environmental histories of Montreal* (pp. 168–86). Pittsburgh, PA: University of Pittsburgh Press.

Power, A. (2001). Social exclusion and urban sprawl: Is the rescue of cities possible? *Regional Studies, 35*(8), 731–42. https://doi.org/10.1080/00343400120084713

Ranmert, W. (1997). New rules of sociological method: Rethinking technology studies. *British Journal of Sociology, 48*(2), 171–91. https://doi.org/10.2307/591747

Reese, L.A., & Rosenfeld, R.A. (Eds.). (2012). *Comparative civic culture: The role of local culture in urban policy-making*. Farnham, Surrey, UK: Ashgate.

Rivadulla, M.J.Á., & Bocarejo, D. (2014). Beautifying the slum: Cable car fetishism in Cazucá, Colombia. *International Journal of Urban and Regional Research, 38*(6), 2025–41. https://doi.org/10.1111/1468-2427.12201

Roy, A. (2011a). Slumdog cities: Rethinking subaltern urbanism. *International Journal of Urban and Regional Research, 35*(2), 223–38. https://doi.org/10.1111/j.1468-2427.2011.01051.x

Roy, A. (2011b). Urbanisms, worlding practices and the theory of planning. *Planning Theory, 10*(1), 6–15. https://doi.org/10.1177/1473095210386065

Rubio, F.D., & Fogué, U. (2013). Technifying public space and publicizing infrastructures: Exploring new urban political ecologies through the Square of General Vara del Rey. *International Journal of Urban and Regional Research, 37*(3), 1035–52. https://doi.org/10.1111/1468-2427.12052

Schlosberg, D. (2007). *Defining environmental justice: Theories, movements, and nature.* New York: Oxford University Press.

Siemiatycki, M. (2006). Message in a metro: Building urban rail infrastructure and image in Delhi, India. *International Journal of Urban and Regional Research, 30*(2), 277–92. https://doi.org/10.1111/j.1468-2427.2006.00664.x

Siemiatycki, M. (2013). The global production of transportation public–private partnerships. *International Journal of Urban and Regional Research, 37*(4), 1254–72. https://doi.org/10.1111/j.1468-2427.2012.01126.x

Simone, A.M. (2004). People as infrastructure: Intersecting fragments in Johannesburg. *Public Culture, 16*(3), 407–29. https://doi.org/10.1215/08992363-16-3-407

Simone, A.M. (2010). *City life from Jakarta to Dakar: Movements at the crossroads*. New York: Routledge.

Soja, E.W. (2010). *Seeking spatial justice*. Minneapolis, MN: University of Minnesota Press.

Stanilov, K, & Scheer, B.C. (2004). *Suburban form: An international perspective*. New York: Routledge.

Star, S.L. (1999). The ethnography of infrastructure. *American Behavioral Scientist, 43*(3), 377–91. https://doi.org/10.1177/00027649921955326

Suchman, L., Blomberg, J., Orr, J.E., & Trigg, R. (1999). Reconstructing technologies as social practice. *American Behavioral Scientist, 43*(3), 392–408. https://doi.org/10.1177/00027649921955335

Swyngedouw, E. (2004). *Social power and the urbanization of water: Flows of power*. Oxford: Oxford University Press.

Takaki, E. (2012). *Bodies in movement: Body experiences in the city*. (Doctoral dissertation in Urbanism). Universidade Federal do Rio de Janeiro, Rio de Janeiro.

Tomalty, R. (2014). *Innovative infrastructure mechanisms for smart growth*. Beaconsfield, QC: Canadian Electronic Library.

Tomalty, R., & Mallach, A. (2015). *America's urban future: Lessons from north of the border*. Washington, DC: Island Press.

Torrance, M. (2009). Reconceptualizing urban governance through a new paradigm for urban infrastructure networks. *Journal of Economic Geography, 9*(6), 805–22. https://doi.org/10.1093/jeg/lbn048

Varley, A. (2013). Postcolonialising informality? *Environment and Planning D, 31*(1), 4–22. https://doi.org/10.1068/d14410

Walks, R.A. (2006). The causes of city-suburban political polarization? A Canadian case study. *Annals of the Association of American Geographers, 96*(2), 390–414. https://doi.org/10.1111/j.1467-8306.2006.00483.x

Webb, G.R. (2007). The popular culture of disaster: Exploring a new dimension of disaster research. In H. Rodriguez, E.L. Quarantelli, & R.R. Dynes (Eds.), *Handbook of disaster research* (pp. 430–40). New York: Springer.

Wellman, K., & Spiller, M. (Eds.) (2012). *Urban infrastructure finance and management*. Chichester, UK: Wiley.

Wesselink, A., Buchanan, K.S., Georgiadou, Y., & Turnhout, E. (2013). Technical knowledge, discursive spaces and politics at the science–policy interface. *Environmental Science & Policy, 30*(1), 22–39. https://doi.org/10.1016/j.envsci.2012.12.008

Wilson, D.C., Vellis, C., & Cheeseman, C. (2006). Role of informal sector recycling in waste management in developing countries. *Habitat International, 30*(4), 797–808. https://doi.org/10.1016/j.habitatint.2005.09.005

Wylie, P.J. (1996). Infrastructure and Canadian economic growth 1946–1991. *Canadian Journal of Economics, 29*(2), S350–5. https://doi.org/10.2307/136015

Yigitcanlar, T. (Ed.). (2010). *Sustainable urban and regional infrastructure development: Technologies, applications and management*. Hershey, PA: IGI Global.

Young, D., Burke Wood, P., & Keil, R. (Eds.). (2011). *In-between infrastructure: Urban connectivity in an age of vulnerability*. Kelowna, BC: Praxis(e) Press.

Young, D., & Keil, R. (2014). Locating the urban in-between: Tracking the urban politics of infrastructure in Toronto. *International Journal of Urban and Regional Research, 38*(5), 1589–1608. https://doi.org/10.1111/1468-2427.12146

Yu, X.J., & Ng, C.N. (2007). Spatial and temporal dynamics of urban sprawl along two urban–rural transects: A case study of Guangzhou, China. *Landscape and Urban Planning, 79*(1), 96–109. https://doi.org/10.1016/j.landurbplan.2006.03.008

Zimmerman, R. (2001). Social implications of infrastructure network interactions. *Journal of Urban Technology, 3*, 97–119. https://doi.org/10.1080/106307301753430764

SECTION 1

Situating Suburban Infrastructures

2 In What Sense Suburban Infrastructure?

JEAN-PAUL D. ADDIE

What, if anything, is held in common across infrastructures as diverse as waste, roads, and trains? And between urban contexts as different as Jakarta, Mumbai, Kampala, Newcastle, and Ramallah?

(Graham & McFarlane, 2015: 12–13)

(Moving Beyond) Suburban Infrastructure as a Chaotic Concept

Suburbs, as the contributions across this volume attest, display a wide and varied abundance of infrastructure. Hard infrastructures, including highways, rail tracks, airports, intermodal yards, oil refineries, power plants, power and fibre optic cables, sewers, and sanitation systems, are crucial – if often inconspicuous – constitutive elements of variegated global suburban landscapes. At the same time, multifaceted social (or soft) infrastructures – the formal institutions and informal practices employed by various actors from national governments to street vendors – fundamentally condition the capacities of people living in, and moving through, suburban places (Simone, 2004). These social and technical systems underpin the growth and experience of "the suburban" by mediating resource flows to, and across, the urban periphery. The provision, maintenance, and governance of transportation, energy, water, and waste systems (or lack thereof) established the conditions for the historical expansion of urban spatial forms and the integration/marginalization of peripheral communities into the wider urban fabric (Gandy, 2003; Law, 2012; Warner, 1978). Contemporary constellations of global suburbia – from large-scale capital intensive developments and swiftly expanding informal settlements to declining inner-ring communities – continue to evolve in a symbiotic, fundamentally politicized, relationship with their infrastructure networks (McFarlane & Rutherford, 2008). As processes of suburbanization occupy a central position in the rapid and ongoing urbanization of the planet, suburban space is a crucial frontier of infrastructural innovation and stress that will deeply shape the future potentialities and challenges of cities, suburbs, and an urbanizing world more broadly (Keil, 2013).

A robust literature now utilizes infrastructure as a critical object of analysis to think through the politics, social relations, and everyday experience of urban life (see, for example, Angelo & Hentschel, 2015; Graham & McFarlane, 2015; McFarlane & Rutherford, 2008; Young, Wood, & Keil, 2011). However, the extended and networked nature of infrastructure systems, their sheer diversity and modes of configuration, and their contingent embedding in varying geographic contexts present conceptual and methodological challenges for critical and comparative urban studies. Reflecting on the case studies curated in their edited volume *Infrastructural Lives* (and the questions posed in the epigraph at the beginning of this chapter), Graham and McFarlane (2015: 13) observe "a tendency for infrastructure studies to focus on particular infrastructures ... [with] little held in common beyond infrastructure itself as a set of material processes." Such concerns raise a further epistemological question for analyses of global suburban infrastructure: Is there anything analytically distinct about *suburban infrastructure*, or the social, technical, and political regimes that singularize the "suburban moment" in its production, governance, or use? Suburban expressions of the urban process are highly pluralized, contextual, interconnected, and endogenous. If the sheer variety (in nature, form, and temporal development) of global suburbs inhibits the construction of universal or all-encompassing definitions of "the suburbs" (Charmes & Keil, 2015; Harris, 2010; Phelps, Wood, & Valler, 2010; Walks, 2013), strong theorization is necessary to prevent "chaotic conceptions" robbing "suburban infrastructure" of its analytical significance (see Sayer, 1992).

This theorization is an important and profoundly political task. Our understanding of the relationship between infrastructures and suburban space (and comparability between instances) is critical to addressing questions of politics, governance, and the applicability of mobile infrastructure policy frameworks, not only between metropolises (Chennai, Stockholm, Toronto, and so forth) and technologies (energy, transport, governance, and suchlike) but within the heterogeneous internal structures of these urban agglomerations. Investment in the core infrastructures of developed and emerging urban societies may be heralded as state spatial strategies to enhance the territorial competitiveness and resilience of metropolitan regions and national economies, but access to infrastructure and the experience of its failures are highly uneven and unequal. Infrastructures invoke dialectics of inclusion/ access and exclusion/marginality. Gridlock, blackouts, crumbling bridges, and leaking pipelines are now a commonplace feature of suburban life, but one whose experience and impacts are dependent upon individuals' and groups' differentiated spatial relations and position relative to dominant power geometries (Graham, 2010). We therefore need to problematize assumptions about our knowledge and experience of, and engagement with, suburban infrastructure to realize its potential analytic utility amidst the maelstrom of contemporary urban growth.

The aim of this chapter is to develop an analytically meaningful framework to analyse suburban infrastructure by paying concerted attention to how infrastructures relate to the production and experience of dynamic and highly variegated suburban environments. My approach is built around two conceptual triads: the first unpacks the modalities of infrastructures as they exist in, for, and of suburbs (broadly understood as the landscapes of extended urbanization); the second discloses the political-economic processes (suburbanization), lived experience (suburbanism), and dynamics of mediation internalized by particular suburban infrastructures. I am not concerned with the tasks of ensuring definitional rigour or bounding what does and does not constitute "suburban infrastructure." Rather, I seek to identify adaptable conceptual and methodological innovations from the distinct *relations* between the suburban and any number of hard and soft infrastructures facilitating social processes and relations across space. The conceptual framework presented in the following is intended to be open and adaptable to the specific geographical context, infrastructures, and conceptual languages (from edge and in-between cities to banlieues and favelas; exopolis to post-suburbia) under empirical investigation. In doing so, it opens avenues to cut through the fuzziness presented by a cacophony of suburban and infrastructural signifiers to (1) realize greater conceptual clarity when discussing the suburbanity of infrastructures and their associated actors, economies, and cultures; (2) facilitate and promote comparative analysis across diverse global suburban contexts; and (3) develop tools to analytically foreground the dialectical relations internalized in the concrete forms, configurations, and governance of suburban infrastructures. I concretize this argument by briefly unpacking the politics of rail infrastructure in the Chicago region, focusing on the changing modalities of suburban infrastructure surrounding the 2007–2009 acquisition of the Elgin, Joliet and Eastern Railroad by the Canadian National Railway.

Theorizing Suburban Infrastructure: A Framework for Analysis

I begin from the proposition that the suburbanity of infrastructure derives from more than its location in a suburban environment. As technical, social, political, cultural, and economic entities, infrastructures invoke a multifaceted and interconnected amalgam of sociospatial relations; consequently, not all infrastructure located in suburbs is suburban, and not all suburban infrastructure is to be found in suburbs themselves. After all, there is nothing necessarily suburban about an airport, a power station, or a fibre optic cable, but an airport may be governed by a regional authority strongly influenced by suburban municipalities; a power station may provide the necessary electricity to support non-central urban growth; fibre optic cables might enable the enhanced securitization of gated communities on the edge of the city. In order to grapple with the specificity of suburban infrastructure, we need to unpack the imbricated ways in which nominally understood infrastructures may be (1) physically embedded in suburban landscapes; (2) produced and performed through place-based suburban governance and sociospatial dynamics; and (3) supportive of suburbanization and suburban ways of life. In other words, we can consider a tripartite division between suburban infrastructures as artefacts and systems *in*, *of*, and *for* suburbs:

- *Infrastructures in suburbs* are principally suburban as a consequence of their physical location in a suburban environment. Such infrastructures may be embedded in suburban places, but the flows they territorialize and their primary functional logics are not contingent on this suburban positioning. Rather, higher order restructuring aligns them to alternative scales of mobility and political economies conditioned elsewhere. Here, we can consider the constituent infrastructural elements facilitating the suburbanization of global distribution and logistics industries – intermodal terminals, international cargo airports, major trucking highways, extended landscapes of warehousing and distribution facilities – as a case in point (see Keil & Young, 2008). These infrastructural artefacts are clearly attuned to processes of globalization rather than essentially suburban in nature. Yet, their physical presence and the imperatives of global competitiveness guiding their planning, operation, and governance do significantly shape the lived experience, development trajectories, and spatial imaginaries of the suburbs that house them, whether by opening economic development opportunities (Cidell, 2011) or by exposing communities to negative externalities, risks, vulnerabilities, and disruptions (Cowen, 2014).

- *Infrastructures of suburbs*, by contrast, are chiefly determined by suburban institutions, communities, landscapes, and governmentalities. They can arise through formal channels structured by local governance, funding, maintenance, and operation. For example, suburban municipal ownership – whether directly or through special taxing districts – can create particular infrastructure systems (for example, regional transport authorities, municipal water boards, forest preserves) that mobilize claims of power and authority over territories both near and far. Infrastructures of suburbs may also be developed through the informal arrangements and practices of users of suburban space – for instance, communities responding to deficiencies in "infrastructure deserts," as McFarlane, Desai, and Graham (2014) discuss with regard to sanitation systems in Mumbai's informal settlements. We can approach the infrastructure of suburbs both through the production, lived experience, or appropriation of networked space and through discourses that construct suburbs in relation to infrastructures normatively understood as "suburban" (for example, automobility as a suburban way of life [Walks, 2015]; homeownership, privatism, and neoliberal spatial polity [Peck, 2011]).
- *Infrastructures for suburbs*, finally, are the material and social elements shaping the resource flows necessary to support suburban growth and ways of life. Processes of suburbanization are enabled through extended infrastructure networks that reach beyond suburbs as a territorially or morphologically defined spatial form. Infrastructures for suburbs tie suburban space and society to central cities through systems supporting traditional economic and land-use patterns and through new infrastructural arrangements that condition the functional integration of polycentric urban regions. At the same time, the metabolic demands of suburbs (for water, waste management, energy, and so on) construct distant geographic landscapes as infrastructural prerequisites for suburban development and reproduction (see Swyngedouw, 2006). Infrastructures for suburbs may thus be framed, following Brenner (2014: 5), as the "operational landscapes" of global suburbanization.

This initial schema, I suggest, is particularly useful in two regards. First, it extends our engagement with the complex spatiality of suburban infrastructure beyond the territorial confines of the suburbs themselves. The distinct topological relations and propinquity disclosed in each categorization illuminates the necessity of incorporating multiple scales of analysis into any examination of suburban infrastructure. Artefacts understood as "infrastructures in suburbs" might be aligned to the broad scale of urban development but still play a vital role in shaping the identity, functionality, and politics of individual suburbs by bounding, enclosing, or dividing space – physically demarcating the "wrong side of the tracks." Second, it draws attention to questions of ownership, governance, and the intent of social and political action. Since individual artefacts and specific systems may internalize multiple scales of urban development and rhythms of mobility, they can invoke distinct and competing political claims (for example, around issues of NIMBYism versus the demands of regional competitiveness). As a result, infrastructures in, of, and for suburbs cannot be considered as mutually exclusive. Rather, they provide a conceptual framework to examine the uses, relations, and ambiguities emergent across the sociotechnical palimpsest of global suburbia.

Considering suburban infrastructures as *things* (broadly considered) relative to suburban space, though, only offers a partial viewpoint – one that does not adequately account for the (sub)urban *processes* giving rise to an ephemeral and transitory amalgam of highly differentiated landscapes (Keil, 2013: 9). Refocusing our attention on the processes internalized in particular infrastructural configurations exposes generative moments of social action and spaces of political practice. Suburbanization is then revealed to be an active and contested moment in the overall process of urban transformation. Here, we can draw a second set of distinctions between the political-economic, experiential, and mediatory dimensions of suburban infrastructure. Again, these categories are not mutually exclusive or ontologically separate. Instead, they are operationalized through a relational three-dimensional dialectic that offers distinct epistemological vantage points onto the contradictory structuring imperatives, governance, experience, and politics of suburban infrastructure (after Lefebvre, 1991: 39):

- *Infrastructures of suburbanization* promote and support increases in non–central-city population and economic activity, and the spatial expansion of urban constellations. The central focus here is the role of infrastructures in the suburbanization of capital and the political-economic process that facilitates capital production, consumption, and circulation that underlie the form and function of suburban space. These include both hard artefacts (for example, pipelines, water systems, and transportation lines) and soft structures (for example, mortgage regulatory frameworks and "innovations" in financialization) that engender urban spatial expansion and establish the grounds to support particular spatial fixes. This categorization therefore draws our attention to the governance modalities of capital and the state – often through the work of the development industry (see Hayden, 2003) – as contextualized within broader trends and urbanization regimes.

- *Infrastructures of suburbanism(s)* are appropriated and repurposed through suburban spatial practice to construct qualitatively differentiated expressions of suburbanism as a way of life, experienced not just in place but as a place. Since infrastructures require the co-production of the subjects who make use of them – in fluid and unpredictable ways (Höhne, 2015) – infrastructures of suburbanism are integral to both the suburbanization of consciousness and the suburbanization of everyday life. They are further generative of governmentalities of authoritarian privatism or emancipation, to the extent that they interpolate inequalities, power relations, or the commodification of suburban space (Ekers, Hamel, & Keil, 2012). In this sense, infrastructures of suburbanism can be understood as the infrastructural components attached to formation and social reproduction of suburban lifestyles, and the construction of peripheral urban locales as distinct spaces of habitation, work, and play (see Walks, 2013 for a detailed discussion).
- *Mediatory infrastructures* articulate suburban constellations within the multiscalar dynamics of contemporary urbanization. Drawing from Lefebvre's (2003: 80) theorization of the urban as a "mixed" or "mediatory" level (not scale), suburban mediatory infrastructures perform the role of connecting and resolving abstract yet essential social relations and the concrete spaces and practices of everyday life. They are sociomaterial practices that bridge between "two epistemological moments within an ontological unity: one we experience – [sub]*urbanism* [the lived experience of suburban space] – the other we don't – [sub]*urbanization* [as a political-economic process] – but we know it really exists nonetheless" (Merrifield, 2002: 160). Mediatory infrastructures shape our knowledge and experience of broad social dynamics and relations. They open analytic avenues

to identify forces, spaces, and relations that might transcend the dialectical tensions between suburbanization (exchange-value) and suburbanism (use-value), and, in doing so, highlight the transformative capacity of infrastructures to puncture new centralities (that can be multiple, fragmented, and overlaid) into seemingly rote and homogeneous landscapes. The mediatory processes internalized in suburban infrastructures may also expose the ways in which suburban space is physically, discursively, and politically embroiled into the wider spatial and temporal dynamics of urban development. For example, suburban municipalities might draw on national infrastructure funds (such as those rolled out following the 2008 financial crisis) to improve local transportation systems and local economic competitiveness, or, conversely, be folded into policy and political discourses articulated at broader scales, as Cochrane, Colenutt, and Field (2015) argue in the case of housing in southeast England.

Individually, the epistemological vantage points offered by examining infrastructures as in, of, and for suburbs, and as internalizing suburbanization, suburbanism, and mediation, enable us to begin to unpack suburban infrastructures as complex concretions of spatially and temporally specific uses and social relations. We can abstract further insights by considering these triads in light of each other (following Harvey, 2006). The resulting nine-cell matrix, shown in table 2.1, discloses the intersections of distinct modalities, materialities, and social relations embedded within particular suburban infrastructures. As the suburban moment is perceived, conceived, and lived in partial and fragmented ways by different people at different moments, juxtaposing the multiple dimensions of suburban infrastructure presents alternative epistemological lenses to disclose the dialectical relations and points of tension emerging at the suburban-infrastructure nexus. Tensions can be temporal as well as spatial, both within and across cells. We can then theorize, for instance, transitions between prevailing infrastructures of suburbanization (for example, from "distributive" to "parasitic" urbanization [Beauregard, 2006]) or trace the infrastructural preconditions supporting emergent and competing ways of suburban living (see Walks, 2013). The content of the cells within this matrix are not exhaustive, and their specific content will depend on the particular theorization (of suburbs and infrastructure) and empirical case under investigation. In this light, it is useful to consider a concrete example of how infrastructure relates to suburban space and social practice.

Table 2.1 Matrix of suburban infrastructures

	Infrastructures of Suburbanization	Infrastructures of Suburbanism	Mediatory Infrastructures
Infrastructures in Suburbs	**Higher order infrastructures as they facilitate suburban expansion:** Splintered premium networks, bypasses (uneven development); National electricity grid, power cables, fibre optics, etc.; Infrastructures produced, maintained, and governed by higher order agencies/scales, but facilitating suburban expansion; Residual elements of previous spatial fixes, remnant space of Fordism	**Higher order infrastructures as they shape suburban life:** Post-suburban growth/mobility hubs (urbanity via densification); Car parks and big box retail power centres promoting new consumption practices; Residential university campuses; Greenbelts; Residual elements of previous spatial fixes (path dependent social practice); Infrastructures as alienating, (dis)connecting; Sites of risk, vulnerability, and opportunity	**Higher order infrastructures integrating suburbs into broader networks (and vice versa):** National highway networks; Airports; Trunk rail lines; Global logistics centres and intermodal terminals; Infrastructures as symbolic markers; Corporate headquarters/science parks/office campuses; Acts of bounding, enclosure, separation (within context of post-metropolitan, regional, and postcolonial urbanization)
Infrastructures of Suburbs	**Place-based infrastructures supporting suburban growth:** Streets, sewers, bus routes, etc., developed, maintained, and governed by local authorities; Claims over territory and growth-oriented politics; Special taxing districts; TIFs, tax breaks, and financial incentives for developers; Localized housing development (physical form) and planning codes (regulatory institutions); Rezoning	**Place-based infrastructures as they shape everyday spatial practice:** Suburban community and advocacy groups; Appropriation and reimagining of (formal and informal) built forms and institutions by suburban inhabitants; Implementation of informal sanitation systems in peripheral urban areas of the global south; Desire lines; Carpooling; Wired connectivity as community; Suburbanity as perceived, lived by suburban inhabitants; Gerrymandering	**Use of place-based infrastructures as spaces of mediation, centrality, difference:** Adapting strip malls for transnational cultural networking and events; Utilization of remnant spaces of Fordism for new, just-in-time practices (new territorialities and topologies); Position of suburban institutions in urban/global governance mosaic; Local partnerships to access national government financing; Inter-suburban economic competitiveness, attempts to locally capture global capital

(Continued)

Table 2.1 Matrix of suburban infrastructures (*Continued*)

	Infrastructures of Suburbanization	Infrastructures of Suburbanism	Mediatory Infrastructures
Infrastructures for Suburbs	**Sites and spaces of extended (sub)urbanization:** Reservoirs and pipelines in non-local watersheds; Auto-manufacturing centres, subsidies for cheap oil/gas; Institutions of financialization, mortgage companies; Private property rights and legal arrangements; Regional or national planning bodies and strategies; Federal/state support for homeownership, construction of new sustainable housing stock	**Extended infrastructures structuring suburban ways of life:** The development of political movements to address peripheralization, automobilities, etc., at multiple scales; Lobbying around the "war on cars"; Struggles over appropriate forms of transport, service provision; Regional commuter-sheds; Google buses; Commodification of distant resources (oil fields, rainforests) in order to meet demands of suburban lifestyles; Hollywood and US commercial film industry representations	**Extended infrastructures of suburban (dis)connectivity:** Suburbanity as relational; Integration into global flows for suburban capital; Mechanisms articulating suburban labour markets into wider networks; Topological connectivity; Co-constituted suburbs and the spaces they support; Expressway off-ramps; Resource wars; Global financial and regulatory agreements (coordinated through the IMF, OECD, EU, etc.); Potentiality of the "Right to the Suburbs"

Source: Addie, 2016

Reading Suburban Infrastructure: The Case of the Elgin, Joliet and Eastern Railroad

In October 2007, Canadian National Railway (CN) submitted an application to the US Surface Transportation Board (STB) to purchase the Elgin, Joliet and Eastern Railroad (EJ&E), a beltline railroad located approximately 40 miles (64 kilometres) from downtown Chicago. The EJ&E's tracks bisect the spectrum of the region's suburban fabric: passing at-grade through the predominantly affluent, white municipalities to the northwest of Chicago (including Lake Zurich and Barrington); the increasingly diverse satellite cities of Waukegan, Elgin, Naperville, and Joliet; the exurban fringes of DuPage, Kane, and Kendall counties; and the lower income and largely black industrial suburbs south of Chicago and in northwest Indiana. CN's application exposed the inherent tensions between Chicago's function as a regionalized global port and multifaceted space of habitation. Actors operating in and over this diverse suburban terrain related to the railroad in divergent ways. The contested politics of infrastructure they subsequently mobilized provides a constructive lens to illuminate the complexities and challenges of analysing and adequately theorizing suburban infrastructure.

Suburbanizing the Infrastructure of Globalization

CN's primary goal in acquiring the EJ&E was to use the railroad as a bypass to reroute intercontinental intermodal freight trains from the highly congested tracks converging on North America's historical rail hub. The deal, which was approved on 24 December 2008 and became operationally effective on 1 February 2009, firmly embedded the EJ&E within CN's continental network and attuned it to the scalar logics and economies of globalization. Its material holdings were incorporated into a more efficient and economically competitive cargo corridor linking the oil-rich Alberta tar sands to refineries and ports along the Gulf of Mexico. The purchase enabled CN to relocate switching operations to Indiana and convert their Gateway Yard in south suburban Harvey, Illinois, to a fully intermodal facility expected to accommodate an increase in containers handled from 350,000 to over 2 million per year. Moreover, the governance and use of the line would be dictated from Montreal rather than Gary, Indiana, in a move that further distanced it from the suburban communities through which it passed.

The CN takeover also meshed with a wider push towards railroad rationalization and modernization in Chicago that firmly recalibrated the EJ&E as an "infrastructure in suburbs" in the first instance. By the mid-2000s, the City of Chicago, the State of Illinois, and the Association of American Railroads had established the Chicago Region Environmental and Transportation Efficiency (CREATE) Program (CREATE, 2005) through a landmark multimodal public–private partnership. The $3.2 billion CREATE initiative forwarded a set of projects eliminating rail junctions and grade crossings within Chicago's municipal borders that, along with CN's EJ&E rerouting, would help reduce the negative externalities of freight activity in the heart of the global city. CREATE garnered the support of a broad coalition of regional interests, from business elites and public officials (including suburban representatives through the Metropolitan Mayors Caucus) to urban community groups who welcomed its proposed improvements to public safety, air pollution, and commuter rail service. In addition, the Chicago Metropolitan Agency for Planning (2010) advocated for a national vision and federal freight program, which would support improvements for regional goods movement and integrate freight needs into infrastructure prioritization as part of a commitment to improve freight policy.

Oppositional Suburbanisms

The picture was less rosy for the suburbs facing a dramatic increase in freight movement along the EJ&E right of way (from five to over twenty trains per day). In response, a coalition of suburban communities (including municipal and county officials from northeastern Illinois and northwestern Indiana) organized as The Regional Answer to Canadian National (TRAC) to oppose the EJ&E acquisition and ensure that both CN and the STB adequately addressed the adverse effects of the deal on their local interests. While TRAC contested claims regarding the beneficial economic and employment contribution the acquisition would have for the region, their chief concerns reflected the changing impact of the railroad as an infrastructure of suburbanism. These concerns manifested around issues of noise and air pollution, increased delays at the 133 at-grade crossings along the line, public safety and health risks, disruption to commuter rail service, and depreciated property values. For Cidell (2015), this reactionary conservativism, particularly among affluent communities northwest of Chicago, represented something more than residents dealing with the negative impacts to their everyday spatial practices. Rather, the suburbanization of freight rail cracked the social and political fallacy of a disconnected and autonomous mode of suburbanism:

> The noise and emissions of CN trains would be a reminder that they [suburbanites who had fled from the hassle and congestion of the city] are still part of an urban area, with all the economic and social inequality that entails, as much as they have tried to avoid it. Moreover, a decline in property values might make it possible for previously excluded people to afford the same sanctuary, reducing its exclusivity and desirability. (Cidell, 2015: 145)

Yet, public officials from the booming satellite town and edge cities adjacent to the EJ&E also had reason to object to the impact of the CN's use of the EJ&E as an infrastructure of suburbanization, which threatened their urbanizing development agendas. Delays at railroad crossings and rising levels of air and noise pollution are not conducive to enhancing the economic attractiveness of aspiring suburban municipalities. Importantly, elevating global and trans-continental freight movement over local mobility regimes jeopardized plans for the Suburban Transit Access Route (STAR Line), a proposed regional commuter rail service connecting major regional employment centres and satellite towns from Joliet, Aurora, and Elgin to O'Hare International Airport using sections of the EJ&E, which had received popular backing across the Chicago region.

Reclaiming Suburban Infrastructure

In contrast to TRAC's vehement opposition, the EJ&E takeover posed a more complex question in Chicago's south suburbs. These communities faced comparable traffic delays, safety concerns, and disruptions to their everyday lives, but CN's investment in local intermodal facilities presented an opportunity to transform their economic prospects (as an infrastructure of suburbanization) and quality of life (as an infrastructure of suburbanism). Here, the South Suburban Mayors and Managers Association (SSMMA) – an intergovernmental agency providing technical assistance and collaborative services to forty-two municipalities in southern Cook and Will counties – in partnership with the Center for Neighborhood Technology (CNT), the Metropolitan Planning Council, and the Delta Institute developed proposals to reimagine and reposition the economically distressed south suburbs of

Chicago as a green manufacturing cluster (Center for Neighborhood Technology [CNT], 2010). The Southland Chicago Green TIME Zone seeks to leverage the region's extant transportation infrastructure and manufacturing facilities to generate 13,400 new jobs, $2.3 billion in new income, and $232 million in new tax income for the Chicago region (CNT, 2010: 14). At its core, discursively and physically, sits CN's expanded Gateway Terminal and the increased cargo flowing along the EJ&E right of way. The proposal's targeted industrial core, Logistics Park Calumet, contains in excess of 1,300 acres of vacant or underutilized land within four miles (6.4 kilometres) of Gateway Terminal (CNT, 2010: 6).

In order to produce desirable neighborhoods, high skilled employment opportunities, and environmental improvements, the Green TIME Zone forwards a tripartite strategy integrating (1) transit-oriented development (TOD); (2) cargo-oriented development (COD); and (3) green manufacturing. Each of these mechanisms is dependent upon its own extended infrastructures of, and for, the south suburbs. TOD needs coordinated zoning practices and land banking to direct growth and curtail entrenched modes of sporadic sprawling suburbanization. Corridor development and community stabilization therefore relies on high levels of political engagement and collaboration between municipalities to address perceptions of a zero-sum competition for inward investment among suburban municipalities. COD seeks to capture the economic benefits and positive externalities of intermodal freight movement by attracting companies looking to take advantage of reduced shipping costs and greater reliability. The development of both hard and soft mediatory infrastructures is necessary for a project like Logistics Park Calumet to effectively integrate local markets into broader international accumulation regimes. These include investments in the built environment to remove spatial barriers to accumulation and the creation of institutional and regulatory infrastructures capable of handling foreign trade (customs inspection stations, and so on) while ensuring firms comply with environmental remediation standards. Green manufacturing, in turn, requires the formation of an economic development infrastructure to support regional supplier integration and production capacity, and to provide workforce training to refocus the local skills base towards alternative energy production, as well as marketing strategies to promote the Green TIME Zone cluster internationally. All, however, depend upon the mobilization of a cross-sectoral and multi-governmental financial infrastructure. To this end, Scott Bernstein, co-founder and president of CNT, notes:

> The [Green TIME Zone] has also created a fund to help finance the land acquisition and pre-development infrastructure costs associated with cargo-oriented and transit-oriented redevelopment; Cook County used a HUD Section 108 loan guarantee against future Community Development Block Grant apportionments for a cargo- and transit-oriented land bank; and the Illinois General Assembly approved a new kind of tax increment financing [TIF] for the target area against the income tax anticipated from new jobs created, in contrast with the typical TIF against anticipated property tax receipts. (Bernstein, 2013: n.p.)

The diverse fiscal mechanisms and scales being mobilized here clearly demonstrate that any attempt to locally reclaim this logistics and distribution landscape is dependent upon the contingent nexus of globally oriented rail infrastructure and existing, but underutilized, industrial capacity (infrastructure in suburbs) and a far-reaching constellation of place-based infrastructures supporting suburban growth (infrastructure for suburbs) to realize the foundations for sustainable local economic development (infrastructure of suburbs).

The EJ&E as Suburban Infrastructure

Following the conceptual framework laid out earlier, we can represent what is distinct about the EJ&E as suburban infrastructure in table 2.2. While the EJ&E takeover bolstered both CN's global logistics network and discourses of resilient regional competitiveness in Chicago, the use and governance of the EJ&E became codified, politically and discursively, as a suburban issue. Suburban communities would be the ones experiencing the disruptions caused by the suburbanization of global logistics activities, but they could also be the beneficiaries of new modes of suburbanization catalyzed by CN's investment, the mediatory dimensions of suburban infrastructure reconfiguring the development potential, ground rents, and economic base of declining industrial inner suburbs. The competing visions and practices of global freight movement, local mobility, and the lived experience of suburban space account for the notion of suburban infrastructure as highly imbricated across space and between scales. Those looking to appropriate the EJ&E for local economic development purposes, consequently, have

Table 2.2 The EJ&E Railroad as suburban infrastructure

	Infrastructures of Suburbanization	Infrastructures of Suburbanism(s)	Mediatory Infrastructures
Infrastructures in Suburbs	Physical rail and road infrastructure, intermodal yards, warehouses; Investment in Gateway Terminal opening employment opportunities and local revenue streams; Built environment of extended metropolitan growth; Potential sites for industrial development; Transformations in built form, urban morphology, increasing ground rents for industrial activity	Traffic delays at at-grade crossings shaping commuting patterns from Barrington to NW Indiana; Negative impacts on quality of life, noise and air pollution, impact on property prices; Questions of safety and risk of freight movements; Positive impacts on quality of life, potential for economic development, improvement in housing stock, and opportunities for industrial retraining	CN as continental rail infrastructure integrating local intermodal and distribution facilities and global economy; Rail infrastructure providing goods and flows that support the Chicago region's growth, attractiveness, and quality of life; Embedding EJ&E purchase within broader regional railroad politics (CREATE), removing unwanted freight activity from city to the suburbs, psychological connection/ fragmentation of regional space
Infrastructures of Suburbs	Local attempts to claim the development trajectory of the south suburbs (SSMMA and the Green TIME Zone); Municipal tax breaks for green manufacturing companies; Housing development (physical form) and planning codes (regulatory institutions) to support TOD and COD; Opposition to modes of suburban	New sociospatial centralities around housing and employment centres; Potential of the STAR Line and new commuting practices utilizing the EJ&E right of way; Privatist responses to CN purchase; Community mobilization, proactive responses around potential regeneration effects, defensive retrenchment around	Green manufacturing as local economic development strategy to repurpose local skills and industrial expertise in a sustainable global economy; Localized institutional partnerships to facilitate access to federal government and global capital (financial infrastructure needed to support the Green TIME Zone); Logistics Park Calumet; Competition between suburbs for stake

	growth that do not boost local economic activity	preserving extant suburban ways of life (The Regional Answer to Canadian National)	in global logistics activities, retailing, manufacturing, etc.
Infrastructures for Suburbs	Global capital and financing to support the CN purchase; Political support from state and national governments; US Departments of Transport, HUD funding for Green TIME Zone; National think tanks (CNT) and mobile policy lessons supporting the case for green manufacturing districts, TOD, COD, etc.; Arguments for a national freight program	Regional support for STAR Line to foster non-radial, non-automobile suburban transit options; Financial mechanisms financing TOD, subsidizing development industry and continued sprawl, financial infrastructure supporting individual homeownership as they shape everyday spatial practice	Relational construction of Southland Chicago as suburban space structured by global economic and logistics activity; Contested politics of scale surrounding the right to utilize global infrastructure "in" suburbs for local benefit; Concrete articulation and experience of globalization, time-space compression, economic opportunity, and the disciplinary logics of capitalism

Source: Addie, 2016

needed to walk a fine line between the fractious politics of their local stakeholders and the disciplinary logics of globalization that had render the CN sale a fait accompli. Analytically, the ability to dialectically read across the cells of the matrix in a non-hierarchical manner proves particularly important here, as it not only prohibits focusing on a single modality but also demonstrates how different social groups can hold widely differing perspectives on, and support differing politics of, suburban infrastructure. There is no singular "suburban" moment and perspective. Rather, there is an ongoing negotiation over the production, knowledge, and appropriation of both infrastructural artefacts and the urban process more broadly. As a result, the EJ&E, as a suburban infrastructure, is revealed to be an active element in the production and articulation of contested suburban spaces and social practices, but one whose internalized processes and relations are understood to be readily comparable to other suburban infrastructural constellations and political struggles over their provision, governance, and use.

Conclusion

Through this chapter, I have argued that what is held in common between diverse suburban environments and distinct infrastructural systems are the (sub)urban relations internalized within particular suburban infrastructures. In theorizing these relations through a complementary framework of *infrastructures in/of/for suburbs* and *infrastructures of suburbanization/suburbanism/mediation*, I have outlined an approach to recognize and engage the unpredictable and overdetermined nature of both suburban infrastructure and suburban space. This approach is not to suggest we arrive at a normative, essential, or readily transferable definition of "suburban infrastructure." Particular infrastructures are characteristically multifaceted and multiscalar entities constructed by complex governance regimes, contested by diverse stakeholders, and generative of distinct social norms. Their concrete articulations are highly varied and experienced in divergent ways by different people. As a result, we require flexible conceptual and comparative tools capable of adapting to the distinct ways in which infrastructures are constructed as problems and potential solutions within the polycentric milieu of global suburbanization. Focusing on the relations between infrastructures and their suburban moment directs investigations towards common and transferable abstractions, founded upon sociospatial relations rather than on the contingent attributes of artefacts and systems in isolation.

This framework forms the basis for robust comparative theory across pipelines, sanitation systems, cultural norms, and governance institutions and edge cities, post-suburbs, in-between spaces, or ethnoburbs in the global north or south. Taking seriously the question, In what sense suburban infrastructure? also foregrounds the political dimensions of such analysis within and across cases. Unpacking the unequal power relations and differential knowledges of suburban infrastructure through the framework presented earlier elevates issues of scale and centrality in the study of suburban infrastructures. In doing so, we are pushed to consider how actors operating across multiple scales articulate and operationalize claims to suburban infrastructures in practice, and how we might reimagine them to claim the right to suburbs.

The chapter has provided a framework for the entire volume by exposing the complexity of the relationship between infrastructures and suburbs. It challenges the superficial assumption whereby, because they are in suburbs, infrastructures are necessarily intended to serve these suburbs. As the present chapter has demonstrated and subsequent chapters will confirm, the purpose of infrastructures located in suburbs can be of a metropolitan scope or indeed be part of continental or global networks. In these circumstances, the nature of their impact on their suburban settings is uncertain. In certain instances, they can be sources of negative externalities, while in other situations they can generate benefits for their suburban sectors, as when they constitute sources of employment. Three subsequent chapters mirror this ambiguous relation between infrastructures located in suburbs and the suburban realm, brought to light in the present chapter. Chapter 12 discusses the broad impact of the Toronto region greenbelt, well beyond its effect on abutting suburbs. The greenbelt is situated in suburban Toronto, but one of its purposes is to contain metropolitan expansion and foster the intensification of the region, including that of the suburban realm. The subsequent chapter, chapter 13, describes the contemporary Chinese suburban infrastructure context, which is characterized by infrastructure development driven more by macroeconomic objectives than the actual needs of the suburbs. Chapter 15 documents the largely unintended social consequences of public transportation improvement.

NOTE

The argument presented in this chapter draws in part from Addie (2016).

REFERENCES

Addie, J.-P.D. (2016). Theorising suburban infrastructure: A framework for critical and comparative analysis. *Transactions of the Institute of British Geographers, 41*(3), 273–85. https://doi.org/10.1111/tran.12121

Angelo, H., & Hentschel, C. (2015). Interactions with infrastructure as windows into social worlds: A method for critical urban studies. *City: Analysis of urban trends, culture, theory, policy, action, 19*(2–3), 306–12. https://doi.org/10.1080/13604813.2015.1015275

Beauregard, R.A. (2006). *When America became suburban*. Minneapolis, MN: University of Minnesota Press.

Bernstein, S. (2013). Industry, infrastructure and intermodalism – Still mixed up on special districts? *Placemakers*. Retrieved from http://www.placemakers.com/2013/07/15/industry-infrastructure-and-intermodalism-still-mixed-up-on-special-districts/

Brenner, N. (Ed.). (2014). *Implosions/explosions: Towards a study of planetary urbanization*. Berlin: Jovis Verlag.

Center for Neighborhood Technology (CNT). (2010). *Chicago Southland's Green TIME Zone*. Chicago: Author.

Charmes, E., & Keil, R. (2015). The politics of post-suburban densification in Canada and France. *International Journal of Urban and Regional Research, 39*(3), 581–602. https://doi.org/10.1111/1468-2427.12194

Chicago Metropolitan Agency for Planning (CMAP). (2010). *Go to 2040 plan*. Chicago: Author.

Chicago Region Environmental and Transportation Efficiency (CREATE). (2005). *Chicago region environmental and transportation efficiency program*. Springfield, IL: Federal Highway Administration, Illinois Department of Transportation.

Cidell, J. (2011). Distribution centers among the rooftops: The global logistics network meets the suburban spatial imaginary. *International Journal of Urban and Regional Research, 35*(4), 832–51. https://doi.org/10.1111/j.1468-2427.2010.00973.x

Cidell, J. (2015). Uncanny trains: Cities, suburbs, and the appropriate place and use of transportation infrastructure. In J. Cidell & D. Prytherch (Eds.), *Transport, mobility, and the production of urban space* (pp. 134–51). New York: Routledge.

Cochrane, A., Colenutt, B., & Field, M. (2015). Living on the edge: Building a sub/urban region. *Built Environment, 41*(4), 567–78. https://doi.org/10.2148/benv.41.4.567

Cowen, D. (2014). *The deadly life of logistics: Mapping violence in global trade*. Minneapolis, MN: University of Minnesota Press.

Ekers, M., Hamel, P., & Keil, R. (2012). Governing suburbia: Modalities and mechanisms of suburban governance. *Regional Studies, 46*(3), 405–22. https://doi.org/10.1080/00343404.2012.658036

Gandy, M. (2003). *Concrete and clay: Reworking nature in New York City.* Cambridge, MA: MIT Press.

Graham, S. (Ed.). (2010). *Disrupted cities: When infrastructure fails.* New York: Routledge.

Graham, S., & McFarlane, C. (Eds.). (2015). *Infrastructural lives: Urban infrastructure in context.* New York: Routledge.

Harris, R. (2010). Meaningful types in a world of suburbs. In M. Clapson & R. Hutchinson (Eds.), *Suburbanization in global society* (pp. 15–48). Bingley, UK: Emerald Group Publishing.

Harvey, D. (2006). Space as a keyword. In N. Castree & D. Gregory (Eds.), *David Harvey: A critical reader* (pp. 270–94). Malden, MA: Blackwell.

Hayden, D. (2003). *Building suburbia: Green fields and urban growth, 1820–2000.* New York: Pantheon.

Höhne, S. (2015). The birth of the urban passenger: Infrastructural subjectivity and the opening of the New York City subway. *City: Analysis of urban trends, culture, theory, policy, action, 19*(2–3), 313–21. https://doi.org/10.1080/13604813.2015.1015276

Keil, R. (Ed.). (2013). *Suburban constellations: Governance, land and infrastructure in the 21st century.* Berlin: Jovis Verlag.

Keil, R., & Young, D. (2008). Transportation: The bottleneck of regional competitiveness in Toronto. *Environment and Planning C: Government and Policy, 26*(4), 728–51. https://doi.org/10.1068/c68m

Law, M.J. (2012). "The car indispensable": The hidden influence of the car in inter-war suburban London. *Journal of Historical Geography, 38*(4), 424–33. https://doi.org/10.1016/j.jhg.2012.04.005

Lefebvre, H. (1991). *The production of space.* Oxford, UK: Blackwell.

Lefebvre, H. (2003). *The urban revolution.* Minneapolis, MN: University of Minnesota Press.

McFarlane, C., Desai, R., & Graham, S. (2014). Informal urban sanitation: Everyday life, poverty, and comparison. *Annals of the Association of American Geographers, 104*(5), 989–1011. https://doi.org/10.1080/00045608.2014.923718

McFarlane, C., & Rutherford, J. (2008). Political infrastructures: Governing and experiencing the fabric of the city. *International Journal of Urban and Regional Research, 32*(2), 363–74. https://doi.org/10.1111/j.1468-2427.2008.00792.x

Merrifield, A. (2002). *Dialectical urbanism: Social struggles in the capitalist city.* New York: Monthly Review Press.

Peck, J. (2011). Neoliberal suburbanism: Frontier space. *Urban Geography, 32*(6), 884–919. https://doi.org/10.2747/0272-3638.32.6.884

Phelps, N.A., Wood, A., & Valler, D. (2010). A post-suburban world? An outline of a research agenda. *Environment & Planning A, 42*(2), 366–83. https://doi.org/10.1068/a427

Sayer, A. (1992). *Method in social science*. New York: Routledge.

Simone, A.M. (2004). People as infrastructure: Intersecting fragments in Johannesburg. *Public Culture, 16*(3), 407–29. https://doi.org/10.1215/08992363-16-3-407

Swyngedouw, E. (2006). Circulations and metabolisms: (Hybrid) natures and (cyborg) cities. *Science as Culture, 15*(2), 105–21. https://doi.org/10.1080/09505430600707970

Walks, R.A. (2013). Suburbanism as a way of life, slight return. *Urban Studies, 50*(8), 1471–88. https://doi.org/10.1177/0042098012462610

Walks, R.A. (2015). Stopping the "war on the car": Neoliberalism, Fordism, and the politics of automobility in Toronto. *Mobilities, 10*(3), 402–22. https://doi.org/10.1080/17450101.2014.880563

Warner, S.B. (1978). *Streetcar suburbs: The process of growth in Boston, 1870–1900*. Cambridge, MA: Havard University Press.

Young, D., Wood, P., & Keil, R. (Eds.). (2011). *In-between infrastructure: Urban connectivity in an age of vulnerability* Kelowna, BC: Praxis (e)Press.

3 Rescaling the Suburban: New Directions in the Relationship between Governance and Infrastructure

JANICE MORPHET

Introduction

The disruption inherent in the separate governance arrangements between cities and their suburbs is at the heart of their spatial differentiation. Cities and the suburbs are the "other" to each other. Yet, their prosperity and social existence are co-dependent. This mutuality allows for operational solutions and "work arounds," but their differentiation has remained central to their identity – until now.

Governments are now challenging city/suburban separation and directing public policy to reframe these long-standing differences in national interests. Cities are sustainable and, despite their existing densities, can accommodate more people, businesses, and services. Cities survive when they are able to cope with change. The financial crisis has shown that the sunk costs of infrastructure investment in cities cannot be abandoned for new investment on greenfield sites at the urban periphery. Infrastructure investment is important where it provides additionality to what already exists, and countries are ranked on their ability to create the governance and operational climate that supports infrastructure investment within and between cities (World Economic Forum [WEF], 2015).

After years of disruptive decline, cities are also returning to their role in generating dynamic economic clusters, agglomerations, specializations, and hubs that are again promoting growth and networks. Cities concerned with attracting investment and jobs are anxious to compare their incentives with those of competitors (Greater London Authority [GLA], 2015). In Western countries, the economic pull of cities that initially stimulated their growth is being exerted again. The flight to the suburbs was supported by unequal economic and culturally differentiated roles in the household. Now, time is at a premium in two-career households, which cannot afford to waste time commuting. These contemporary households, as well as active retirees who prefer to live in dynamic communities, want "edgy" rather than "edge" lifestyles. Some suburbs and out-of-town businesses are no longer viable. What happens when the suburbs are the location of choice for the economically marginalized, who travel long distances to work and have no time or money for community-focused lives?

Cities and suburbs are being driven to change how they relate to one another in a multitude of ways. Government policies are increasingly influenced by international organizations such as the United Nations (UN), the Organisation for Economic Co-operation and Development (OECD), and the European Union (EU). There is a clear need for cities to accommodate demographic shifts towards urban centres, while the long-standing culture of suburban separation is harder to change. Top-down reforms in administrative boundaries can be mired in opposition campaigns from those whose identity is defined by the existing separation. Changes have to be the result of independent evaluations or nudged through incentives to create outwardly cooperative forms of governance that can hold and retain the differences within them. Cities and suburbs working together can frame a political agenda based on the mutual benefits of improved access and infrastructure investment in order to promote megaprojects or tackle "not-spots." In this new relationship, the role of infrastructure is critical to the arguments for, and the acceptability of, change. This chapter considers the drivers for new directions in public policy and the way in which governments are responding to them at the local level.

Changing Relationships between Cities and Their Suburbs

The examination of the relationships between the urban core and its suburban periphery demonstrates that these connections are frequently beset by political conflict and opposing objectives (Phelps, Vento, & Roitman, 2015). Cities are concerned with growth, not least as they demand more space for both wealth-generating activities and the needs of the people employed to support the economic system. In all parts of the world, successful cities are magnets for people, and housing availability frequently lags behind the pressures of in-migration (Accetturo, Manaresi, Mocetti, & Olivieri, 2014). In contrast, suburban areas are characterized as areas that are in conflict with city objectives (Matthews, Bramley, & Hastings, 2015). The suburban is the escape from the city, where lower residential density equates to a perception of higher standards of living. People have traditionally been willing to commute longer distances to achieve the benefits of suburban living while continuing to work in the city (Headicar, 2015). The opposition between the city and its suburbs is frequently compounded by different governance regimes. Cities develop to the edge of their administrative boundaries and then rely on the suburban populations for their economic survival. Suburban political regimes survive only through their oppositional stances to the city's outward pressures.

However, these patterns of city and suburban relationships are now coming under threat for a variety of reasons. First, many younger people no longer find cars to be as culturally acceptable as their parents did. Fewer young people are obtaining driver's licences, and many retain their early reliance on public transport or bicycles into adulthood (Goodwin, 2012). As Filion (2013) points out, this trend may not reduce car ownership, and the suburbs will likely rely on car-based access in the short to medium term, but suburbs may change in the future as successive generations make different locational choices for long-term living.

Second, suburban patterns of commuting survive best where there is an unequal economic income relationship between partners in households, where the highest paid household member bears a longer commute and the lowest paid partner is more likely to find part-time or local work in order to support the domestic demands of the household and children (Wheatley, 2012). The emergence of dual-career households with more equal incomes and job status means that neither partner can fully perform this domestic role at a distance with ease. As income levels in dual-career families equalize, journey to work reduces for both partners (Pickup, 1978). The mechanism for finding more time to manage job and home has been through a reduction in commuting. Households with more equal jobs are likely to be found working nearer the city centre and remaining there if there have children. Couples who minimize their travelling time have contributed to a revival of inner-city housing, services, and jobs (Karsten, 2003). This reversal of travel and living patterns has also been accommodated through the gentrification of the original inner suburbs, which were abandoned earlier for suburban flight (Boterman & Karsten, 2014; Goodsell, 2013; Rérat & Lees, 2011; Skaburskis, 2012). Land formerly used for employment has been repurposed to build more houses and apartments. These new city-centre apartments are not only for the young. In cities such as Liverpool and Toronto, these apartments are also purchased by retirees returning from the suburbs.

Changing household and domestic patterns have also meant that the suburbs may no longer be so attractive. Lower densities, the development of "edgetowns" (Garreau, 2011), and ever longer car commutes to the city are changing the role of the suburbs. Suburbs are under pressure to change through intensification, increasing economic development, and providing new service hubs to support the expansion of public transport. Elsewhere, the role of suburbs is changing entirely: in San Francisco, Silicon Valley workers live in the city and are bused out each day to their work campuses on the suburban periphery, while in other areas abandoned retail malls are being adaptively reused for other purposes such as education (Stabiner, 2011).

Thus, the suburbs are in transition (Grant, 2013), with the certainties of their traditional role now being eroded by changes that are out of their control, such as cultural shifts and changes in urban organizational patterns. While Nelson (2013) argues that these changes are most apparent in American suburbs, which tend to be less densely developed than those in other countries, the factors inducing these changes are visible elsewhere. Once characterized by their recognizable economic homogeneity, the suburbs are now equally characterized by extremes of poverty and wealth. Indeed, as Forsyth (2013) points out, there are now more poor people living in the US suburbs than in core cities.

Rescaling the Suburban – Policy Provenance

The changing dynamic between the city and its suburbs has been the focus of wider public policy changes that can be evidenced by three shifts in the understanding of the relationship between place governance and economic growth (Department for Communities and Local Government [DCLG], 2008c; Her Majesty's Treasury [HMT], 2003).

The first change is the adoption of more sustainable principles and their application through world treaties and agreements. This modification has resulted in pressures to reduce energy use, transport emissions, and material consumption worldwide. The principles of producing and consuming locally, as agreed in the 1992 UN Conference on Environment and Development (also known as the Rio de Janeiro Earth Summit), have privileged cities over other spatial forms.

Prior to this agreement, many large cities were in population and economic decline, with businesses and residents migrating to lower density suburbs for out-of-town retailing and campus workplaces. Now, cities have re-established their economic advantages of agglomeration and reinforced cultural changes in households. For example, improvements in public transport have encouraged urban growth through densification and reduced private car use (Goodwin, 2012; Metz, 2013). However, the sometimes complex governance of the supply and management of public transport can prevent these new economies from thriving and reduce city advantages of efficiency, as found in the wider Chicago Tri-State metropolitan area (Merk, 2014).

The second development is a growing concern among countries regarding their economic growth and gross domestic product (GDP). In the past, external trade between nations has been the primary focus of governments to drive economic growth. While the World Trade Organization (WTO) was set up to negotiate trade terms between countries, regional organizations such as the European Union (EU), the Association of Southeast Asian Nations (ASEAN), the North American Free Trade Agreement (NAFTA) superseded in 2018 by the United States-Mexico-Canada Agreement (USMCA), and the Southern Common Market (*Mercado Comum do Sul;* MERCOSUR) have also been created to promote trade between geographic groups of states at advantageous terms. However, Krugman (1991; 2011) found that, while trade between countries remains important to their economies, internal trade within countries contributes equally to the GDP. He states that governments focusing only on external trade are more likely to underperform economically. Further, Krugman argues, internal trade is most successfully supported through major urban areas and cities, which offer the most efficient economies through the benefits of agglomeration. The most successful cities are those that have a critical economic mass and enjoy infrastructure efficiencies, such as public transport, which can efficiently link urban and suburban labour markets.

Critics of this approach (Brenner 1999; 2004) have suggested that to realign the state to support the needs of the economy is placing the interests of business above the people. Further arguments assert that these newly created larger substate spaces are managed through non-democratic means such as partnerships or stakeholder boards, which again favour the needs of business. Others argue that these fuzzy, economically based governance arrangements have been the means through which central government can centralize control over local spaces and assert decision-making power that should otherwise be taken locally (Allmendinger & Haughton, 2010; Porter & de Roo, 2012). Jessop (1997) has suggested that these new governance systems are spatiotemporal "fixes," although Pemberton and Morphet (2014) have categorized them as a process of transitional territorialism rather than an end state.

Despite these criticisms, the realignment of internal administrative boundaries to mirror economic spaces has taken hold of the policy agenda of international economic organizations including the OECD, International Monetary Fund, World Bank, and World Economic Forum. The policy has been developed to focus on the role of cities and their areas of economic activity within a given country's economy. Rather than defining a city by its administrative boundary, there is now a move to redefine cities by their journey-to-work areas and the relationship of cities with others within polycentric clusters (Dijkstra & Poelman, 2012; Organisation for Economic Co-operation and Development [OECD], 2013). These commuting boundaries are then proxies for labour market areas (DCLG, 2008b; 2008a). Including the wider city commuting area into a consideration of the city immediately starts a new relationship between the cities and their suburbs. These adjusted boundaries may be termed "functional economic areas" (FEAs) or sometimes "functional urban areas" (FUAs). Consideration and evaluation of FEAs suggest that cities and their peripheries would be better managed if their administrative boundaries were aligned to their economic boundaries rather than being in opposition, enabling issues such as transport and housing to be managed in a more sustainable and efficient way. This policy agenda is framed to combine the cities and their suburbs in new formal relationships, while maintaining some of the territorial integrity and character of both in these new entities.

Although numerous objections to the neoliberal aspect of rescaling administrative spaces to align with functional economic areas do exist, this policy approach has been adopted across the OECD (Gurria, 2014) as well as within the EU (Dijkstra & Poelman, 2012). The OECD has promoted the integration of territory and governance through FEAs in two specific ways. First, the OECD has found that the productivity of FEAs is enhanced where there is an alignment in the administrative boundaries and those of the FEA (Ahrend, Farchy, Kaplanis, & Lembcke, 2014). A study of five member countries (Germany, Mexico, Spain, the United Kingdom, and the United States) found that productivity increased with city size, and that cities with more fragmented governance within their FEAs had lower productivity than those with a unified government for the FEA. Furthermore, the study revealed that cities benefited where they had access to other agglomerations. In another study, it was found that decreasing travel time for populations within FEAs led to a net gain in productivity for the area as a whole, and that halving travel time gave the equivalent of a 0.2 to 0.4 percentage points increase in annual per capita growth (Ahrend & Schumann, 2014).

The second avenue OECD has found to promote integrated FEAs is through a series of studies on the economies of cities and their leadership. One OECD study of economic leadership in Amsterdam, Hamburg, Manchester, and Stockholm (OECD, 2015) found that the fragmented nature of cities and their multiplicity of stakeholder interests has become more complex and requires greater clarity in leadership styles and messages for the cities to be successful. The plethora of public sector organizations that manage the wider city area are regarded as difficult to comprehend by investors and businesses, and their fractured governance reduces the scale of their impact on central government. The OECD concluded that streamlining and aligning governance across an FEA would bring a leadership dividend greater than the existing forms of governance.

The OECD has undertaken nearly one hundred territorial reviews of specific urban areas in their member countries since 2001, including the Chicago Tri-State metropolitan area in 2012, Toronto in 2009, and Newcastle in 2004. Each of these studies demonstrates why the governance of the FEA is hampering its economic growth and its contribution to the country's economy. These reviews also reinforce the model of FEA-wide governance by providing examples of comparable areas that have integrated governance. In the most recent study of Rotterdam–The Hague, the Dutch government was directed towards governance models integrating the city and its suburbs, as in Manchester, Barcelona, and Singapore, and advised to adopt a strategic planning approach similar to that used for London.

This use of comparisons within the OECD member countries to exemplify policy messages and practices is also reinforced through benchmarking between FEAs. The OECD benchmark criteria are not only confined to size of population and economy, but also include other factors such as FEA governance models. The scale of independence and devolution from central government control, and the degree to which there is vertical and horizontal alignment in the objectives for the area as agreed by governance bodies and stakeholders, are also shown as being contributory factors to their economic success (Charbit, 2011). In addition, the

OECD has worked with the EU to identify a new definition of functional economic areas (Dijkstra & Poelman, 2012), which is now used to examine and compare cities across its membership. By 2014, 50 per cent of the OECD's population lived in 275 FEAs, and these contributed more than 50 per cent of the OECD's GDP and employment (Brezzi & Veneri, 2014). The role of FEAs in developing polycentric groups that include multiple cities is also exemplified as a way of maximizing economic growth and GDP. By 2014, over 50 per cent of the OECD's member countries were reforming substate governance in this direction (Gurria, 2014).

However, this focus on further developing the economic capacity of cities is constrained by limits to agglomeration created by transport congestion for people and goods (Hamza, Frangenheim, Charles, & Miller, 2014). The potential for economic growth in cities also depends on access to a widening labour market and supporting this market through other infrastructure, including utilities. As such, increased investment in improving existing infrastructure within the whole of the city's economic area, including for those living within the suburban labour market area, is critical. Where cities are currently bounded by areas of social exclusion on their suburban peripheries, this infrastructure investment becomes more pressing.

The third driver for city redefinition reflects a concern of Western countries about the relative scale and growth of cities within the BRIC (Brazil, Russia, India, China) countries, where economic challenges are of growing concern. The redefinition of cities in OECD member countries to include both city FEAs and polycentric urban groupings will increase the number of cities that can be defined in comparable ways to those in the BRIC countries. Also, where city boundaries include the suburbs, this redefinition may also encourage infilling and growth in ways that support more infrastructure investment and a growing labour market. In Europe, only London and Paris can be defined as global megacities; and overall the role of cities has been prioritized through a new urban agenda that is being developed to respond to these perceived challenges (Commission of European Communities [CEC], 2015). This new agenda will include a system of urban benchmarking using social, sustainable, and economic criteria.

Statecraft and Scalecraft

The international pressures to reform and rescale internal substate government and its boundaries are considerable, although each senior level administration will approach these external pressures in different ways. An understanding of the approaches of central governments in achieving these changes is informed by the literature on statecraft together with its emerging alliance with scalecraft as a means of analysing substate governance rescaling (Fraser, 2010; Pemberton & Searle, 2016). This rescaling is located in different cultural contexts, experiencing its intermediate slippery stages or "transitional territorialism" (Pemberton & Morphet, 2014) to realignment.

Statecraft refers to how the state implements international agreements (Bulpitt, 1986). Its characteristics have been described as a conscious gaming strategy, where the application of international agreements is used to achieve domestic objectives (Buller & James, 2012). Statecraft also frames and narrates the way in which policies are implemented. The dominant national culture can also be characterized by singularity, that is, any state may differ from other states. Statecraft suggests that while central governments are considering the implementation of substate rescaling, they will also be considering harnessing other political or government objectives to the policies in a process of "gold-plating." At the very least, central governments will be incorporating the international policy into a domestic agenda that aligns with objectives, cultural norms, and practices within the state.

While statecraft offers one avenue to consider for the application of international policy agendas and their implementation (Kingdon, 1984), the more recent focus on functional economic areas and their contribution to GDP suggests that there is also a need to consider scalecraft as part of this approach (Ahrend, Farchy, Kaplanis, & Lembcke, 2014). Scalecraft may not be determined by the geographies it encompasses, but rather recognizes that geographies are political constructs in support of wider state purposes. As Fraser (2010) points out, scalecraft is concerned with the destruction of spatial entities as well as the creation of new ones. New governance scales may be created in response to evidence or changed external political priorities, such as the global economy. They may be constructed to create greater citizen involvement and meet considerations of democratic engagement. New scales may be chosen simply because they are different from what has gone before and, in their creation, provide an opportunity to reprioritize policies or meet other objectives.

Scalecraft can be considered as a component or tool of statecraft, but this sub-classification does not take into account the strength of the institutions that may be created within scalecraft initiatives. As Richards

(2015) points out, the establishment of new governance scales in the United Kingdom, such as those in London and other places where mayors are directly elected for functional economic areas, may create new economic opportunities as well as challenges. As these roles mature and rescaled governance takes hold, substate structures can arise that are more equal to central government (Richards, 2015). The economic power of these rescaled structures can grow, both in their own larger economies as well as in central governments' reliance on them to drive national economies. They may act as mechanisms to develop more individual and locally determined priorities, where the mayor or leader is not a supplicant to government but has the power to either implement these solutions directly or hold the national government to ransom in order to achieve local decisions.

Power at the substate level is dependent upon the integration of relationships between cities and suburbs. Their relationships, reinforced through increased infrastructure investment as an outward manifestation of the benefits of combining cities and suburbs, may be used as a means to justify rescaling. However, in the short term, there are likely to be a range of statecraft/scalecraft intersections where policies meet unevenly and result in differing outcomes. Thus, in the long term, rescaling may mean that the relative power relations between central and substate areas change.

Rescaling the Suburban – Applying an International Policy

The development of a new international policy for rescaling the state and the way that different governments respond through the application of statecraft/scalecraft has led to a variety of substate governance modes in practice. The application of new city/suburban governance institutions is evident across the OECD area. As Gurria (2014) noted, over 50 per cent of OECD member states have started to implement these changes. In particular, it is possible to identify new institutions in Australia, New Zealand, Ireland, the United Kingdom, and Denmark.

Reforms in these countries focus particularly on areas that have a new administrative recognition of existing economically related but democratically independent spaces. In the creation of a new strategic authority for Auckland, New Zealand, the proposed soft measures of partnership and joint working arrangements were set aside by the government for a more formal structure, with a directly elected mayor and driven specifically by the need to create a pan-FEA infrastructure and transport body (McFarlane, Solomon, & Memon, 2015). In another area of New Zealand, Hawkes Bay, the framing arguments used were also based on efficiency, but this time focused on the organizational rather than the territorial benefits, although the outcomes were the same (Kortt, Dollery, & Drew, 2016). These examples suggest that central governments are using a range of policy arguments tailored to each locality to achieve the same ends. Similar amalgamations across city and suburban areas are being undertaken in Australia, such as the governance reforms in New South Wales, the Fit for the Future review, where new joint organizations of councils have statutory recognition and provide a forum for councils and the state government to work together to deliver regional priorities, including infrastructure.

In Europe, there are numerous examples of substate rescaling. In Paris, the first overarching governing body, the *Métropole du Grand Paris*, was implemented on 1 January 2016. The new institution brings together all of Paris's metropolitan area municipalities to make joint decisions affecting the entire city. The governance of the region had been criticized for being fragmented, with the urban core separated from its suburbs and the suburbs from each other. These new government arrangements have been promoted to help the region compete with its better-integrated rival, London. The new government arrangements were also preceded by a transport plan for the area, and its boundaries were specifically drawn to define the suburban areas and major city airports within the FEA.

In Gothenburg, thirteen municipalities have come together to work on climate change, although the new grouping has been given responsibility for infrastructure planning as well (Lundqvist, 2016). Other European examples include combining local authorities in Switzerland (Pluss, 2015) and a new single urban and suburban authority for Cork (Cork Local Government Committee, 2015) with associated infrastructure and planning responsibilities. The OECD (2015) has also pointed to FEA government reforms defined by single administration approaches to infrastructure planning and investment in Amsterdam, Hamburg, Manchester, and Stockholm, arguing that these reforms have added economic and social value through more unified leadership of wider and integrated urban and suburban areas (Katz & Noring, 2015).

In the United States, new substate government arrangements are driven through the metro regions approach, though they may be less formally constructed than in Europe. In a comparison between the United States and the European Union, Carbonell and Yaro (2005) identified an absence of spatial strategy and a lack of preparedness for population growth on the part of the United States. By invoking the 1809 and 1909 territorial plans of former presidents Thomas Jefferson and Theodore Roosevelt, respectively, they argue that the future for the United States will rely on filling in the suburbs, making them part of the cities, and improving infrastructure in these new metro city areas. They recommended eight "megalopolis" regions that together would make an American spatial plan, similar to the European Spatial Development Perspective (CEC, 1999), and that could take on the expected city competition emerging from Europe and the BRIC countries (Carbonell & Yaro, 2005). These new megaregions can be defined by the use of a new geography (Lang & Nelson, 2009) and promoted through infrastructure investment (Ross & Woo, 2009).

The execution of these new areas reflects the variations in government models and the differences that occur across state boundaries. The definition of the new US megaregions was undertaken by Hagler (2009) for the Regional Plan Association, who identified infrastructure as a means of enhancing economic competitiveness and reducing social inequality. Katz and Bradley (2013) describe this new approach as collaborative federalism, where both the state and the FEAs move away from hierarchical to mutual relationships.

Hence, policies for rescaling the suburbs are manifest in different ways, but they all have some common features. First, they are new models of government and include democratic accountability. Second, they all have strong political leaderships with directly elected leaders of a new combined local authority. Third, in creating new state spaces, particular care has been taken to retain the integrity of the differences between the urban and suburban, and to find ways to respect these differences in practice. Cities and suburbs are unified by mutual necessity, thus creating a different set of relationships in comparison with the past when either the city or the suburb was dominant.

Strategies for Applying the New Urban/Suburban Policy Doctrine

The application of substate administrative rescaling in each country is complex and will need to align with constitutional models and prevailing cultures. In considering government approaches to the creation of these new FEAs that join cities with their hinterlands, it is possible to identify three models that governments have used to implement the changes. The first is a formal model, through the use of legislation. The second model is quasi-independent, where the government appoints external and independent experts to provide advice on government reform. The third model is the nudged or incentivized approach that encourages local authorities to create larger units of common working. While each of these models is considered in turn in the discussion following, it is noticeable that a common feature in all three approaches to creating new government models is the role of infrastructure in linking the cities to their hinterlands and to other cities. The focus on infrastructure includes attention to the benefits of a larger geographical scale, the need for substate government stability to create confidence in investment, and the efficiencies that are underpinning the narratives of state rescaling in FEAs.

In considering these models in more detail, the formal or legislative reform approach is overtly led by government and is top-down in its implementation. The new scales of government are implemented by changes in state or federal legislation. This approach has been applied in France, Spain, Italy (Bezes & Parrado, 2013), and Denmark (Blom-Hansen, 2010) in a centrally imposed fashion, although in some cases the definition of areas may have included some local choice. Another version of this top-down approach has been applied in Scotland, where legislation has created strategic planning areas that require new forms of governance (Lloyd & Purves, 2009; Morphet, 2011). Although the new governance model appears to be focused on the specific strategic planning outcomes, the inclusion of major infrastructure, environmental capital, and social issues such as poverty in the FEA make the strategic plan governance a de facto new form of substate governance.

The second model is one where independent commissions have been established by state governments to review the most effective government forms for the area. These have been used in Ireland for governance reforms in Cork and Dublin, for example. In Australia, the reform of FEA governance in Sydney has been cloaked in the reform reviews for the whole of New South Wales. The same approach has been used in Wales (Pemberton, 2015). In this approach, independent commissions are appointed to review the current arrangements for local government within a given area. Their remit is set by central government, which offers an arms-length centralist approach. Although there may be some degree of flexibility in the boundaries for the new authorities, their functions and roles set out in the terms of reference for the independent commissions leave the likely outcomes to be as predetermined as in the first model. As in all approaches, the use of comparators in other countries, the concerns about competitiveness, and its expressions through scale are employed to justify the outcomes. However, the use of an independent commission allows government to deflect criticism and provides both time and mechanisms for the proposed changes to be accepted.

The third model is the use of nudged or incentivized approaches to create new FEA governance institutions. This method is the one that has been applied in England, through a layered and developed approach to forming FEA governance alliances through multi-area agreements (2007–2010), local enterprise partnerships (from 2010), and finally combined authorities from 2012. FEAs have been encouraged to apply for governance change through devolution deals that are negotiated with the government. In return for joining new FEA government institutions with directly elected, executive mayors for their whole area, local authorities have been offered a range of powers and funding to encourage their compliance.

Another version of this incentivized approach to work together in FEAs has been used by the EU in the creation of cross-border networks. Here, using the specific European Territorial Co-operation (Interreg) program, funds have been made available to support cross-border working within the FEA. These approaches, which started in 1991, have also now grown to a greater and more institutionalized scale through the megaregion cohesion partnerships that are being adopted across the EU. They started in the Baltic Sea and have been extended to the Danube, the Adriatic, and, most recently, the Alpine region. Here, the governance scales beyond the FEA are creating a context for their operation, and the strategic plans and programs for the megaregions are being agreed upon and adopted by the EU as a whole. This wider contextual government level at mesoregional level is also being used in France and in England – the latter through the "Powerhouse" and "Engine" brands (which are economic development strategies for Northern England).

A Negotiated Settlement

The development of new governance models that stretch across cities into the suburbs has been propelled by their contributions to economic growth and realized through improved infrastructure investment. These new models represent a rebalancing of the relationships between the city and the suburbs, with a shift in the relative power relationships between them. They offer a form of governance that enables both cities and their suburbs to benefit, and addresses their respective fears of growth and decline. These new aligned FEA relationships provide more funding, more investment, and ultimately more power that could allow for greater self-determination.

The advantages for cities in these new alliances with the suburbs include additional capacity and less conflict, although successful relationships may depend on an overt form of mutual respect for their complementary roles. In England, smaller peripheral local authorities have preferred to join their city FEAs rather than remain with their rural neighbours. Past experience with poor hierarchical governance relationships can act as push factors, while cities' welcome and promises of investment can pull willingness for reform. Where there are more formal government reforms in FEAs, the creation of directly elected mayors or new institutions can overcome the uncertainties of governance and the lack of democratic mandate.

These new institutional arrangements can also provide the suburbs with a more stable relationship with their city. Rather than being in opposition, the suburbs can benefit from the city's advantages and gain investment as a result. Cities can also act as a shield against loss of population and, through infrastructure investment, give new life to the suburbs. This process can not only encourage suburban densification around public transport hubs but also can enable a transition to new lifestyles and household patterns. Just as the cities have been able to reinvent their roles, these new arrangements can afford the same opportunities to the suburbs.

Conclusion: What Are the Implications of State Rescaling for Suburban Infrastructure?

Although the case for creating FEAs is based on the function of cities and their contribution to national GDP, the suburbs can also benefit from these new partnerships. The cities have seen the suburbs draining away the economically active, leaving behind a more dependent population. Cities have been anxious about this drain and seen their power to negotiate working arrangements with suburbs decline. The international focus on the role of cities has given them power and confidence, but they have also needed to work with their suburbs in a more equal way.

As we have noted, the suburbs are now also facing change. Their lower densities are under attack as localities need to respond to climate change. They need more public transport and a more densified development to succeed. They need to positively attract city dwellers rather than rely on past suburbanizing practices. As cities grow in attractiveness, the role of suburbs is undermined. The investment in infrastructure that links the suburbs with the city provides more access to jobs for all members of the household, and new transport hubs provide locations for the kinds of support services that households need, including childcare, services, and retail.

While critics of this new economic geography have been concerned that recent changes in governance patterns have been at the behest of neoliberal priorities, cities have always been organized on economic practices and specializations. Primary intentions of the state have focused on economic sectors in the past, but the new approach of scalecraft has potentially created some institutional structures that, in time, may reduce the power of the central state. Indeed, the issue has been raised that "cities [may be] the new countries" (Coughlan, 2016). It is important to recognize that these power shifts are manifesting themselves through infrastructure investment and its subsequent effects on locations. The suburbs are densifying and changing their roles.

Chapter 3 has introduced an infrastructure rescaling theme to this volume. It has identified the advantages of integrating the governance of metropolitan regions and of thereby merging the administration of cities and suburbs. The chapter has highlighted the economic advantages of metropolitan governance and pointed to the enhanced infrastructure coordination capacity resulting from such an institutional rescaling. It has demonstrated the existence of a direct connection between the territories covered by institutions and the efficient delivery and operation of infrastructure systems. To function in an optimal fashion, infrastructure networks often require institutional structures that parallel their scale of operation, hence the importance of metropolitan-wide governance for infrastructures connecting cities and suburbs such as public transportation systems. Another chapter in this volume also concentrates on the impact of institutions on infrastructures. Chapter 6 associates the absence of vital sewerage infrastructure in a city close to Delhi with the absence of agency capable of promoting and funding such an infrastructure. A further infrastructure dimension of institutions reverberates throughout the volume. Because they are essential to infrastructure planning, development, and funding, institutions can be perceived as "enablers" of infrastructures in a manner that recalls how infrastructures support other human activity. In this sense, institutions can be seen as "infrastructures for infrastructures."

NOTE

This chapter is based on a previously published article. Morphet, J. Rescaling the suburban: New directions in the relationship between governance and infrastructure. *Local Economy, 32*(8), 803–17.

REFERENCES

Accetturo, A., Manaresi, F., Mocetti, S., & Olivieri, E. (2014). Don't stand so close to me: The urban impact of immigration. *Regional Science and Urban Economics, 45*, 45–56. https://doi.org/10.1016/j.regsciurbeco.2014.01.001

Ahrend R., Farchy, E., Kaplanis, I., & Lembcke, A. (2014). What makes cities more productive? Evidence on the role of urban governance from five OECD countries. (OECD Regional Development Working Paper 2014/05). Paris: Organization for Economic Cooperation and Development.

Ahrend, R., & Schumann, A. (2014). Does regional economic growth depend on proximity to urban centres? (OECD Regional Development Working Paper 2014/07). Paris: Organization for Economic Cooperation and Development.

Allmendinger, P., & Haughton, G. (2010). Spatial planning, devolution, and new planning spaces. *Environment and Planning C: Government and Policy, 28*(5), 803–18. https://doi.org/10.1068/c09163

Bezes, P., & Parrado, S. (2013). Trajectories of administrative reform: Institutions, timing and choices in France and Spain. *West European Politics, 36*(1), 22–50. https://doi.org/10.1080/01402382.2013.742735

Blom-Hansen, J. (2010). Municipal amalgamations and common pool problems: The Danish local government reform in 2007. *Scandinavian Political Studies, 33*(1), 51–73. https://doi.org/10.1111/j.1467-9477.2009.00239.x

Boterman, W.R., & Karsten, L. (2014). On the spatial dimension of the gender division of paid work in two-parent families: The case of Amsterdam, the Netherlands. *Tijdschrift voor Economische en Sociale Geografie, 105*(1), 107–16. https://doi.org/10.1111/tesg.12073

Brenner, N. (1999). Globalisation as reterritorialisation: The re-scaling of urban governance in the European Union. *Urban Studies, 36*(3), 431–51. https://doi.org/10.1080/0042098993466

Brenner, N. (2004). *New state spaces: Urban governance and the rescaling of statehood*. Oxford: Oxford University Press.

Brezzi, M., & Veneri, P. (2014). Assessing polycentric urban systems in the OECD: Country, regional and metropolitan perspectives. (OECD Regional

Development Working Papers 2014/01). Paris: Organization for Economic Cooperation and Development.

Buller, J., & James, T.S. (2012). Statecraft and the assessment of national political leaders: The case of new labour and Tony Blair. *British Journal of Politics and International Relations*, *14*(4): 534–55. https://doi.org/10.1111/j.1467-856x.2011.00471.x

Bulpitt, J. (1986). The discipline of the new democracy: Mrs Thatcher's domestic statecraft. *Political Studies*, *34*(1), 19–39. https://doi.org/10.1111/j.1467-9248.1986.tb01870.x

Carbonell, A., & Yaro, B. (2005). American spatial development and the new megalopolis. *Land Lines*, *17*(2), 1–4.

Charbit C. (2011). Governance of public policies in decentralized contexts: The multi-level approach. (OECD Regional Level Working Papers 2011/04). Paris: Organization for Economic Cooperation and Development.

Commission of European Communities (CEC). (1999). *The European spatial development perspective*. Brussels: Author.

Commission of European Communities (CEC). (2015). Results of the public consultation on the key features of an EU Urban agenda. (Commission Staff Working Paper SWD 20-15, 109/final). Brussels: Author.

Cork Local Government Committee. (2015). Local government arrangements in Cork. (Report of the Cork Local Government Committee, September). Cork, IE: Author.

Coughlan S. (2016, 21 January). "Are cities the new countries?" *BBC News Online*. http://www.bbc.co.uk/news/education-35305586

Department for Communities and Local Government (DCLG). (2008a). *Planning and optimal geographical levels for economic decision making – the sub-regional role*. London: Author.

Department for Communities and Local Government (DCLG). (2008b). *Review of economic assessment and strategy activity at the local and sub-regional level*. London: Author.

Department for Communities and Local Government (DCLG). (2008c). *Why place matters and the implications of the role of central, regional and local government*. London: Author.

Dijkstra, L., & Poelman, H. (2012). *Cities in Europe: The new EU OECD definition*. Brussels: Commission of European Communities.

Filion, P. (2013). Optimistic and pessimistic perspectives on the evolution of the North American suburb. *Planning Theory and Practice*, *14*(3), 411–13. https://doi.org/10.1080/14649357.2013.808833

Forsyth, A. (2013). Suburbs in global context: The challenges of continued growth and retrofitting. *Planning Theory and Practice*, *14*(3): 403–6. https://doi.org/10.1080/14649357.2013.808833

Fraser, A. (2010). The craft of scalar practices. *Environment and Planning A*, *42*(2), 332–46. https://doi.org/10.1068/a4299

Garreau, J. (2011). *Edge city: Life on the new frontier*. New York: Random House LLC.

Goodsell, T.L. (2013). Familification: Family, neighborhood change, and housing policy. *Housing Studies*, *28*(6), 845–68. https://doi.org/10.1080/02673037.2013.768334

Goodwin, P. (2012). Peak travel, peak car and the future of mobility: Evidence, unresolved issues, and policy implications, and a research agenda. (International Transport Forum Discussion Paper, No. 2012/13). Paris: Organization for Economic Cooperation and Development.

Grant, J.L. (2013). Suburbs in transition. *Planning Theory and Practice*, *14*(3): 391–415. https://doi.org/10.1080/02673037.2013.768334

Greater London Authority (GLA). (2015). *Resources of global city comparison indicators*. Retrieved from http://data.london.gov.uk/dataset/resources-of-global-city-comparison-indicators

Gurria, A. (2014, 6 October). Smart investment in regions and cities. (Speech to EU regions, Open Days). Retrieved from http://www.oecd.org/about/secretary-general/smartinvestmentinregionsandcities.htm

Hagler, Y. (2009). *Defining US megaregions – America 2050*. New York: Regional Plan Association.

Hamza, C., Frangenheim, A., Charles, D., & Miller, S. (2014). *The role of cities in cohesion policy*. Brussels: Commission of European Communities.

Headicar, P. (2015). Homes, jobs and commuting: Development location and travel outcomes. In R.P. Hickman, M. Givoni, D. Bonilla, & D. Bannister, (Eds.), *Handbook on transport and development* (pp. 59–72). Cheltenham, UK: Edward Elgar.

Her Majesty's Treasury (HMT). (2003). *Productivity in the UK 4 – the local dimension*. London: Author.

Jessop, B. (1997). The entrepreneurial city: Re-imaging localities, redesigning economic governance, or restructuring capital. In N. Jewson and S. MacGregor (Eds.), *Transforming cities: Contested governance and social transformation* (pp. 28–41). London: Routledge.

Karsten, L. (2003). Family gentrifiers: Challenging the city as a place simultaneously to build a career and to raise children. *Urban Studies*, *40*(12): 2573–84. https://doi.org/10.1080/0042098032000136228

Katz, B., & Bradley, J. (2013). *The metropolitan revolution: How cities and metros are fixing our broken politics and fragile economy*. Washington. DC: Brookings Institution Press.

Katz, B., & Noring, L. (2015, 17 August). Europe for cities. *Opinion*. Retrieved from https://www.brookings.edu/opinions/europe-for-cities/

Kingdon, J.W. (1984). *Agendas, alternatives, and public policies*. Boston: Little, Brown.

Kortt, M.A., Dollery, B., & Drew, J. (2016). Municipal mergers in New Zealand: An empirical analysis of the proposed amalgamation of Hawke's Bay councils. *Local Government Studies*, *42*(2), 228–47. https://doi.org/10.1080/03003930.2015.1007133

Krugman, P. (1991). *Geography and trade*. Cambridge, MA: MIT Press.

Krugman, P. (2011). The new economic geography, now middle-aged. *Regional Studies*, *45*(1), 1–7. https://doi.org/10.1080/00343404.2011.537127

Lang, R., & Nelson, A. (2009). Megapolitan America: Applying a new geography. In C. Ross (Ed.), *Megaregions: Planning for global competitiveness* (pp. 107–26). London: Island Press.

Lloyd, G., & Purves, G. (2009). Identity and territory: The creation of a national planning framework for Scotland. In S. Davoudi & I. Strange (Eds.), *Conceptions of space and place in strategic spatial planning* (pp. 71–94). London: Routledge.

Lundqvist, L.J. (2016). Planning for climate change adaptation in a multi-level context: The Gothenburg metropolitan area. *European Planning Studies*, *24*(1), 1–20. https://doi.org/10.1080/09654313.2015.1056774

Matthews, P., Bramley, G., & Hastings, A. (2015). Homo economicus in a big society: Understanding middle-class activism and NIMBYism towards new housing developments. *Housing, Theory and Society*, *32*(1), 54–72. https://doi.org/10.1080/14036096.2014.947173

McFarlane, K., Solomon, R., & Memon, A. (2015). Designing institutions for strategic spatial planning: Auckland's governance reforms. *Urban Policy and Research*, *33*(4), 452–71. https://doi.org/10.1080/08111146.2015.1061988

Merk, O. (2014). Metropolitan governance of transport and land use in Chicago. (OECD Regional Development Working Paper, 2014/08). Paris: OECD.

Metz, D. (2013). Peak car and beyond: The fourth era of travel. *Transport Reviews*, *3*(3), 255–70. https://doi.org/10.1080/01441647.2013.800615

Morphet, J. (2011). *Effective practice in spatial planning*. Abingdon, UK: Routledge.

Nelson, A.C. (2013). The resettlement of America's suburbs. *Planning Theory and Practice*, *14*(3), 392–403. https://doi.org/10.1080/02673037.2013.768334

Organisation for Economic Co-operation and Development (OECD). (2013). *Definition of functional urban areas for the OECD metropolitan database*. Paris: Author.

Organisation for Economic Co-operation and Development (OECD). (2015). *Local economic leadership*. Paris: Author.

Pemberton, S. (2015). Statecraft, scalecraft and local government reorganisation in Wales. *Environment and Planning C: Government and Policy*, *34*(7), 1306–23. https://doi.org/10.1177/0263774X15610581

Pemberton, S., & Morphet, J. (2014). The rescaling of economic governance: Insights into the transitional territories of England. *Urban Studies*, *51*(11), 2354–70. https://doi.org/10.1177/0042098013493484

Pemberton, S., & Searle, G. (2016). Statecraft, scalecraft and urban planning: A comparative study of Birmingham, UK, and Brisbane, Australia. *European Planning Studies*, *24*(1), 76–95. https://doi.org/10.1080/09654313.2015.1078297

Phelps, N.A., Vento, A.T., & Roitman, S. (2015). The suburban question: Grassroots politics and place making in Spanish suburbs. *Environment and Planning C: Government and Policy*, *33*(3), 512–32. https://doi.org/10.1068/c13136

Pickup, L. (1978). Women's gender-role and its influence on travel behaviour. *Built Environment*, *10*(1), 61–8.

Pluss, L. (2015). Municipal councillors in metropolitan governance: Assessing the democratic deficit of new regionalism in Switzerland. *European Urban and Regional Studies*, *22*(3), 261–84. https://doi.org/10.1177/0969776412469804

Porter, M.G., & de Roo, G. (Eds.). (2012). *Fuzzy planning: The role of actors in a fuzzy governance environment*. Farnham, Surrey, UK: Ashgate Publishing.

Rérat, P., & Lees, L. (2011). Spatial capital, gentrification and mobility: Evidence from Swiss core cities. *Transactions of the Institute of British Geographers*, *36*(1), 126–42. https://doi.org/10.1111/j.1475-5661.2010.00404.x

Richards, S. (2015, 26 October). David Cameron faces much larger obstacles than the Lords to control the UK. *The Independent*. Retrieved from http://www.independent.co.uk/voices/david-cameron-faces-much-larger-obstacles-than-the-lords-to-control-and-continue-to-govern-the-uk-a6709636.html

Ross, C., & Woo, M. (2009). Identifying megaregions in the United States: Implications for infrastructure. In C. Ross (Ed.), *Megaregions: Planning for global competitiveness* (pp. 53–82). London: Island Press.

Skaburskis, A. (2012). Gentrification and Toronto's changing household characteristics and income distribution. *Journal of Planning Education and Research*, *32*(2), 191–203. https://doi.org/10.1177/0739456x11428325

Stabiner, K. (2011, 21 January). New lives for "dead" suburban malls. *New York Times*. Retrieved from http://newoldage.blogs.nytimes.com/2011/01/21/new-lives-for-dead-suburban-malls/?_r=0

Wheatley, D. (2012). Work-life balance, travel-to-work, and the dual career household. *Personnel Review*, *41*(6), 813–31. https://doi.org/10.1108/00483481211263764

World Economic Forum (WEF). (2015). *Global competitiveness report 2014–2014*. Geneva: Author.

4 Financial Infrastructures of Suburbanism: From Suburbanization to Value Extraction

ALAN WALKS

Introduction

Urban development requires financing. However, the financial infrastructures that undergird the construction of cities, and their relationships with the evolution of suburbanism(s), are typically overlooked. While finance is ages old, the kinds of suburbanization that have arisen in the post–Second World War period have not only accompanied but have also been dependent on the development of very particular financial structures. Furthermore, the evolution of suburbanization processes over time has occurred in tandem with the evolution of financial innovations, from state-backed mortgage insurance, to automobile loans, to mortgage securitization. However, these key financial innovations, originally developed in order to facilitate suburbanization processes, have transformed over time into financial infrastructures that work to extract value from suburbanism.

This chapter begins by briefly theorizing the context of the financial revolution that has occurred alongside the spread of suburbanization. It then analyses the construction of the financial infrastructure, which arose out of the deliberate responses by policymakers to promote certain forms of suburban expansion in the face of the Great Depression. The financial innovations that have emerged out of, and transformed, this original financial infrastructure are then examined, as are its effects on suburban residents. The implications of this history for understanding the role of financial infrastructures and innovations are discussed in the conclusion.

The Financial Origins of the (Sub)Urban Revolution

Henri Lefebvre in 1970 broached the possibility of an urban revolution rooted in the evolution of global capitalism. Not only were cities spreading all over the world, but, according to Lefebvre, capitalism was in the throes of a transformation away from industrialization as the generator of value to a system in which accumulation is rooted in urbanization (Lefebvre, 2003 [1970]). Furthermore, he foretold a time when the world would be entirely urbanized, in the sense of having urban ways of life and urban political-economic logics dominating all spaces, whether mercantile, industrial, agricultural, or oceanic. The logic of capitalism, according to Lefebvre, will increasingly govern in accordance with the dominance of urbanization as the main driver of accumulation. However, Lefebvre did not spell out exactly how accumulation was to be derived directly from urbanization as opposed to industry, and it is not clear if he suspected this process would eventually involve financial innovation or predominant forms of suburbanization.

The extent to which cities and, indeed, the entire global population and economy have changed since the late 1800s (what Arrighi [1994] refers to as the "Long Twentieth Century") is evident in some basic statistics. Since 1901, the world population has grown from 1.6 billion to over 7 billion, while the proportion living in cities has grown from 13 per cent (220 million people) to over 50 per cent (3.6 billion people in 2011) (Walks, 2015: 3). It is the post–Second World War period that has witnessed the largest growth in absolute terms, and by 2011 the proportion living in cities was 43 per cent greater than the whole global population in 1950 (Walks, 2015: 3). The vast majority of the increase in population since the war, at least in developed nations, has been accommodated through the building of new suburban communities at the fringes of cities. In turn, the suburbs have acted as a huge absorber of capital investment.

Lefebvre's theorization of urban revolution derives from the productive dialectic tensions between simultaneous primary tendencies within urbanism of centrality and dispersal, leading to various kinds of "implosion-explosion" (political, economic, cultural, and social). While he did not formalize it, his theorizing allows for the construction of a dialectical theory of suburbanism, producing six dimensions of urbanism-suburbanism, which in combination produce an infinite number of possible suburbanisms (Walks, 2013b). However, the actual history of urbanization has favoured certain kinds of suburbanisms over others. The most dominant and salient forms of suburbanism, at least among developed nations, are characterized by high levels of home ownership, automobile use, and domesticity at relatively low densities in predominantly single-family houses (Jackson, 1985; Moos & Mendez, 2015). Meanwhile, much of the (sub)urban landscapes built in the post-war period are not only dominated by automobile infrastructure, but are also monofunctional in providing few other transport options (Freund & Martin, 1993). Hence automobility has become the primary means of mobility in the suburbs of most developed nation metropolitan areas. Whereas motor vehicle travel totaled 2.8 trillion passengers per kilometre in 1950, just fifty years later this travel came to 32 trillion, a more than 1,000 per cent increase (Walks, 2015). The "Long Twentieth Century," it could be argued, has been both a "suburban century" and a "century of automobility" (Volti, 1996; Walks, 2015).

The shift from urban economies based around the "forces of centred urbanism" (Rae, 2003) to regional economies in which economic, social, and political activities are linked across large spaces within and among metropolitan areas has involved nothing short of a (financial) revolution. David Harvey (2012) argues that all financial crises under capitalism have roots in urbanization processes. Once capital investment switches from the primary circuit (manufacturing and production) to the secondary circuit (urban built environment), there is a growing tendency to treat land as a "purely financial asset" (Harvey, 2006 [1982]: 347) and in the process create financial securities collateralized by land and housing (Marx's "fictitious capital"). The influx of investment into urban residential and office development leads to overbuilding and financial bubbles, often evidenced at their peak by high numbers of tall buildings (Harvey, 2012: 34). It is often in response to financial crises that state policy becomes most innovative. The 1920s was the time in the United States when automobility grew fastest, facilitating the construction of (then) far-flung suburbs, with much of this development afforded by rapidly rising availability of credit and a stock market boom (Jackson, 1985; Volti, 1996).

While the Great Depression of the 1930s formally began with the popping of the stock market bubble, the crash was preceded by a huge real estate bubble, which, when it deflated, brought new housing construction to a halt. It was the build-up of delinquent mortgages during the 1930s that largely hobbled the solvency of US banks, provoking bank runs and extending the depression (Gaffney, 2009; Simpson, 1933). By the early 1930s, half of all US mortgages were in default, and in 1932 roughly 250,000 mortgages were in foreclosure (Aalbers, 2011: 83). As the Great Depression deepened, it became evident to US policymakers, as well as policymakers in other nations, that they needed to devise a plan to save the banks and underwater homeowners, encourage the flow of credit back into mortgage markets, and restart housing construction. President Hoover organized a conference on home building in 1931, and all four of the recommendations from the proceedings were eventually accepted (Jackson, 1985). These recommendations were to use state policy to find ways of reducing mortgage interest rates, to encourage long-term amortized mortgages, to reduce home construction costs, and to aid private companies in housing lower income households. It is thus out of this process that a number of key financial innovations central to the construction of the post-war suburbs were born: amortizing mortgages, state-backed mortgage insurance, and the originate-and-distribute model of mortgage issuance, which over time led to securitization.

The Financial Infrastructure of Post-War Suburbia

Cities of all kinds – falling anywhere on the multidimensional axes of urban-suburban – must be financed before they can be built. Much of this finance comes from the way that demand for space is provided on the consumption side. In the post-war period, residential mortgages secured by private property undergirded most demand for housing, while other kinds of loans – both secured and unsecured – fuelled demand for other household items like automobiles and furniture. The latter, such as automobile loans, largely took place outside the purview of the state during the early post-war period, although Fordist economic regulation on the demand and supply side kept automobile companies producing at high levels of employment, ensuring that lending for these kinds of consumer goods remained highly profitable (see Aglietta, 1979; Paterson, 2007). However, all of this financing would not have been possible without the establishment of a very large, state-backed system of mortgage finance, which underwrote and directly shaped contemporary forms of suburbanization throughout the post-war period.

Before the 1930s, most mortgages were funded privately or else through insurance companies or (in certain countries) building societies and savings and loan banks. There was no state involvement in mortgage lending, and mortgages typically had loan-to-value ratios usually less than 50 per cent and not exceeding 66 per cent, structured as interest-only loans with terms between two and six years (Jackson, 1985; Woodard, 1959). In the United States, for instance, most mortgages in the 1920s were bullet or balloon payment loans, and only 20 per cent of US mortgages were amortizing (Green, 2014). The United States allowed commercial banks to originate mortgages from 1913 onwards, but this practice was the exception. However, by adding to the profusion of mortgage credit, it likely augmented the 1920s real estate bubble and subsequent bust, as well as the reticence of banks to get back into the mortgage market in the 1930s (Aalbers, 2011).

Amortizing Mortgages and Mortgage Insurance

During the Great Depression, in order to save banks and homeowners from the rising foreclosure crisis, first the Hoover, and then the Roosevelt, administrations in the United States initiated a myriad of programs and institutions. One of the most important from the perspective of financial innovation was the Home Owners Loan Corporation (HOLC), established in 1933. The HOLC purchased and refinanced defaulted mortgages on new terms at relatively low interest rates, and over time was able to fine-tune a new kind of mortgage that private lenders would then be compelled to copy. This new type was the long-term self-amortizing mortgage through which borrowers could pay back the loan with uniform monthly payments combining principal and interest over an extended time frame (originally twenty years, later extended to thirty years) at fixed rates of interest, with loan-to-value ratios based on standardized appraisals (Jackson, 1985). The reform of mortgage terms produced significant relief for illiquid homeowners, who, faced with the credit crunch of the early 1930s, were having difficulty refinancing their mortgages. Indeed, roughly 40 per cent of eligible mortgagees received aid from HOLC programs in the early 1930s (Jackson, 1985).

In 1934, the same (Roosevelt) administration created the Federal Housing Administration (FHA) to insure private mortgages issued by banks in order to encourage banks and other financial institutions to get into the mortgage lending business (with the insurance covering the lender, not the borrower). FHA mortgage insurance covered 100 per cent of the outstanding mortgage balance in the face of default, thus transferring all default risk from lenders to the federal government. By establishing qualifying criteria upon which it would provide this insurance (regarding amortization terms, loan-to-value ratios, interest rates, and gross debt service levels), the FHA was able to standardize the terms of the vast majority of mortgages issued by private lenders going forward. This system worked very well, such that, by the 1950s when housing markets began once again to boom, private lenders were only interested in originating mortgages that conformed to HOLC/FHA guidelines.

The US model was adopted by other national governments around the globe. For instance, in Canada the Dominion Housing Act (DHA) was passed in 1935 and began issuing joint long-term self-amortizing mortgage loans with private lenders, a variant on what the HOLC and FHA were doing in the United States (Bélec, 1997; Miron, 1988). The subsequent National Housing Acts (NHA) of 1938, 1944, and 1947 extended amortization terms to twenty and then twenty-five years, and provided for 80 per cent loan-to-value financing (Woodard, 1959). The Canadian (originally "Central") Mortgage and Housing Corporation (CMHC), a Crown corporation of the federal government, was formed in 1946 with the dual role of funding and coordinating social housing activity with the provinces, while simultaneously aiding the profitability of private business, particularly "private mortgage companies" (Bacher, 1993: 179). In addition to setting guidelines on building standards and funding new social housing construction, the CMHC took over the issuance of joint mortgage loans, and in 1954 followed the US lead in switching to mortgage insurance for private residential mortgages, with an intention to stimulate new private sector housing activity without it being a drain on the federal treasury (Bacher, 1993; Poapst, 1993; Walks & Simone, 2016). In Japan, a variant on this type of agency is found in the state-run Japan Housing Loan Corporation (JHLC), formed in 1950. Like the CMHC, the JHLC was given a mandate to fund new housing construction, regulate building standards, and act as the mortgage lender of last resort (Seko, 1994). Similarly, in Australia the Housing Loan Insurance Corporation (HLIC) was established in 1965, modelled on Canada's CMHC, with a mandate to build social housing insured private mortgages, until it was privatized in 1998. In Europe, a number of different models arose to insure both mortgage lenders and, in some cases, borrowers against default (Elsinga, Premius, & Cao, 2009). One of the first schemes was set up in the Netherlands in 1956; although transformed into an independent private institution in 1995, their National Mortgage Guarantee (NHG) served as the model for similar state programs in Belgium, Sweden, and other northern European countries in the 1990s (Elsinga et al., 2009).

Because state mortgage insurance institutions like these were developed to serve particular public policy objectives, they contained provisions meant to finance new building and to protect taxpayers. Until the entry of the state into mortgage finance, mortgage credit was very difficult to acquire for purchase of unserviced land, for new buildings, and for housing at the urban fringe. Many new suburban areas were effectively redlined, compelling would-be residents to self-finance and self-build (Harris, 1996; 2004; Harris & Forrester, 2003). Since lenders used the land as collateral, and well-serviced land in higher-status neighbourhoods and areas near the core was easiest to sell in the case of non-payment and foreclosure, until the 1930s it was far easier to acquire a mortgage on an existing property, and most mortgages went to the elite. State support for mortgages, of course, was meant not only to promote homeownership among middle and lower income households, but also to augment job creation efforts and stimulate construction, so the various national agencies applied additional criteria to their operations, often in order to encourage lenders to provide mortgages for what were then considered riskier investments.

This policy meant that in many nations most state-backed mortgage insurance was reserved for new housing in the suburbs. In the United States, for instance, most FHA title II mortgage insurance programs, representing 70 per cent of all FHA-insured mortgages, were restricted to new homes, as were some other programs (Green, 2014; Hoagland & Stone, 1969). In Canada, mortgage insurance for mortgages on residential owner-occupied housing could only be offered on new home purchases (Poapst, 1993; Woodard, 1959). Because banks entered mortgage markets mostly in order to access the risk-free guarantees provided by state mortgage insurance, households found it far easier to obtain a mortgage for new suburban dwellings, and they could also borrow significantly more for such purchases. Already by the 1940s, criticism of the idea of state-backed mortgage insurance highlighted how it altered normal lending patterns, led to a profusion of mortgage credit, and provoked riskier lending:

> The insured mortgage is the type of loan which is made without reference to the judgement and prudence of the lender. It is made according to a statutory formula and under the influence and solace of the device of the federal guarantee. The bulk of today's mortgage lending, at present terms, ratios, and rates, would not be made by mortgage lenders if they had to assume the full burden of the risks involved. (Rathje, 1948, cited in Hoagland & Stone, 1969: 529)

Such mortgage insurance programs had very clear effects, not only promoting suburbanization but also shaping the kind of dwellings and social composition that would be found in the new post-war suburbs. Houses purchased through such programs were disproportionately concentrated in fringe suburban communities (Bélec, 2015; Jackson,

1985). In turn, it was mostly wealthy households purchasing with significant down payments that could access mortgages to purchase older housing, typically (still) through insurance companies or private arrangements, thus helping to maintain the housing markets perhaps only in existing older elite areas. But demand for housing in most inner-city neighbourhoods dropped considerably, and it was mainly landlords wanting to purchase old housing for conversion to rental, as well as ethnic communities relying on their own communities for private mortgages, that continued to provide demand in older areas (for examples in Canada, see Murdie, 1986; 1991). It was not until the passage of the 1968 Fair Housing Act in the United States, and the 1969 Bank Act in Canada, that restrictions on state mortgage insurance for the purchase of existing housing were lifted. It is thus not a coincidence that incipient gentrification within the inner cities of the largest metropolitan areas started in the late 1960s (for Canadian cities, see Walks & Maaranen, 2008a; 2008b), ushering in very literally a "back to the city movement by capital, not people" (Smith, 1979). Still, throughout the 1970s, the typical bid-rent geography of property values remained inverted, with houses in fringe US suburbs commanding higher prices than equivalent properties in the inner cities (Wheaton, 1977).

State housing administrations also funded social housing in various ways, which also had a significant effect on the urban landscape. In the United States and Canada, mortgage insurance was extended to public housing and non-profit housing agencies. Whereas much new social housing in European cities (and some Canadian cities) was located at or near the fringes in new suburbs, in the United States the vast majority of such units were built in older inner-city neighbourhoods, often those concentrating the local African American population (Harloe, 1995).

Redlining

In the United States, the HOLC and FHA also adopted redlining practices that discouraged lending within inner cities. They produced "residential security" maps in which the risk to lenders was colour-coded in four categories. Through such maps, HOLC/FHA guidelines sought to push public and private lenders to concentrate mortgages in areas where demand would be positive "in good times and bad," which were coloured green, while discouraging lending in areas that have "little or no value today due to the colored element now controlling the district," shaded red (Jackson, 1985: 197–200). While these guidelines may have largely reflected what private lenders were already doing, they nonetheless institutionalized race-based redlining practices and made US federal housing intervention officially racist until the mid-1960s (Aalbers, 2011). This component of the financial architecture not only worked to support suburbanization while accelerating inner-city decline, but it also helped entrench intra-urban racial segregation and ensured that post-war suburbanism in the United States took on a particularly racialized (white) form that continued on for decades (Jackson, 1985; Massey & Denton, 1993). Such policies helped naturalize the image of US suburbs as the reward for hard work and mask the role of state policies, and, by concentrating white middle-class families into middle-class suburbs, contributed to the extension of "colour-blind" policies assumedly based on meritocracy, including those related to public education choice and busing (Lassiter, 2004). The way that suburbanization was financed in the United States was therefore a direct factor in producing the peculiar form of "advanced marginality" that Wacquant (2007) associates with the US experience of racialization and segregation. In European nations, state mortgage insurance generally worked against redlining practices, but in their absence, redlining by private lenders was also detectable (Aalbers, 2011).

Secondary Mortgage Markets and Mortgage Securitization

Mortgage securitization is also ultimately a product of the US response to the Great Depression. The Federal National Mortgage Association (FNMA, or Fannie Mae) was initiated in 1938 in order to provide liquidity to mortgage markets by issuing bonds and using the proceeds to purchase mortgages from private lenders. This policy took the mortgages off banks' balance sheets, allowing them to continue lending. It encouraged banks to originate mortgages for the sole purpose of selling them to other investors, and thus was termed the "originate-and-distribute" model of mortgage lending. Like FHA mortgage insurance, this model encouraged lenders to issue amortizing mortgages that conformed to FHA standards. Until 1968, when Fannie Mae was privatized, its bonds were officially fully backed by the US federal government.[1] Federal backing made Fannie Mae's bonds safer than privately issued bonds, increased investor demand, and brought interest rates down for the FHA-conforming mortgages that Fannie Mae purchased. Together, this system effectively used the power of the state to reduce private financing and transaction costs for new housing construction and purchase, in turn significantly promoting home ownership. While ostensibly meant to support the private "market," for many years these state institutions dominated the mortgage market. The early peak came in 1943 when 80 per cent of all new units were financed by FHA-insured loans. Through the 1950s, roughly one quarter of all housing units were funded with FHA-insured loans, many of them purchased by Fannie Mae (Hoagland & Stone, 1969: 528).

Fannie Mae's initial charter included the option of innovating with various types of securities backed by the mortgages it purchased, as well as stimulating a secondary mortgage market through which mortgages could be traded among different investors (Buser & Mcconnell, 2014). However, it was not until the late 1960s that the first true mortgage-backed securities (MBSs) were issued. In 1968, the US federal government privatized Fannie Mae, turning it into a government-sponsored enterprise (GSE) with its own shareholders. At the same time, it created both its "competitor" GSE, the Federal Home Loan Mortgage Corporation (FHLMC or Freddie Mac), in order to finance savings and loan institutions, as well as the Government National Mortgage Association (GNMA, or Ginnie Mae), which remained an agency of the federal government. It was the latter (Ginnie Mae) that created the first true MBS in 1968, followed in short order in 1971 by Freddie Mac and later in 1981 by Fannie Mae (Belsky, 2014). Over the 1970s and 1980s, MBSs underwent significant experimentation within the United States, mostly through the activities of these mortgage agencies (FNMA, FHLMC, and GNMA). Securitization and the establishment of secondary mortgage markets grew in parallel with the deposit-funded model of mortgage funding evident among US savings and loans lenders and among banks elsewhere (this model is termed the "originate-and-hold" model).

Mortgage securitization acts as a way of transforming illiquid spatially fixed material assets like land and housing into liquid financial assets that can be traded in international investment circuits (Gotham, 2009). In turn, mortgage securitization allows investors based elsewhere in the world to invest in particular local housing markets. It involves the pooling of a large number of individual mortgages into a single fund, allowing investors to purchase portions of this pool instead of the individual mortgages. By pooling these loans, investment risks from each individual loan are also averaged and smoothed, allowing the securities to be traded and priced based on collective repayment rates and average interest rates. Mortgage securitization also addresses the problems related to maturity mismatch, which plagues lenders who have to borrow short to lend long; with securitization, lenders can sell a portion of the claims represented by the MBSs well before maturity and thus react to changes in economic conditions and interest rates (Buser & Mcconnell, 2014).

These initial mortgage securitization schemes effectively used capital markets to fund housing, increasing the amount of investment available to fund (sub)urban expansion. It was mostly large institutional funds that purchased MBSs, and, with the expansion of funded pension plans from the 1960s onward, this trend meant that increasingly many retirement nest eggs depended on the orderly expansion of housing markets and prices (Toporowski, 2000). Because mortgage securitization was first developed in the United States and because, after the end of the Bretton Woods fixed exchange rate regime in the early 1970s, international investors often sought out "safe" investments in US financial markets, housing markets in the United States particularly attracted large flows of financial investment into housing.

Securitization spread to other nations beginning in the late 1980s. First the United Kingdom in 1985, and then Canada in 1987, sought to mirror the apparent success of the US model by adopting their own mortgage securitization programs (Karley & Whitehead, 2002; Walks, 2014; Walks & Clifford, 2015), leading to building booms in the late 1980s in both nations. In Europe, lenders also used capital markets to fund mortgages, but traditionally through a different kind of security: covered bonds. In the latter case, lenders issue private bonds backed by the mortgages they issue, giving bondholders claims on the underlying collateral in case of default and making the bonds potentially less risky in the face of any widespread financial crisis (Kim & Cho, 2014). While these bonds were also largely purchased by institutional investors, it was not until the creation of the eurozone (EZ) and the adoption of the euro in 1999 that investment in European covered bonds became sufficiently globalized, at which point many European countries had also adopted securitization (discussed at a later point). In combination with state-provided mortgage insurance, early securitization schemes facilitated the smooth and continuous flow of funding into new housing construction, promoting particular forms of suburbanization in exactly those places where housing demand was strongest: either fast-growing cities in suburban nations like the United States, Canada, and Australia or in countries like Japan and the Netherlands, where significant portions of cities had been destroyed during the war. During this period, the financial architecture that was developed facilitated predominant forms of post-war suburbanization. In terms of the typology laid out by Jean-Paul Addie in this volume, this financial architecture at the time largely developed *for the suburbs*, both as an infrastructure of suburbanization and as a key infrastructure of post-war suburbanism.

Evolution of Financial Infrastructure: Suburbanism and the Extraction of Value

The considerable advantage enjoyed by the new suburbs as a result of the post-war financial architecture began to wane between the late 1960s and the 1990s. In addition to the expansion of mortgage insurance to existing properties in older areas, in the United States the 1968 Fair Housing Act prohibited discrimination in mortgage lending, while the 1975 Federal Home Mortgage Disclosure Act forced lenders to report on the mortgages they issued by socio-demographics and place of residence, and the 1977 Community Reinvestment Act required lenders to meet the needs of previously marginalized inner-city communities (Aalbers, 2011). Not only did this reshaping break down the biases towards a racialized (white) middle-class form of suburbia, it also changed the financial logic involved with investment in such forms of suburbanism. But even in nations lacking similar histories or pieces of legislation, gentrification took hold in growing cities, particularly those characterized by strong and growing financial sectors (Lees, Slater, & Wyly, 2008). Financial innovation was not only to drive employment growth in the main centres of finance, but also evolved an architecture that has worked to extract value through investment in the built environment. As discussed later, this new architecture became particularly evident in the United States and elsewhere during and after the global financial crisis (GFC).

Already by the mid-1990s, the scholarly literature was evoking a post-suburban world characterized by suburban decline, rising social diversity, deindustrialization, and infrastructure deficits (Essex & Brown, 1997; Lucy & Phillips, 1997). Although many newer suburbs continued to flourish, often by explicitly targeting wealthy individuals through low-tax and low-spending regimes promoted by home-rule governance traditions in the United States (Peck, 2011), much planning scholarship, public policy, and public opinion began to question the dominant form and potential sustainability of post-war suburbanization (Nicolaides, 2006). This debate occurred at the same time that profit rates began to decline across the globe, leading to a number of financial crises, including the 1997 Asian Financial Crisis (Brenner, 2002; McNally, 2009). As investment moved away from developing nations and back into asset-backed securities in the United States and other developed countries, a number of financial innovations were put into practice that turned mortgage finance systems and loan securitization chains into technologies of value extraction (Lapavitsas, 2009; McNally, 2009).

Once again, it was in the United States that such a financial architecture was mainly innovated. The first private-label MBS (these are MBSs typically not conforming to FHA guidelines, for selling to private, non-agency investors) was issued in 1977 by the Bank of America. But it was only after the establishment and acceptance of collateralized mortgage obligations (CMOs) and collateralized debt obligations (CDOs) in the 1990s that private structured finance evolved as a significant profit-making industry. This evolution was driven by a number of factors. First, the Basel Accord, passed in 1988 and adopted incrementally over the 1990s by various nations, provided guidelines on risk-weighting for lending that allowed lenders to hold less capital against securitizations than for other types of loans, while simultaneously encouraging investors to hold MBSs over other kinds of securities (Major, 2012). Second, declining profits and interest rates in the aftermath of the Asian Financial Crisis, and again after the 911 terrorist attacks in the United States (in part, on behalf of the US Federal Reserve led by Alan Greenspan), provoked a search for yield among institutional investors, leading the major Wall Street banks to use securitization to engineer higher-return financial assets to meet this demand (Ashton, 2009; Lysandrou, 2011). Third, the systematic use of statistical risk modelling procedures, credit scoring systems, and automated underwriting methods allowed risk to be quantified in internationally accepted ways, facilitating the global trading and pricing of such financial assets (Belsky, 2014; Buser & Mcconnell, 2014; Engelen et al., 2011).

A key technology involved with the private-label system is the simultaneous pooling and "tranching" (division) of MBSs into CMOs and (with other kinds of debt) CDOs. These new forms funnel payments to investors through the "waterfall" structure of payment allocation.[2] This process encouraged rating agencies to rate senior tranches of MBSs with triple-A ratings, thus making them acceptable to the most conservative large investors, such as pension funds. Another innovation concerned the shift from simple pass-through securities, in which anyone who purchased a portion of them received their share of whatever interest was paid into the pool, to derivative MBSs funded by asset-backed commercial paper (ABCP) based on the master trust structure with bullet or balloon payments at maturity, making them more desirable for purchase by large institutional investors around the globe.[3] A third innovation involved the creation of CDO-squared, in which very large CDO structures were built from the pooling of smaller CDOs, and the production of synthetic CDOs, in which the value of the CDO is partially produced not through claims on the flows of actual payments but instead by financial engineering through the use of credit default swaps and other derivatives (Engel & McCoy, 2011; Engelen et al., 2011). Once these "innovations" were incorporated into the world of structured finance, lenders had an interest in churning out as many loans as they could, as they profited through extreme forms of leverage (collateralizing one form of debt with another) and a profusion of fees, while the risks were held off balance sheet or offloaded onto others entirely (Engel & McCoy, 2011; Engelen et al., 2011). Because lenders were not bearing the risks, they were willing to lend far more, and in more reckless ways, than if they had to keep the loans on their books.

Although largely innovated in the United States, such products spread across the globe in varying amounts during the 1990s and early 2000s. Many nations strove to adopt an "asset-based welfare" approach to public policy (in which households were encouraged to amass wealth through private sector investment, particularly in housing, rather than relying on public services for welfare), as this approach relied on the private sector and appeared to reduce public liabilities (Doling & Ronald, 2010). France and Sweden first adopted mortgage securitization in 1990, Spain and Australia in 1993, Finland in 1994, the Netherlands and Belgium in 1996, and Italy in the early 2000s (Bailey, Davies, & Smith, 2004; Brinkhuis & van Eldonk, 2008; Kim & Cho, 2014; Van den Eynde, 1998). This funding mechanism was in addition to funding mortgages on capital markets via covered bonds, which continued to be favoured in some European nations (including France, Germany, Italy, and Denmark). The United Kingdom, and then Canada, reflecting ongoing innovation in the United States, switched to new MBS schemes based on the master trust structure in the late 1990s and 2001 respectively, which significantly increased demand for these securities (Wainwright, 2009; Walks & Clifford, 2015).

The shift to funding mortgages through international capital markets led to a massive influx of investment into urban housing sectors, particularly to speculators purchasing property to rent out. By 2011 or so, 69 per cent of mortgages in Sweden and Ireland, 53 per cent in the Netherlands, and 44 per cent in the United Kingdom had been funded through capital market securities (either MBSs or covered bonds), similar in scope to the 64 per cent in the United States (Kim & Cho, 2014: 100–3). Although some nations, including Canada, mostly maintained a public-label securitization system, this prudence did not prevent those systems from turning predatory. In Canada, for instance, the fact that banks were no longer holding the mortgages encouraged them to continually increase the amount they were willing to lend (Walks, 2014), leading to ever growing record household debt levels (Walks, 2013a). When the global financial crisis (GFC) hit, the federal government responded with programs that further encouraged banks to increase mortgage issuance, producing not only a housing bubble but also imperiling the balance sheets of both households and the federal government going forward (Walks, 2014; Walks & Clifford, 2015).

In the United States, private-label MBSs and CDOs were most associated with "subprime" lending. It typically involved mortgage loans issued to those with lower credit scores, often on non-standard terms that essentially forced borrowers to refinance ("flip") the loan often, providing additional fees to lenders. The latter included "teaser" interest rates that adjust to much higher rates after an initial period (usually, two years) and balloon payments through which the entire mortgage comes due after a short period (typically ten years). Very high interest rates, high prepayment penalties, and negative amortization were also common features of subprime loans, often in combination (Engel & McCoy, 2011; Immergluck, 2009; Squires, 2009). These kinds of terms were justified by the market completion model, by which greater variation in loan terms allows lenders to use risk-based pricing to create products that meet both borrower and investor demand (Ashton, 2009; Wyly, Moos, Kabahizi, & Hammel, 2009). The US federal government helped fuel the rise of subprime lending by requiring the GSEs to purchase triple-A rated private MBSs from 2004 onward (a time when the ratings agencies like Moody's, Standard and Poor's, and Fitch were willing to rate anything highly, even securities "structured by cows" [Smith, 2008]), as well as by pre-empting state-level anti-predatory legislation that would have limited many of the more predatory terms (Immergluck, 2009; Schwartz, 2009).

The rapid rise of predatory mortgage lending in the United States disproportionately victimized African American and Latino households, as well as lower income households, who often did not understand the terms of the loans they were agreeing to (Engel and McCoy 2011; Immergluck, 2009; Wyly et al., 2009). In a number of older deindustrializing cities, such as Detroit and Cleveland, as well as some newer cities such as Atlanta, the collapse of the subprime lending boom led to a foreclosure crisis most acutely felt in the inner cities, particularly in poorer neighbourhoods disproportionately occupied by African Americans (Immergluck, 2010; 2011).

However, a majority of the neighbourhoods badly affected by rising foreclosures were found in the suburbs (Immergluck, 2010; 2011). Indeed, in a significant number of US metropolitan areas, and in virtually all the "hot market" cities experiencing growth, the suburbs bore the brunt of the foreclosure crisis (Immergluck, 2010; 2011). As Schafran and Wegmann (2012) demonstrate in relation to the San Francisco Bay area, suburban effects related to foreclosure overlap with geography, race, and ethnicity. It was the newer suburbs disproportionately occupied by African American and Latino households that witnessed the most rapid increases in house prices in advance of the GFC, but the most rapid declines afterward. Meanwhile, whiter and wealthier areas near the core enjoyed the most rapid increase in relative prices after the GFC, widening the price gap between the gentrifying inner city and the declining suburbs (Schafran & Wegmann, 2012).

This pattern through which suburban areas disproportionately gained, and then lost, significant market value before and after the GFC, leaving homeowners in negative equity, is also evident among the cities of other nations that experienced a housing bubble and then market collapse. For instance, in Ireland, the owners of new units in suburban and exurban areas (as well as rural second homes) experienced the greatest loss, which is where unfinished estates turned into modern-day ruins (Kitchin, O'Callaghan, & Gleeson, 2014). In Spain, it was mainly fringe neighbourhoods of the large metropolitan areas, as well as seaside towns, that witnessed the most rapid increase in foreclosures and abandonment (De Weerdt & Garcia, 2016), leaving many apartment blocks empty "in suburbs rich and poor" (Moran, 2012; see also White, 2013). In retrospect, the mortgage lending boom and bust, brought on by the global proliferation of MBSs and CDOs/CMOs, worked not only to extract significant value from unsuspecting homebuyers (what David Harvey [2005] has called "accumulation by dispossession") but also became a method for extracting value from the suburbs, and the suburbanization process, on a global scale.

Automobile lending has also undergone significant transformation, again largely originating in the US experience. Until the mid-1980s, motor vehicle finance was provided directly by lenders associated with dealers, and, like the originate-and-hold model of mortgage lending, loans would remain on the initial lenders' balance sheets for the life of the loan. Typically, automobile loans would not be issued for more than four-year terms, which reduced lender risk and typically meant that purchasers remained with negative equity for roughly the first two and a half years (depending on the interest rate payable on the loan), providing roughly eight to fourteen years of debt-free ownership once the loan was paid off.

This method of financing changed in 1986 when the General Motors Acceptance Corporation (GMAC, the credit arm of General Motors) issued the first asset-backed securities (ABSs) collateralized by automobile loans (Zweig, 2002). The initial automobile loan ABSs were pass-through securities, in which investors in ABSs received their share of principal and interest payments. These, like their MBS equivalents, are subject to prepayment risk, particularly during the first three years of the loan (Roever, McElravey, & Schultz, 1998). Automobile leases have also been securitized since the mid-1980s. They are the easiest to securitize, since they involve short-term contracted payments that typically do not change over the life of the contract, thus enabling investors to largely avoid prepayment risk (Schorin, 1998). However, beginning in the early 2000s, the finance arms of the large automakers began experimenting with their own non-amortizing bullet bond structures in order to appeal to international investors, with the intention of "in many ways, replacing corporate bonds" (Kothari, 2006: 401). In the United States, credit rating downgrades to the big three automakers throughout the early 2000s made it costlier for them to raise funds directly through capital markets, further encouraging them to resort to ABS markets instead (Kothari, 2006: 402). By 2001, 31 per cent of all ABSs in the United States were originated by one of the big three automakers (Ford, General Motors, and Daimler-Chrysler), and between 1995 and 2004, the outstanding balance of US automobile loan ABSs virtually quadrupled (Kothari, 2006: 402–3). The securitization of automobile loans spread globally beginning in the late 1990s. Ford Credit – the lending arm of the Ford Motor Company – issued the first automobile loan ABSs in the United Kingdom in 1997, and the practice quickly spread to countries in Asia (Kothari, 2006).

The use of securitization to fund automobile loans, and the rising profitability of the credit arms of the major automobile manufacturers, is a significant factor in the "financialization" of the profit structures of non-financial corporations (Krippner, 2005). After a lull during the GFC and subsequent recession, origination of automobile loan ABSs once again picked up, and in 2015 a record $102 billion of new automobile loan ABSs were issued in the United States, bringing the total outstanding automobile debt close to US$1 trillion (Scully, 2016). Rising profits have attracted new issuers funded by private equity firms, which have funnelled investment into "securitizations of subprime car loans," triggering probes by the US Justice Department into possible lending abuse (McLaughlin, Schoenberg, & Robinson, 2015). As in the subprime mortgage debacle, because originators do not shoulder the risk but instead transfer it to investors, they are willing to issue larger loans for longer and more predatory terms. Loan originators have hence innovated with longer amortization terms and higher loan-to-value ratios, creating the eight-year automobile loan as the new standard amortization length (Weisbaum, 2013). While longer terms and lower interest rates allow for lower monthly payments, under typical eight-year terms, borrowers will now remain in negative equity for at least six years (Shecter, 2014). And, in some countries (such as Canada), lenders are providing loans for greater than the value of the vehicles being purchased (Marr & Shecter, 2014). While this practice allows households to consolidate other debts into automobile loans, potentially at more favourable interest rates (since automobile loans charge lower interest rates than credit cards or other unsecured loans), it also means that borrowers can never settle their debt by selling the vehicle, and it significantly extends the time over which interest payments are made and in turn total interest payments. The securitization of automobile loans has thus also helped turn automobile debt into another form of value extraction.

Since the post-war suburbs disproportionately concentrate homeowners, and since much of the post-war suburbs remain automobile dependent, this lending practice similarly means that the auto-mobile suburbs themselves have become a kind of technology extracting value from residents and funnelling it to international investors. In Jean-Paul Addie's scheme, the financial architecture has stopped being a benevolent force for the suburbs. Instead, financial innovation, increasingly collateralized by auto-mobile suburbanisms, has become in its own right a form of suburban infrastructure, but one with less-than-benevolent outcomes for current or future suburban residents.

Conclusion

Often overlooked in discussions of urban infrastructure are the financial structures that make growth and development possible. This oversight is particularly true of the financial infrastructures undergirding demand for space on behalf of various agents. This article has focused on a subset of these, namely the structural foundations and transformations involving mortgage lending and automobile loans. The initial financial infrastructure improvised during the early post-war period provided the basis for massive investment into the suburbanization process, and in turn facilitated new ways of life and new constructions of suburbanism that likely would not have existed without it. Using the Jean-Paul Addie categorization, this innovation was a financial infrastructure *for the suburbs*, and also represented one of the key infrastructures *of* post-war suburbanism.

However, declining profitability and changes occurring within capitalism, not least related to declining returns associated with certain kinds of suburbanization (including industrial suburbanisms), induced a transformation of this financial infrastructure. While still undergirding the suburbanization process, financial innovations involving the securitization of mortgages and automobile loans transformed them into vehicles for extracting value *from the suburbs*. As now the bulk of structured securities are collateralized by the flows of payments from households with little choice about whether to consume and thus must continue making their payments (particularly in automobile-dependent suburbs with no access to public transit, where it is very difficult to live without an automobile), it could be said that the dominant financial infrastructure has grown into a kind of suburban infrastructure in its own right. However, it is a form of suburban infrastructure that is no longer geared towards the promotion of the suburban, but increasingly to the extraction and dispersion of suburban value (including its diversion into investments in gentrification). Going forward, it will be important to once again reform, democratize, and transform this financial infrastructure to make it work in the public interest in order to replenish and foster progressive and diverse new suburbanisms.

NOTES

1 Even after privatization in 1968, most investors assumed that Fannie Mae and the other main GSE (Federal Home Loan Mortgage Corporation,

or Freddie Mac) enjoyed an implicit government guarantee and so saw Fannie Mae's bonds as safer than private bonds, increasing demand for them and driving down mortgage interest rates on conforming mortgages. This assumption turned out to be true when the global financial crisis hit in 2008.

2 In this structure, the senior tranches, containing the highest-rated MBSs, are paid first, followed by the mezzanine tranches. Only when the owners of those tranches have been fully paid is the lowest-rated (equity) tranche paid. Often the equity tranche would be held by the lender or the financial institution that created the CMO in order to instill confidence among investors in the senior tranches (Belsky, 2014; Fligstein & Goldstein, 2010).

3 A pool of mortgages operating with a pass-through structure involves interest payments that decline over time, and also entails prepayment risk, making them less than ideal for institutional investors who prefer bonds that pay consistent returns. When structured as regular bullet bonds, however, investors can hold the bond to maturity, receiving interest payments during that time and the original principal at maturity (see Wainwright, 2009: 380). However, this practice required complex funding structures, often through short-term funding (including ABCP), which were vulnerable to a credit crunch. It was problems with such short-term funding that led to the run on the UK bank Northern Rock (Marshall et al., 2012) and to the collapse of the ABCP market in 2007 (Loxley, 2009).

REFERENCES

Aalbers, M. (2011). *Place, exclusion and mortgage markets*. London: Wiley-Blackwell.

Aglietta, M. (1979). *A theory of capitalist regulation*. London: New Left Books.

Arrighi, G. (1994). *The long twentieth century: Money, power and the origins of our times*. London: Verso.

Ashton, P. (2009). An appetite for yield: The anatomy of the subprime mortgage crisis. *Environment and Planning A*, *41*(6), 1420–41. https://doi.org/10.1068/a40328

Bacher, J. (1993). *Keeping to the market place: The evolution of Canadian housing policy*. Montreal: McGill-Queen's University Press.

Bailey, K., Davies, M., & Smith, L.D. (2004). *Asset securitization in Australia*. Canberra: Reserve Bank of Australia. Retrieved from http://www.rba.gov.au/publications/fsr/2004/sep/pdf/0904-1.pdf

Bélec, J. (1997). The Dominion Housing Act. *Urban History Review*, *25*(2), 53–62. https://doi.org/10.7202/1016071ar

Bélec, J. (2015). Underwriting suburbanization: The National Housing Act and the Canadian city. *Canadian Geographer*, *59*(3), 341–53. https://doi.org/10.1111/cag.12175

Belsky, E.S. (2014). Mortgage markets: The United States. In H.K. Baker & P. Chinloy (Eds.), *Public real estate markets and investments* (pp. 79–98). Oxford: Oxford University Press.

Brenner, R. (2002). *The bubble and the boom: The US in the world economy*. London: Verso.

Brinkhuis, D., & van Eldonk, R. (2008). Still going strong: The Netherlands remains the jurisdiction of choice for structured finance transactions. In *Global securitisation and structured finance 2008* (pp. 215–20). London: Global White Page Ltd. Retrieved from http://www.globalsecuritisation.com/08_GBP/GBP_GSSF08_215_220_Nthds.pdf

Buser, S.A., & Mcconnell, J.J. (2014). Development of the market for mortgage-backed securities. In H.K. Baker & P. Chinloy (Eds.), *Public real estate markets and investments* (pp. 121–36). Oxford: Oxford University Press.

De Weerdt, J., & Garcia, M. (2016). Housing crisis: The Platform of Mortgage Victims (PAH) movement in Barcelona and innovations in governance. *Journal of Housing and the Built Environment*, *31*(3), 471–93. https://doi.org/10.1007/s10901-015-9465-2

Doling, J., & Ronald, R. (2010). Home ownership and asset-based welfare. *Journal of Housing and the Built Environment*, *25*(2), 165–73. https://doi.org/10.1007/s10901-009-9177-6

Elsinga, M., Priemus, H., & Cao, L. (2009). The government mortgage guarantee as an instrument in housing policy: Self-supporting instrument or subsidy? *Housing Studies*, *24*(1), 67–80. https://doi.org/10.1080/02673030802561422

Engel, K.C., & McCoy, P.A. (2011). *The subprime virus: Reckless credit, regulatory failure, and next steps*. New York: Oxford University Press.

Engelen, E., Erturk, I., Froud, J., Johal, S., Leaver, A., Moran, M., Nilsson, A., & Williams, K. (2011). *After the great complacence: Financial crisis and the politics of reform*. Oxford: Oxford University Press.

Essex, S.J., & Brown, G.P. (1997). The emergence of post-suburban landscapes on the north coast of New South Wales: A case study of contested space. *International Journal of Urban and Regional Research*, *21*(2), 259–87. https://doi.org/10.1111/1468-2427.00072

Fligstein, N., & Goldstein, A. (2010). Anatomy of the mortgage securitization crisis. In M. Lounsbury & P.M. Hirsch (Eds.), *Markets on trial: The economic sociology of the U.S. financial crisis, part A* (pp. 29–70). Bingley, UK: Emerald Group Publishing.

Freund, P., & Martin, G. (1993). *The ecology of the automobile*. Montreal: Black Rose Books.

Gaffney, M. (2009). Money, credit, and crisis. *American Journal of Economics and Sociology*, *68*(4), 983–1038. https://doi.org/10.1111/j.1536-7150.2009.00659.x

Gotham, K.F. (2009). Creating liquidity out of spatial fixity: The secondary circuit of capital and the subprime mortgage crisis. *International Journal of Urban and Regional Research*, *33*(2), 355–71. https://doi.org/10.1111/j.1468-2427.2009.00874.x

Green, R.K. (2014). *Introduction to mortgages and mortgage backed securities*. Amsterdam: Academic Press.

Harloe, M. (1995). *The people's home? Social rented housing in Europe and America*. Oxford, UK: Blackwell.

Harris, R. (1996). *Unplanned suburbs: Toronto's American tragedy, 1900–1950*. Baltimore, MD: Johns Hopkins University Press.

Harris, R. (2004). *Creeping conformity: How Canada became suburban, 1900–1960*. Toronto, ON: University of Toronto Press.

Harris, R., & Forrester, D. (2003). The suburban origins of redlining: A Canadian case study, 1935–1954. *Urban Studies*, *40*(13), 2661–86. https://doi.org/10.1080/0042098032000146830

Harvey, D. (2005). *A brief history of neoliberalism*. Oxford: Oxford University Press.

Harvey, D. (2006 [1982]). *The limits to capital*. London: Verso.

Harvey, D. (2012). *Rebel cities: From the right to the city to the urban revolution*. London: Verso.

Hoagland, H.E., & Stone, L.D. (1969). *Real estate finance*. Homewood, IL: Richard D. Irwin Inc.

Immergluck, D. (2009). *Foreclosed: High-risk lending, deregulation, and the undermining of America's mortgage market*. Ithaca, NY: Cornell University Press.

Immergluck, D. (2010). Neighborhoods in the wake of the debacle: Intra-metropolitan patterns of foreclosed properties. *Urban Affairs Review*, *46*(1), 3–36. https://doi.org/10.1177/1078087410375404

Immergluck, D. (2011). The local wreckage of global capital: The subprime crisis, federal policy, and high-foreclosure neighbourhoods in the US. *International Journal of Urban and Regional Research*, *35*(1), 130–46. https://doi.org/10.1111/j.1468-2427.2010.00991.x

Jackson, K.T. (1985). *Crabgrass frontier: The suburbanization of the United States*. Oxford: Oxford University Press.

Karley, N.K., & Whitehead, C. (2002). The mortgage-backed securities market in the U.K.: Developments over the last few years. *Housing Finance International*, *17*(2), 31–6.

Kim, K.-H., & Cho, M. (2014). Mortgage markets: International. In H.K. Baker & P. Chinloy (Eds.), *Public real estate markets and investments* (pp. 97–120). Oxford: Oxford University Press.

Kitchin, R., O'Callaghan, C., & Gleeson, J. (2014). The new ruins of Ireland? Unfinished estates in the post-Celtic Tiger era. *International Journal of Urban and Regional Research*, *38*(3), 1069–80. https://doi.org/10.1111/1468-2427.12118

Kothari, V. (2006). *Securitization: The financial instrument of the future*. Hoboken, NJ: Wiley & Sons.

Krippner, G. (2005). Financialization of the American economy. *Socio-Economic Review*, *3*(2), 173–208. https://doi.org/10.1093/ser/mwi008

Lapavitsas, C. (2009). Financialized capitalism: Crisis and financial expropriation. *Historical Materialism*, *17*(2), 114–48. https://doi.org/10.1163/156920609x436153

Lassiter, M.D. (2004). The suburban origins of "colour-blind" conservatism: Middle-class consciousness in the Charlotte busing crisis. *Journal of Urban History*, *30*(4), 549–82. https://doi.org/10.1177/0096144204263812

Lees, L., Slater, T., & Wyly, E. (2008). *Gentrification*. London: Routledge.

Lefebvre, H. (2003 [1970]). *The urban revolution*. N. Bononno (Trans.). Minneapolis, MN: University of Minnesota Press.

Loxley, J. (2009). Financial dimensions: Origins and state responses. In J. Guard & W. Antony (Eds.), *Bankruptcies and bailouts* (pp. 62–75). Halifax, NS: Fernwood Press.

Lucy, W.H., & Phillips, D.L. (1997). The post-suburban era comes to Richmond: City decline, suburban transition, and exurban growth. *Landscape and Urban Planning*, *36*(4), 259–75. https://doi.org/10.1016/s0169-2046(96)00358-1

Lysandrou, P. (2011). Global inequality as one of the root causes of the financial crisis: A suggested explanation. *Economy and Society*, *40*(3), 323–44. https://doi.org/10.1080/03085147.2011.576848

Major, A. (2012). Neoliberalism and the new international financial architecture. *Review of International Political Economy*, *19*(4), 536–61. https://doi.org/10.1080/09692290.2011.603663

Marr, G., & Shecter, B. (2014, 22 November). How Canada's auto loan bubble has become a ticking time bomb. *National Post*. Retrieved from http://business.financialpost.com/2014/11/22/how-canadas-auto-loan-bubble-has-become-a-ticking-time-bomb/

Marshall, J.N., Pike, A., Pollard, J.S., Tomaney, J., Dawley, S., & Gray, J. (2012). Placing the run on Northern Rock. *Journal of Economic Geography*, *12*(1), 157–81. https://doi.org/10.1093/jeg/lbq055

Massey, D.S., & Denton, N.A. (1993). *American apartheid: Segregation and the making of the underclass*. Cambridge, MA: Harvard University Press.

McLaughlin, D., Schoenberg, T., & Robinson, M. (2015, 24 February). Auto loan securitization probed by U.S., states, Yates says. *Bloomberg News*.

Retrieved from http://www.bloomberg.com/news/articles/2015-02-24/justice-department-probing-auto-loan-securitization-yates-says

McNally, D. (2009). From financial crisis to world-slump: Accumulation, financialisation, and the global slowdown. *Historical Materialism*, *17*(2), 35–83. https://doi.org/10.1163/156920609x436117

Miron, J.R. (1988). *Housing in postwar Canada*. Montreal: McGill-Queen's University Press.

Moos, M., & Mendez, P. (2015). Suburban ways of living and the geography of income: How homeownership, single-family dwellings and automobile use define the metropolitan social space. *Urban Studies*, *52*(10), 1864–82. https://doi.org/10.1177/0042098014538679

Moran, L. (2012, 12 February). Spain's ghost towns: Built during the boom years but now lying empty... as jobless total tops five million. *Daily Mail*. Retrieved from http://www.dailymail.co.uk/news/article-2102074/Spain-haunted-ghost-towns-built-boom-years-unemployment-tops-5million.html

Murdie, R. (1986). Residential mortgage lending in metropolitan Toronto: A case study of the resale market. *The Canadian Geographer*, *30*(2), 98–110. https://doi.org/10.1111/j.1541-0064.1986.tb01035.x

Murdie, R. (1991). Local strategies in resale home financing in the Toronto housing market. *Urban Studies*, *28*(3), 465–83. https://doi.org/10.1080/00420989120080451

Nicolaides, B. (2006). How hell moved from the city to the suburbs. In K.M. Kruse & T.J. Sugrue (Eds.), *The new suburban history* (pp. 80–90). Chicago, IL: University of Chicago Press.

Paterson, M. (2007). *Automobile politics: Ecology and cultural political economy*. Cambridge: Cambridge University Press.

Peck, J. (2011). Neoliberal suburbanism: Frontier space. *Urban Geography*, *32*(6), 884–919. https://doi.org/10.2747/0272-3638.32.6.884

Poapst, J.V. (1993). Financing of post-war housing. In J.R. Miron (Ed.), *House, home and community: Progress in housing Canadians, 1945–1986* (pp. 94–109). Ottawa, ON: Canada Mortgage and Housing Corporation.

Rae, D. (2003). *City: Urbanism and its end*. New Haven, CT: Yale University Press.

Roever, W.A., McElravey, J.M., & Schultz, G.M. (1998). Auto loan asset-backed securities. In F.J. Fabozzi (Ed.), *Handbook of structured financial products* (pp. 123–38). Hoboken, NJ: Wiley & Sons.

Schafran, A., & Wegmann, J. (2012). Restructuring, race, and real estate: Changing home values and the new California metropolis, 1989–2010. *Urban Geography*, *33*(5), 630–54. https://doi.org/10.2747/0272-3638.33.5.630

Schorin, C.N. (1998). Automobile lease ABS. In F.J. Fabozzi (Ed.), *Handbook of structured financial products* (pp. 139–52). Hoboken, NJ: Wiley & Sons.

Schwartz, H. (2009). *Subprime nation: American power, global capital, and the housing bubble*. Ithaca, NY: Cornell University Press.

Scully, M. (2016, 14 January). Deutsche Bank said to probe subprime auto securitizations. *Bloomberg News*. Retrieved from http://www.bloomberg.com/news/articles/2016-01-14/deutsche-bank-said-to-probe-sales-of-subprime-auto-securities

Seko, M. (1994). Housing finance in Japan. In Y. Noguchi & J. Poterba (Eds.), *Housing markets in the US and Japan* (pp. 49–64). Chicago, IL: National Bureau of Economic Research (NBER)/University of Chicago Press. Retrieved from http://www.nber.org/chapters/c8821.pdf

Shecter, B. (2014, 23 October). Canadians may be buying "too much car": Moody's sounds alarm on bank auto loans. *National Post*. Retrieved from http://business.financialpost.com/2014/10/23/canadians-may-be-buying-too-much-car-moodys-sounds-alarm-on-bank-auto-loans/

Simpson, H.D. (1933). Real estate speculation and the depression. *American Economic Review*, *23*(1), 163–71.

Smith, N. (1979). Toward a theory of gentrification: A back to the city movement by capital, not people. *Journal of the American Planning Association*, *45*(4), 538–48. https://doi.org/10.1080/01944367908977002

Smith, Y. (2008, 22 October). S&P: "We'd do a deal structured by cows" and other rating agency dirty linen. *Naked Capitalism* (blog). Retrieved from http://www.nakedcapitalism.com/2008/10/s-wed-do-deal-structured-by-cows-and.html

Squires, G.D. (2009). Inequality and access to financial services. In J. Niemi, I. Ramsay, & W.C. Whitford (Eds.), *Consumer credit, debt, and bankruptcy: Comparative and international perspectives* (pp. 11–31). Oxford: Hart Publishing.

Toporowski, J. (2000). *The end of finance: The theory of capital market inflation, financial derivatives, and pension fund capitalism*. London: Routledge.

Van den Eynde, P. (1998). Securitization in Europe: Overview and recent developments. In F.J. Faboozi (Ed.), *Handbook of structured financial products* (pp. 9–20). Hoboken, NJ: Wiley & Sons.

Volti, R. (1996). A century of automobility. *Technology and Culture*, *37*(4), 663–85. https://doi.org/10.2307/3107094

Wacquant, L. (2007). *Urban outcasts: A comparative sociology of advanced marginality*. Cambridge: Polity Press.

Wainwright, T. (2009). Laying the foundations for a crisis: Mapping the historico-geographical construction of residential mortgage backed securitization in the UK. *International Journal of Urban and Regional Research*, *33*(2), 372–88. https://doi.org/10.1111/j.1468-2427.2009.00876.x

Walks, A. (2013a). Mapping the urban debtscape: The geography of household debt in Canadian cities. *Urban Geography*, *34*(2), 153–87. https://doi.org/10.1080/02723638.2013.778647

Walks, A. (2013b). Suburbanism as a way of life, slight return. *Urban Studies, 50*(8), 1471–88. https://doi.org/10.1177/0042098012462610

Walks, A. (2014). Canada's housing bubble story: Mortgage securitization, the state, and the global financial crisis. *International Journal of Urban and Regional Research, 38*(1), 256–84. https://doi.org/10.1111/j.1468-2427.2012.01184.x

Walks, A. (2015). Driving cities: Automobility, neoliberalism, & urban transformation. In Walks, A. (Ed.), *The urban political economy and ecology of automobility: Driving cities, driving inequality, driving politics* (pp. 3–20). London: Routledge.

Walks, A., & Clifford, B. (2015). The political economy of mortgage securitization and the neoliberalization of housing policy in Canada. *Environment and Planning A, 47*(8), 1624–42. https://doi.org/10.1068/a130226p

Walks, A., & Maaranen, R. (2008a). Gentrification, social mix, and social polarization: Testing the linkages in large Canadian cities. *Urban Geography, 29*(4), 293–326. https://doi.org/10.2747/0272-3638.29.4.293

Walks, A., & Maaranen, R. (2008b). *The timing, patterning and forms of gentrification and neighbourhood upgrading in Montreal, Toronto, and Vancouver 1961 to 2001.* (Cities Centre Research Paper 211). Toronto, ON: University of Toronto.

Walks, A., & Simone, D. (2016). Neoliberalization through housing finance, the displacement of risk, and Canadian housing policy: Challenging Minsky's financial instability hypothesis. *Research in Political Economy, 31,* 49–77. https://doi.org/10.1108/s0161-723020160000031004

Weisbaum, H. (2013, 18 April). Car loans stretch out to 8 years, costing borrowers more. *CNBC News*. Retrieved from http://www.cnbc.com/id/100654029

Wheaton, W.C. (1977). Income and urban residence: An analysis of consumer demand for location. *The American Economic Review, 67*(4), 620–31.

White, S. (2013, 6 May). Bankers whisper: Spain's bailout bill could rise. *Reuters*. Retrieved from http://www.reuters.com/article/us-spain-banks-idUSBRE94509020130506

Woodard, H. (1959). *Canadian mortgages*. Toronto, ON: Collins.

Wyly, E., Moos, M., Kabahizi, E., & Hammel, D. (2009). Cartographies of race and class: Mapping the class-monopoly rents of American subprime mortgage capital. *International Journal of Urban and Regional Research, 33*(2), 343–54. https://doi.org/10.1111/j.1468-2427.2009.00870.x

Zweig, P.L. (2002). Asset-backed securities. *The Concise Encyclopedia of Economics*. Retrieved from http://www.econlib.org/library/Enc1/AssetBackedSecurities.html

SECTION 2

Suburban Infrastructures in Crisis

5 Phases of Neoliberal Infrastructure: Test Zones of Post-Soviet Europe

FREDERICK PETERS

Urban–suburban networks of water infrastructure in Eastern Europe, in their hydromorphological and social historical contexts, signal the character of the transformations of the social forces determining municipal infrastructure in the region, but also speak to gradual shifts in phases of "actually existing neoliberalism" (Brenner & Theodore, 2002). Post-communist Eastern Europe in many ways tested the viability of neoliberal strategies concerning privatization of public assets such as municipal infrastructure. Now, as we are approaching three decades of post-communism or, as the European Bank for Reconstruction and Development (EBRD) referred to it, "transition" – to a presumed liberal democratic model of capitalism – analysis of the experiences and their outcomes in the region allows for some nuanced assessment of those strategies.

Three main arguments are laid out in this chapter. First, the restructuring of ownership models in municipal water infrastructure in Eastern Europe was a manifestation of the attempts at neoliberal restructuring of water services globally. We will explore this phenomenon though the state–market negotiation of public–private partnerships (PPPs) as seen in the Eastern European cities of Gdańsk and Tallinn. These two cities were, in their respective countries, Poland and Estonia, the first ones to fully adopt this trajectory. Second, while this restructuring was founded on both European Union (EU) and local regulations, as well as supported by mandates of international financial institutions such as the EBRD, it also reflected local political upheaval and contradictions inherent in neoliberal strategies, such as the need for state management of "free" markets. Third, ownership of assets such as water services infrastructure should be recognized as being along a continuum between public and private, with high degrees of social forces struggle in the management of that relationship. We can talk of PPP phases, versions (V) to use a computer software metaphor. Like software, versions have glitches; fixes try to address them and can create others. Political determination of the negotiation of water infrastructure suggests the possible renegotiation of those relationships to increase democratic legitimation, making another version of the public–private partnership into a new phase also possible. Periodization is required to capture the nuances of shifting neoliberal strategies and practical responses to investment pressures over the past thirty years.

Water infrastructure – pipes and treatment plants, pumping stations, and the like – or, similarly, power grids, transport, or other communication networks have traditionally been understood in technical terms, the domain of experts with instrumental perspectives on how to define adequate service provision (Swyngedouw, 2003; 2004). The restructuring of ownership models to include private capital in partnerships with public ownership was critiqued in the scholarship of geographers such as Swyngedouw (2004), Kaika (2005), and Bakker (2010). These researchers portrayed this restructuring as central to later twentieth century capitalist expansion projects. At the same time, scholars such as Fine (1999), Leys (2001), Saad-Filho and Johnson (2005), and Neil Smith (1984; 2006) have exposed the relationship between the privatization of infrastructure and rising social inequity. But neoliberalism is not a unitary strategy, as Castree (2006) demonstrated. There are many facets to this phenomenon. Market supremacy is a hallmark of neoliberalism, but how that played out was always locally determined, and not simply – as in the cases of post-communist countries such as the Baltic countries and Poland, which I focus on here – in centres of financial and political power such as the EBRD halls at Bishopsgate, London, and the EU offices in Brussels and Strasbourg. As technical requirements for acceptable levels of organic and inorganic compounds that may be consumed as drinking water or expelled as wastewater became a significant health and environmental concern, the need to reorganize water services became a highly charged political matter. In some places, how to pay for water infrastructure upgrading raised questions of social equity and justice.

Defining phases in the evolution of water system ownership from initial privatization initiatives can help capture impacts on social equity. I focus on Gdańsk, Poland, and Tallinn, Estonia, comparing these two versions of public–private partnership, as these cities were first in their respective countries to bring the private sector into their water infrastructure. Building on Matti Siemiatycki's (2012; 2013) sense of periodization of public–private partnerships (PPPs), I suggest that it makes sense for the two cases to define the initial period, 1990s to 2005, as V1.0. V2.0 covers 2005 to 2015, a time span distinguished from the initial period by efforts to fix the "bug" of legitimacy through claims of renewed public oversight over the original privatization, made with more or less conviction. This period does not yet conform to Karl Polanyi's (1944) description of a pendulum swing back towards the public sector as a correction for capitalism's excesses. V3.0 of PPPs in water management is potentially on the horizon, as the shift would bring in a new version of social partnership, which could address present infrastructure ownership and management issues. Maintaining the precarious balance of risk and finance between the public sector and corporations will give rise to political struggles among water system operators, the corporations and financial institutions, public officials, critics in the local media, and the affected public who pay for water services. The emergence of what we could call V3.0 of water system infrastructure management is inevitable, as the precariousness of present ownership and management arrangements cannot last forever. But the form such a V3.0 would take is still uncertain. Therein lies the potential for higher levels of social equity in the way water services are addressed and financed, in other words, a politics of democratic infrastructure management.

Research Process

The research into the specific cases of Gdańsk and Tallinn was conducted in two major phases. Initial field work occurred in 2008, the year of the subprime collapse, and was limited to corporate research through databases accessed online at the Ashridge Business School library and meetings with senior EBRD officials responsible for investments in the region. Through a snowball methodology, each person I met suggested further respondents and assisted greatly with understanding the bank's perspective. Field work in Gdańsk and Tallinn was delayed until 2009; having cultivated email correspondence with academics located in the two cities and written to the private companies, I arrived on site to reach out to previously contacted respondents. All the interviews I conducted were followed up with fact checking.

Background: Peripheral Infrastructure and Central Planning

Water infrastructure governance was restructured in the post-Soviet period in Eastern Europe. In Poland, Estonia, and the other countries that aimed to join the EU in 2004, it happened surprisingly quickly.

Countries dependent on the Soviet Union were in a financial crisis throughout the 1970s and 1980s, and this crisis played out as water quality deterioration became intolerable. Oil price hikes in the 1970s and reduction in oil deliveries from the Soviet Union in the early 1980s enhanced dependence on domestic coal, low-quality lignite, for electricity and heat across Eastern Bloc countries. The economic crisis of state socialism in the 1980s therefore led to increased environmental degradation. Environmental laws pre-date the move to liberal democracy throughout the Soviet Bloc, even if their enforcement often conflicted with other priorities in terms of environmental damage mitigation. There was "a surprising correspondence between the neo-liberals after 1989 and communist politicians before 1989 in the ways in which they understood and acted in regard to the environment" (Pavlínek & Pickles, 2004: 243).

However, as Eastern European nations transitioned to open-market economies and democratic forms of government, they needed to confront their legacy of environmental degradation. They were requested to harmonize their environmental controls and legal structures with their counterparts in Western Europe (eventually codified in the Water Framework Directive [WFD] of 2000, part of the *Acquis communautaire*) in order to receive assistance.

Water infrastructure, while a matter of civil engineering embedded in local hydromorphic conditions, is also expected to meet socially negotiated targets. The restructuring of post-communist legal forms is grounded in environmental and social engineering. The WFD is the result of, and resulted in, a new code for water services infrastructure (European Parliament and Council, 2000). Water regulations in the EU were rationalized through the 1990s and amalgamated into one piece of legislation. The WFD subsumed prior EU environmental laws and applied to Poland, the Baltic nations, and all those countries that intended to eventually join the EU – the "accession" countries. The WFD evolved through a power struggle between the elected European Parliament, the appointed European Commission, and the appointed Council of Ministers (Kaika & Page, 2003a: 325). The complex process that led to the WFD was a primary example of disagreement over the valuation of water that reflects the contradictions in the way the environment is "managed" in Europe, above all in relation to water infrastructure investment.

European institutional power over environmental issues shifted. First, legislative power moved from the appointed Council of Ministers towards the elected European Parliament at a time when the former championed a decidedly more market-oriented version of the WFD. The European Parliament leaned towards a heavier handed vision of the treaty, where the directives would be mandatory for all member states; it also sought input from environmentalists, including the World Wildlife Fund and the European Environmental Bureau. At the same time, a second power shift brought on by this process was the increased influence of unelected bodies on legislation. Lobby groups – private firms in the water industry and their representatives, as well as environmental non-governmental organizations – gained great influence. Kaika and Page (2003a: 316) see the significance of this shift in terms of a transition from government to governance, which they define as a neoliberal, non-democratic form of administration.

Under the EU, management of Western European rivers, such as the Rhine or the Maas, had been addressed by cross-border agreements. But under the new directive, there was a scalar shift from national to river basin management agencies. Groundwater and coastal waters were, like river basins, subject to environmental objectives established within the legislation (European Parliament and Council, 2000: 11). Contrary to the 1992 Dublin Conference (International Conference on Water and the Environment [ICWE], 1992), the directive was explicit in defining water as a renewable resource. The directive rejected the notion of water as primarily an economic good. But it also raised the notion of "full-cost recovery," without defining the scope of which costs were legitimate and which were not (Kaika & Page, 2003b: 332). The WFD differentiated between "full-cost recovery" as a principle in water management accounting for the environmental costs of water treatment and the different uses of water, and the principle of "polluter pays" to mitigate environmental degradation. This differentiation left a wide berth for practical implementation (European Parliament and Council, 2000: art. 9). In theory, public consultation was enshrined in the WFD: paragraph 14 of the preface emphasizes that the success of the directive "relies on close cooperation and coherent action at Community, Member State and local level as well as on information, consultation and involvement of the public, including users" (European Parliament and Council, 2000: 2). In practice, this statement did not amount to much. For instance, opening up water pricing decisions was a struggle for municipalities that were partnered with private water company operators. In the case of Tallinn, the private operator – after years of struggle with the municipality of Tallinn – turned to the EU in 2010 with a formal complaint about national Estonian legislation aimed at curbing price hikes. To date, the dispute remains unsettled (Tallinna Vesi, 2016).

The WFD enshrined legal contestation and addressed environmental degradation in the region. At the same time, it tended to promote a strategy for the valorization of water under an undeclared pro-privatization philosophy, based on the view that the provision and treatment of water under new rules required intensive capital influx. This thinking was precisely the kind of regulatory response that Polanyi (1944) suggested laissez-faire capitalism engenders. Polanyi wrote about the formation of countermovements to unbridled "free markets," which aimed to use legislation to minimize the social suffering that capitalism engendered. But he also noted that such movements only exacerbated capitalism's contradictions, including environmentalist capital accumulation strategies, which James O'Connor (1988) called the "second contradiction of capitalism."

Local legal and environmental rules put in place to conform to EU membership requirements had financial implications for municipal governments, which now controlled their water infrastructures but did not yet have the capacity to both govern and finance them. From being central state–controlled entities, water management organizations were transformed into municipally owned companies. These companies, now subject to significant investment requirements in order to meet environmental regulations, became, in certain cases such as in Gdańsk and Tallinn, objects of interest for international financial organizations and multinational corporations. In this fashion, a market was created for water services delivery, thus leading to the circumstances for the first negotiations for PPPs in the water services sector, V1.0.

The Backstory: Historical Periodization

Tallinn's history parallels the early history of Gdańsk. Like many Hanseatic coastal trading towns in the region, they both began as medieval port cities, ruled by different administrations over time. They were ethnically diverse, and social striations tended to reflect the presence of different ethnic groups and divisions between working, merchant, and political classes. The city administrations did not control their own city, or if they did, in exceptional cases, it was not for long. The cities were mostly under the aegis of some greater nearby power. Their histories, and the histories of their respective countries, Estonia and Poland, reflect their subjugation to the imperial ambitions of militaristic powers. Seen from a water infrastructure developmental history, their different experiences stem from variations in local hydromorphology or water sources. Kowalik and Suligowski (2001) trace the technical history of Gdańsk's development from the fourteenth century canal building (the Radunia canal) for drinking water based on Roman gravity-fed aqueduct technology.

Urban modernization in the European context is defined in part by the building of urban water systems. Industrialization and the concomitant urban population expansion of the mid-nineteenth century brought disease and stench, which for about forty years led to significant investment in water service infrastructure across the continent. By 1900, every major city in Europe, from London to St Petersburg, had made some sort of significant water infrastructure investment. A historical perspective suggests that substantial returns on such investment could be achieved in the early modernizing context, as studies of the Berlin water infrastructure in the 1870s onward show (Dinckal & Mohajeri, 2001).

The building of centralized systems for water treatment in Gdańsk was initially started in the 1870s (the Heubude/Stogi plant), with significant additions built in the 1930s (the Zaspa plant) and the 1970s (the Wschód plant, completed in 2000) (Swinarski, 1999). These projects were mostly paid for by the municipality, but the construction of Zaspa began in a period of transition from Weimar to Third Reich Germany, and Wschód began under the Polish Soviet Socialist Republic (SSR) and was completed with EU and Scandinavian financial assistance.

Hanni (1999) chronicles the Tallinn water system development and modernization, where water was supplied by rivers and the nearby Lake Ülemiste. A treatment plant for drinking water was completed in the 1920s using artesian wells dug from the 1840s. Well digging was expanded throughout the mid-twentieth century, and many wells were dug during the German occupation in the 1940s and later used for industry in the Estonian SSR period. Many of these wells remained unaccounted for in the privatization of the system in the 2000s. Sewerage in Tallinn remained, until the 1970s, a system of pipes under the streets, which dumped raw sewage into the sea. A treatment plant was not completed in Tallinn until 1984 (four years after Olympic sailing took place in Tallinn Bay). In both cities, portions of their current water systems continue to use parts of these older systems.

Periodizing Neoliberal Infrastructure

The Soviet system collapsed in domino fashion between 1989 and 1992 in the eastern Baltic region, followed by a significant wave of neoliberalism – from about 1991 to 2005 – encouraged by international financial institutions and EU governance bodies. This period marked the beginning of neoliberal infrastructure in Eastern Europe. I posit that, in the 2005–2015 period, a new version of neoliberalism developed – a new public–private partnership, a PPP V2.0. These developments can be explored through that period's evolution, as Gdańsk and Tallinn tried to address the "mistakes" made in the initial privatization of their water systems. An evolution also took place in the methods used to manage conflicts in the relationships between private and public bodies and between populations and the water systems these bodies managed. This evolution constitutes what I am calling a potential PPP V3.0.

Neoliberal restructuring has in many ways exposed the "black box" of corporate water management, be it as municipal entities or market-driven public–private partnerships. Empirical studies looking into the cases of Gdańsk, Poland, and Tallinn, Estonia, ground a discussion of what neoliberal infrastructure means, but also allow for a dialogue about what a new mode of water system management could look like. Both cities share a twentieth century experience of multiple regime changes, accompanied by hardship for the population. Both cities also share the post-Soviet experience of rapid radical restructuring of their economies in the early 1990s, externally and internally driven by a neoliberal turn towards the privatization and devaluation of the concept of "public." With post-communist restructuring, local municipal governments now had control of their water infrastructures but did not yet have the capacity to govern and finance them. This situation led to a high degree of international financial interest in Gdańsk and Tallinn.

Gdańsk

The first period of privatization in Poland can be seen as the re-municipalization of the water system administration by devolving its control from the national level in Warsaw to the municipality. But this transition also involved an internationalization of the financing for its reconstruction through the EU, the EBRD, and by involving the French SAUR water company. In the 1990s, Gdańsk's municipally controlled water system was transformed into an "asset" and partially privatized, with 51 per cent in French hands, as SAUR Neptune Gdańsk (SNG). V1.0 of the privatization agreement was thus a simple majority ownership by a private firm. Warsaw's central control was reduced, and the city was allowed to (and in fact had to) negotiate with outside entities more directly – other cities, as well as corporate and international financial entities – thereby taking over responsibilities that had previously been handled by the central state. As Poland ended Soviet-style communist rule and turned westward to re-engage with Europe, Gdańsk once more became an internationalized city, looking for economic growth by reintegrating into the Baltic region and Europe.

Privatization V2.0 began in Gdańsk in 2005. Evolving from the original privatization arrangement, three city staff (answerable to politicians) now had the engineering and finance departments of SNG (answerable to the SAUR head office in Paris) as their counterparts in the relationship. At first, it was a tense marriage between a large international concern and a politically splintered municipality, which brought uneven capacity to the relationship. Then the two entities came to a new arrangement (D. Skuras, personal communication, 12 October 2009). On 1 January 2005, the city took ownership of the water infrastructure via an asset holding company, Gdańsk Water and Canal Infrastructure Corporation (GIWK, *Gdańska Infrastruktura Wodociągowo-Kanalizacyjna Sp z o.o.*), leasing the infrastructure to SNG. European Commission money supported GIWK projects to improve sewerage and carry out pipeline work, which SNG had half-heartedly undertaken (Gdańsk Water and Canal Infrastructure Corporation [GIWK], 2010). GIWK solved the city's management difficulties by clarifying the respective responsibilities of the city and the water company, and thereby addressing the confusion in the public/private distribution of responsibilities stemming from the 1992 agreement with SAUR.

The creation of GIWK answered the need to repair the damage made during the previous PPP. Dimitris Skuras, director of the Department of Public Utilities of the City of Gdańsk, identified the problem the city had in the past with SNG: insufficient staffing and managerial capacity (D. Skuras, personal communication, 12 October 2009). City officials saw GIWK's role as making that relationship more even.

The EBRD's own donor report of 2007 praised the "innovative" arrangement in Gdańsk's water system – the partnership between a private sector operator and a publicly owned asset holding company, SNG and GIWK. This arrangement extended, in their words, "the benefits of grant financing to a public-sector entity in investments where there is also a private sector operator" (European Bank for Reconstruction and Development [EBRD], 2007). The EBRD perceives this partnership as a model for structural change in Eastern Europe, where private sector involvement is facilitated and public–private partnerships coexist with EU funding, since EU Cohesion Fund money is granted to government bodies, not to private corporations. It is a situation where the market is clearly supported by the municipal state, even as GIWK seeks to protect city interests by repairing the uneven capacities of contract enforcement that came from the original privatization deals (1992, and renegotiated in 1996).

Skuras mentioned that EU and EBRD officials had met with him and GIWK staff to regulate financial matters and ensure that all was in order. No one had asked him before if the arrangements "worked" in the interests of the city and water infrastructure management. It did work, he said.

What was, in the early 1990s, an awkward marriage centred on the running of water services in Gdańsk became, by 2005, an "arrangement" secured by the EBRD and European Commission financing for GIWK, so that SNG was no longer responsible for water infrastructure investment. The burden of investment (for pipes across the city, and so on) had been lifted from SNG and placed onto GIWK, an entity eligible for European Commission (and EBRD) support. These funding arrangements amounted to indirect public money (on a local and international EU level) supporting private sector profit, given that SNG was 51 per cent privately owned. Thomas Maier, EBRD director for municipal and environmental infrastructure, said that these arrangements will not only "modernise the city's water network. They will also *facilitate the involvement of the private sector by improving the efficiency and mitigating the risk* [emphasis added], thus demonstrating how public-private partnerships can coexist with EU grants" (EBRD, 2006).

The loan to GIWK (the 2005 funding for GIWK was approximately EUR 133 million, with 75 per cent of it coming from the EU Cohesion Fund and EUR 12 million coming from EBRD) was the first EBRD loan to the water system of Gdańsk. The EBRD showed early interest in the region, but the SAUR deal did not require specific EBRD money in 1992 and had the support and urging of the French government behind it. EU Cohesion Fund grants, the EBRD, and other quasi-public funds made the utility (SNG) an attractive investment for the private sector. In Gdańsk, by 2005, flushing the toilet and, of course, paying the household water bill became of international interest.

The large water companies began divesting themselves of assets outside their core activities to focus on their water utilities. Private equity groups and institutional investors stepped into the spaces these companies vacated. The construction company Bouygues sold SAUR, including its 51 per cent holdings of SNG and other operations in France and

Russia, to the French private equity firm PAI in 2005. The controlling interest has since been sold again, and is held by a consortium of French capitalists. After the sale to PAI, private equity interest in water utilities was reflected in the 2007 joint purchase of France's SAUR water by a French nationalized bank (the Caisse des dépôts et consignations), AXA Investment Managers, and AXA Private Equity in partnership with Séché Environnement, a French waste management company. This purchase affected SAUR Neptune Gdańsk.

A main difference from their prior involvement is that these consortiums are looking for safe investment sites, rather than solely taking over infrastructures from which short-of-cash governments are divesting, hence the rise of institutional investors and hedge funds. Macquarie Group for instance, has been active in the buy-up of the German RWE (involved in the energy, heating, and disposal sectors) holdings and has attempted to buy up SAUR's out-of-France holdings. Veolia, one of the two biggest French water companies, was buying up United Utilities assets in 2010, although not Tallinn's water company, owned in part by them and the EBRD.

Tallinn

Like in Gdańsk, within ten years of independence, conflict between Estonian local elites ran along an ideological divide concerning the role of private infrastructure investment by foreign private capital. Estonians and Poles have had to deal with the communist political legacy since independence from Moscow. Following the same path as Gdańsk, Tallinn's water infrastructure was also privatized after the fall of the Soviet Union. A British company, United Utilities Group Plc of Warrington, England (UU), purchased a controlling interest in Tallinn's waterworks in a consortium with an Italian and a US-based firm to form the International Water United Utilities (IW/UU). Tallinn privatized later than Gdańsk, in 2001, but Tallinn began financing infrastructure much earlier than Gdańsk, brokering the PPP arrangement through the EBRD. The development bank still retains a significant 20 per cent interest. Yet, in Tallinn itself, there remains much disagreement around how this privatization of water services occurred and what were the circumstances surrounding its implementation.

Capital requirements for infrastructure renewal in the 1990s triggered a budgetary crisis in Tallinn, causing the new Estonian ruling elite to see state-owned assets in a new light. V1.0 began unfolding.

This history has direct connections to the formation of the new ruling classes in Estonia; hence the same names appear throughout the history of the privatization of Tallinn's water system. While the Scandinavians (including the Finns) stepped in to provide Estonia with technical and managerial assistance, the municipality had to confront interest in privatization on the part of local financial institutions and the political class. The new independent state and local governments became closely tied with the banks. In Tallinn, politicians tried to align the networked ecology of the city to their ideological trajectories. They sought to tie the city water system to the logic of capital accumulation.

In 1996, under new Estonian commercial laws, the city began the process of corporatizing the system by restructuring the water infrastructure into a shareholder company. In May 1997, the municipal council restructured the Tallinn Waterworks and Sewerage Municipal Enterprise, incorporating it as the joint stock company, AS Tallinna Vesi (ASTV, translated as Tallinn Water Ltd.). It was capitalized at EEK 850 million (DM 10.62 million at the time, EUR 5.43 million), EEK 10 per share. The 85 million shares remained in the possession of the City of Tallinn (De La Motte, 2005). ASTV could, by 1999, deliver high-quality drinking water to the city and adequately treat wastewater from Tallinn, as reported by Toivo Eensalu, former director of the municipal water company from 1984 to 2000 and ASTV from 1997 (T. Eensalu, personal communication, 21 October 2009).

As Eensalu and Vello Ervin, senior bureaucrats of the city, confirm, the initial political discussion around selling off assets of the water works had focused on limiting the percentage of shares on offer to 30 per cent (V. Ervin, personal communication, 22 October 2009). From ASTV management's perspective, partnering with a water company from the West would finance a needed investment in connection pipes and provide management expertise in running the Tallinn water system more efficiently. They reasoned that an investor could offer management and technical assistance.

The decisions made about privatization were taken in a period when the position of mayor was held by five different people over a span of four years. These were years of ethical as much as ideological struggle. Against the recommendation of ASTV management, then mayor Jüri Mõis and council made the political decision in December 2000 to sell a *controlling* interest of 50.4 per cent of ASTV to a consortium of three firms – Italy's Edison SpA, US-based Bechtel, and UK-based United Utilities International – named the International Water United

Utilities (IW/UU) (Lobina, 2001: 7). After the initial privatization, very little further improvement in the water system occurred, according to Eensalu, who reported staff reduction at the laboratory and cleaning facilities. Toivo Eensalu was relieved of his post once IW/UU gained a controlling interest in the company in 2001. He then became the director and later a board member of a neighbouring municipality's water services, Viimsi Vesi Ltd. Considerable controversy surrounded the original privatization arrangements and the subsequent price and service levels.

In the last meeting of the city council before Christmas 2000, thirty items were on the agenda, including five contracts with IW/UU, totalling 371 pages. Tallinn finalized the decision to sell its controlling interest in the water company at that meeting (Hukka, Seppälä, & Teinonen, 2005). City council agreed to the terms of sale with IW/UU, awarding the consortium a fifteen-year contract to manage water and wastewater services for over 400,000 people, effective in 2001. Ervin suggests that city councillors had the impression that the sale of a controlling interest in ASTV would relieve the city of the burden of EEK 3 billion for sewerage expansion (V. Ervin, personal communication, 22 October 2009). It did not.

The purchase of a majority stake in ASTV by the joint venture was financed in part by further loans from the EBRD – EUR 55.0 million to ASTV itself and EUR 10.0 million to IW/UU as an acquisition loan, signed 4 December 2001 (EBRD, 2017). The EBRD subsequently lent another EUR 80 million to ASTV, of which EUR 50 million was to be syndicated to international commercial banks (Hukka, Seppälä, & Teinonen, 2005; EBRD, 2017). This loan added to the DM 22.5 million 1994 loan that had financed the rehabilitation of water and wastewater treatment plants, groundwater wells, and wastewater networks. An EBRD release from 11 November 2002 proudly identified "ASTV [as] one of the first utility companies to be privatized in the Baltic States ... The loan will help ASTV enter a new phase in its continuing development by improving the efficiency of its balance sheet following its privatisation in 2001" (EBRD, 2002). That same press release claimed that the loan would help achieve success in the financial operations of ASTV.

However, financial success was in reality achieved by raising consumer water tariffs by 50 per cent (over inflation) in the period between 2001 and 2005 (Lobina, 2001:7). The EBRD also asserted that the loan would assist ASTV in the achievement of certain goals, but these had been attained some time before privatization – ASTV had already succeeded in meeting Estonian and EU environmental guidelines. This situation was the first incarnation of the PPP arrangement.

PPP V.2.0 began in 2005. ASTV was going "public." It was going to offer an initial public offering on the Estonian Stock Exchange. That June, an article in the *Baltic Times* quoted Toivo Promm, deputy mayor of Tallinn, speaking about the increased transparency a wider shareholder base would bring to the company, reducing political influence: "As the circle of shareholders has considerably widened, decisions of Tallinn Water will no longer depend so much on political winds that might blow" (Baltic Times, 2005). For Promm and the government, there was an elision between the meanings of "public," as if the stock market were a democratic and transparent institution, whereas state ownership, which had been lost in the initial privatization in 2001 and was still a politically controversial topic four years later, was not. It was a deft political "bait and switch" to claim that private shareholding – so-called stock market democracy – would offer more transparency and answerability than public ownership, and it was an effort to elide the five-year political controversy concerning privatization and dismiss it as wind.

United Utilities participated in this process by placing part of its total shareholding into the business, reducing its joint stake held with the EBRD to 35.3 per cent. In the end, the City of Tallinn owned 34.7 per cent and the Nordea Bank (Finland) 11 per cent of the utility. From its initial public offer of EUR 9.25, the stock rose to EUR 13.48 in December 2005 and to EUR 16.25 on 27 March 2007, although it lost 57 per cent of its value between January 2007 and June 2011. It resurged to 2008 levels by October 2012 (EUR 8.85), still below the opening price (Nasdaq, n.d.).

As the Baltic region entered the EU, the aspirations for financial success through PPPs became subject to the anxieties of market-based systems. ASTV successfully met environmental standards, but did so *before* privatization. City officials, company officials, and, of course, investors witnessed the stock value gyrations negatively affecting dividend payouts. Capitalization of the company plummeted between 2006 and 2011 (Nasdaq OMX website). This fluctuation in capitalization evidences the extent to which privatized municipal services are vulnerable to the volatility of market forces. The value of an efficiently managed water system is undeniably high, but when that value is siphoned off by market forces, the public interest can suffer.

Much of the conflict centred on contractual obligations and the management of finance and assets. David Hetherington, chief executive officer at ASTV from 2007 to April 2010, insisted in 2009 that ASTV was in the process of meeting all its contractual obligations with Tallinn. Hetherington commented that the city ought to become a better manager of its contracts rather than make these concerns "political" (D. Hetherington, personal communication, 23 October 2009).

Achieving constructive dialogue between the city and the water company was a problem. The company called in the EU to mediate on 10 December 2010, lodging a complaint regarding a new anti-monopoly bill proposed by the Estonian Parliament. The company claimed the proposed bill would unilaterally alter the agreements it had made with the city concerning the original privatization. The company called this bill a breach of its 2001 contract, and, in the words of a Reuters report, of "European Union fundamental freedoms of establishment and movement of capital" (Reuters, 2011). Certainly, one of the consequences of the Tallinn water privatization case has been the employment of legal services on both sides, appointed with the task of clarifying what contractual obligations entail.

In a fashion often seen in PPP arrangements, the ASTV relationship to the city has been marked by repeated legal and regulatory struggles, affected by the vagaries of the market and multiple corporate and international financial institution agendas. While ASTV achieved high EU standards for environmental responsibility, and dividend payouts included the city as a shareholder of the company, the reputation of ASTV has been tainted by large tariff increases. Political and legal disputes characterized the privatization experiences of Tallinn, to an extent that Gdańsk managed to avoid through the establishment of a more even relationship between the city and its water management company.

Eventually, and as a testimony to the continued international corporate interest in water service investment spanning the 2010s, Veolia Water UK bought into the European and non-regulated UK interests of United Utilities in June of 2010 in a deal that originally included Tallinna Vesi (Hõbemägi, 2010). However, the deal was called off for "undisclosed reasons" wrote a partner from the law firm Sorainen, which handled the deal, in a personal correspondence to the author. But there has been no resolution to the dispute over tariff increases, according to the Q1 2016 Quarterly Report of Tallinna Vesi (AS Tallina Vesi, 2016).

Version 3 of the PPP arrangement in Tallinn remains elusive until either the EU rules on the rights of a state to limit profits on public service utilities, or the city or state government succeeds in returning the water company to democratic public sector control, in contrast to the so-called democracy of stock market shareholder ownership.

Plausible Futures from PPPs Past

Following the restructuring of the post-communist second test case of neoliberalism (after Naomi Klein's [2008] periodization, with Chile's post-1973 neoliberalism as the first period), a high degree of institutional transformation in the expanded EU was required, for instance, in the regulatory framework analysed by Kaika and Page (2003a; 2003b). Devolution of central government water services ownership was accompanied by local regulatory changes to accord with EU regulations. Corporate partnerships were encouraged in the mandates of international financial institutions such as the EBRD. Municipal-level politics in water management also shifted in the face of these changes. The social need approach to infrastructure, however incompatible with prevalent financial and technocratic thinking, can be used to criticize the two early local adopters of private sector partnerships investigated here, Gdańsk and Tallinn.

To summarize, four key points are emphasized. First, the radical restructuring of water services in the Baltic region reflected more than just a restructuring of post-communist economies. It also illustrated the attempted restructuring of the entire water services industry in Europe and elsewhere. The post-communist Baltic should thus be viewed as a region identified by international corporations and finance as an opportunity for capital accumulation strategies, as an expansion territory.

The second point emphasizes the support of the restructuring, chronicled in the chapter, by EU-level environmental regulatory initiatives as well as by international financial institution agendas, although not without internal contradictions and opposition. The concerns voiced by 1980s grass roots environmental movements in communist Eastern Europe became justifications for the neoliberal regulatory and financial agendas of the 1990s.

Point three concerns the convergence of factors responsible for observed outcomes. The complex dynamic interplay between history, governance, finance, and local geographical (hydromorphic) conditions suggests that "ownership" of water infrastructure, the management of it, the construction of new systems, and the adaptation of the existing hydromorphological conditions of a locality occurred along continuums between and among the public and private sectors and public and private financial sources.

The final point notes that the rapid pace of this restructuring and its extent reflected an unprecedented attempt at the implementation of "really existing neoliberalism." This ideology was expressed through multiple scales of governance and regulation changes (local and national governance, as well as international governance with the EU, and pressures from international financial institutions). The Gdańsk and Tallinn narratives have shown the political circumstances that have brought about the described neoliberal transitions and pointed to the plausibility of more democratic forms of infrastructure management.

While water infrastructure – pipes and treatment plants, pumping stations, and so on – or highways, or electricity generation have traditionally been understood in technical terms, which is the domain of experts with instrumental takes on defining adequacy of service provision, austerity budgets in infrastructure governance highlight the political side of "hard" infrastructure decisions. Once infrastructures are shown to be politically determined and meant to satisfy social needs, it becomes possible to think of ways to achieve financial and managerial capacity without giving away infrastructural assets in questionable arrangements.

Many chapters in this volume refer to the impact the shift to neoliberalism has had on suburban infrastructures. They identify this political and economic transition as the main factor responsible for changes in the decision-making that determines the form, distribution, and access to infrastructures, as well as their social consequences. One of the main themes of the book is the growing social inequality resulting from new patterns of infrastructure provision and operation. By chronicling how the passage from communism to neoliberalism has affected the quality and cost of water services in Gdańsk and Tallinn, chapter 5 has presented an extreme case of the transition to neoliberalism. The shift to neoliberalism has been given the most attention in the chapter, which accounts for its presence in the book even though it does not focus principally on suburban areas. The chapter has demonstrated that transitions experienced in its two case studies have been anything but linear. They have involved pendulum effects, an adaptation to changing societal contexts, and reactions to the cost and standards of the services provided by the water systems. Chapter 5 has demonstrated how the impact of neoliberalism on infrastructures is shaped by national and local contexts, and thus differs from one place to another.

REFERENCES

AS Tallina Vesi. (2016). *AS Tallinna Vesi Results of operations – for the 1st quarter of 2016*. Retrieved from https://www.tallinnavesi.ee/wp-content/uploads/2016/03/ASTV_Q1_2016.pdf

Bakker, K. (2010). *Privatizing water: Governance failure and the world's urban water crisis*. Ithaca, NY: Cornell University Press.

Baltic Times. (2005, 1 June). Tallinn Water pulls off successful IPO. *Baltic Times*. Retrieved from https://www.baltictimes.com/news/articles/12812/

Brenner, N., & Theodore, N. (2002). Cities and the geographies of "actually existing neoliberalism," *Antipode*, *34*(3), 349–79. https://doi.org/10.1111/1467-8330.00246

Castree, N. (2006). From neoliberalism to neoliberalisation: Consolations, confusions, and necessary illusions. *Environment and Planning A*, *38*(1), 1–6. https://doi.org/10.1068/a38147

De la Motte, R. (2005). D20: Water time case study – Gdańsk, Poland. Retrieved from http://www.watertime.net/docs/WP2/D20_Gdansk.doc

Dinckal, N., & Mohajeri, S. (Eds.). (2001). *Blickwechsel: Beiträge zur Geschichte der Wasserversorgung und Abwasserentsorgung in Berlin und Istanbul*. Berlin: TU Verlag.

European Bank for Reconstruction and Development (EBRD). (2002). EBRD loan to improve water, waste-water services in Tallinn: Local residents expected to benefit from improved level of service. (Press release). London: Author.

European Bank for Reconstruction and Development (EBRD). (2006, 13 July). EU and EBRD support Gdańsk water infrastructure. *Welcome Europe*. Retrieved from http://www.welcomeurope.com/news-europe/eu-ebrd-support-gda-324-sk-water-infrastructure-9257+9157.html

European Bank for Reconstruction and Development (EBRD). (2007). *EBRD donor report 2007: Making change happen*. Retrieved from www.ebrd.com/publications/donor-report-2007-english.pdf

European Bank for Reconstruction and Development (EBRD). (2017). EBRD investments 1991–2017. Retrieved from www.ebrd.com/documents/corporate-strategy/ebrd-investments-19912017.xlsx

European Parliament and Council. (2000, 22 December). Directive 2000/60/EC of the European Parliament and of the Council of 23 October 2000 establishing a framework for community action in the field of water policy. (EU Water Framework Directive [WFD]). *Official Journal of the European Communities*, *43*(L 37), 1–72. Retrieved from http://ec.europa.eu/environment/water/water-framework/index_en.html

Fine, B. (1999). Privatization theory with lessons from the United Kingdom. In A. Vlachou (Ed.), *Contemporary economic theory: Radical critiques of neoliberalism*. London: MacMillan Press.

Gdańsk Water and Canal Infrastructure Corporation (GIWK). (2010). Another investment in Gdańsk East wastewater treatment plant. Retrieved from http://www.giwk.pl/ and http://www.przetargi.giwk.pl/

Hanni, H. (1999). Water supply and sewerage in Tallinn since medieval times. *European Water Management*, *2*(4), 62–8.

Hõbemägi, T. (2010, 14 June). Veolia Water to become large shareholder of Tallinna Vesi. *Baltic Business News*. Retrieved November 2012 from http://www.balticbusinessnews.com/article/2010/6/14/Veolia_Water_to_acquire_26_5_of_Tallinna_Vesi

Hukka J.J., Seppälä, O.T., Teinonen, R. (2005). D40: Water time case study: Tallinn, Estonia. Retrieved from http://www.watertime.net/wt_cs_cit_ncr.html#Estonia

International Conference on Water and the Environment (ICWE)· (1992).The Dublin statement on water and sustainable development. Retrieved from-http://www.un-documents.net/h2o-dub.htm

Kaika, M. (2005). *City of flows, modernity, nature, and the city*. New York: Routledge.

Kaika, M., & Page, B. (2003a). The EU water framework directive: Part 1. European policy making and the changing topography of lobbying. *European Environment*, *13*(6), 314–27. https://doi.org/10.1002/eet.331

Kaika, M., & Page, B. (2003b). The EU water framework directive: Part 2. Policy innovation and the shifting choreography of governance. *European Environment*, *13*(6), 328–43. https://doi.org/10.1002/eet.332

Kiisler, I. (2005, 9 March). Estonian capital to sell third of water utility to raise cash. Estonian Radio, BBC Monitoring Former Soviet Union. Retrieved November 2012 from http://www.accessmylibrary.com/coms2/summary_0286-19034563_ITM

Klein, N. (2008). *The shock doctrine: The rise of disaster capitalism*. Toronto, ON: Vintage Canada.

Kowalik, P., & Suligowski, Z. (2001). Comparison of water supply and sewerage in Gdansk (Poland) in three different periods. *AMBIO: A Journal of the Human Environment*, *30*(4), 320–3. https://doi.org/10.1639/0044-7447(2001)030[0320:cowsas]2.0.co;2

Leys, C. (2001). *Market driven politics: Neoliberal democracy and the public interest*. London: Verso.

Lobina, E. (2001). Water privatization and restructuring in Central and Eastern Europe, 2001. University of Greenwich, Public Services International Research Unit, London. Retrieved from http://www.psiru.org/reports/water-privatisation-and-restructuring-central-and-eastern-europe-2001.html

Nasdaq. (n.d.). Historical data function. Retrieved from http://www.nasdaqbaltic.com/market/?instrument=EE3100026436&list=2&pg=details&tab=historical&lang=en¤cy=0&date=&start=14.05.2005&end=14.11.2018

O'Connor, J. (1988). Capitalism, nature, socialism: A theoretical introduction. *Capitalism, Nature, Socialism*, *1*(1), 11–38. https://doi.org/10.1080/10455758809358356

Pavlínek, P., & Pickles, J. (2004). Environmental pasts/environmental futures in post-socialist Europe. *Environmental Politics, 13*(1): 237–65. https://doi.org/10.1080/09644010410001685227

Polanyi, K. (1944). *The great transformation*. New York: Rinehart.

Reuters. (2011, 5 May). Tallinna Vesi AS submits complaint regarding competition authority's decision. Retrieved June 2011 from http://www.reuters.com/finance/stocks/TVEAT.TL/key-developments/article/2307365

Saad-Filho, A., & Johnson, D. (2005). *Neoliberalism: A critical reader*. London: Pluto Press.

Siemiatycki, M. (2012). Public-private partnerships: At the intersection of finance and design. In J. Khamsi (Ed.), *Huburbs: Creating transit supportive communities* (pp. 23–8). Toronto, ON: University of Toronto Press.

Siemiatycki, M. (2013, 13 June). Is there a distinctive Canadian PPP model? Reflections on twenty years of practice. Paper presented at Public-Private Partnership Conference Series CBS-Sauder-Monash, 12–14 June 2013, Vancouver, BC.

Smith, N. (1984). *Uneven development: Nature, capital, and the production of space*. Oxford: Blackwell.

Smith, N. (2006). Nature as accumulation strategy. In L. Panitch & C. Leys (Eds.), *Coming to terms with nature* (pp. 16–36). (Socialist Register 2007, 43). London: Merlin Press.

Swinarski, M. (1999). The development of waste water treatment systems in Gdansk in 1871–1998. *European Water Management*, 2(4), 69–76.

Swyngedouw, E. (2003). Dispossessing H_2O: The contested terrain of water privatization. *Capitalism, Nature, Socialism, 16*(1), 81–98. https://doi.org/10.1080/1045575052000335384

Swyngedouw, E. (2004). *Social power and the urbanization of water – Flows of power*. Oxford: Oxford University Press.

Tallinna Vesi. (2016, 29 April). Quarterly report: Results of operations – for the 1st quarter 2016. Retrieved from https://cns.omxgroup.com/cdsPublic/viewDisclosure.action?disclosureId=709056&messageId=887764

6 "Designed to Fail": Technopolitics of Sewage in India's Urban Periphery

SHUBHRA GURURANI

Introduction

> Every day at 5 a.m. sewage is pumped from our internal lines to the main line. You will see there is a manhole, the diesel pump on the other side, and the sewage line that connects with the main line. HUDA maintains the main line, while the developers maintain their own colonies. We all pay maintenance charges to GDL[1] for the services. This has been going on for fifteen years. The reason our lines are not connected to the city line, our developer tells us, is that the main [city] line may backup, which will back up our line, and our homes and apartments, so they manually pump it into the main line. They [the developers] have paid extra development charges too, but there is no proper citywide sewerage. There are four different governmental bodies here, and there is a great deal of uncertainty and lack of clarity about who is responsible for what, who is accountable, and everybody of course blames the other. *The system here is designed to fail.* (AK, private housing colony resident, personal communication, 8 September 2014; emphasis added)

In many of our conversations with the residents of private housing colonies and villages, migrant workers, activists, and planners, we could discern a palpable sense of frustration, anger, and scepticism with municipal authorities and private developers about the management of sewer infrastructure in Gurgaon, a city approximately twenty kilometres southwest of India's capital New Delhi. According to some estimates, in 2012 there were over one hundred housing colonies that were not connected to the Haryana Urban Development Authority (HUDA) master lines.[2] Only two sewage treatment plants were in operation in Behrampur and Dhanwapur villages, which together treated only a small proportion of networked sewage, while the majority was disposed of untreated in local drains. Recently, Mr R.S. Rathee, president of Gurgaon Citizen Council, noted: "In most places, either the local network is not connected with the master network or the levels of pipes do not match. The sewage network has to function on gravity and this does not happen here."[3] In such a scenario, city residents acknowledge that they are forced to make ad hoc arrangements, pay bribes, and mobilize networks of influence to ensure access to basic infrastructure. Unfortunately, Gurgaon is not alone when it comes to a dismal state of infrastructure. In a recent report, "Urban Shit: Where Does It All Go?," Rohilla et al. (2016) note: "Only a third of urban houses in India are connected to sewer systems. The majority of the houses – 38.2 per cent, as per Census 2011 – use toilets connected to septic tanks." In an attempt to examine the changing terrain of infrastructure, I turn to Gurgaon and examine the "technopolitics" of sewage in this urbanizing periphery.

Infrastructural practices here, as Larkin (2013) points out, are not simply "out there," or down there, waiting to be observed or evaluated. Rather, networks of infrastructure, or their absence, "comprise the architecture for circulation, literally providing the undergirding of modern societies, and they generate the ambient environment of everyday life" (328). Taking cues from recent works in anthropology, science and technology studies, and geography that have shown how pipes, pumps, electricity, and roads constitute the multiple domains of modernity and configure subjectivities, social realities, and future imaginations (Anand, 2012; Elyachar, 2012; Simone, 2004), I approach waste infrastructures as social-technical-material systems and examine everyday practices of accessing/managing sewage in an urban periphery. Unlike roads, railways, electricity, and telecommunications, which are visible infrastructures that can showcase emergent sociality and modernity, connect to global networks, and evoke a sense of pride, sewers are not objects of public imagination or pride. Invisible, dirty, and associated with filthy work, sewers do not exactly invoke poetic prose. But, sewers are critical in shaping the social geography of urban spaces. In an attempt to show how the technopolitics of sewers unfolds, I draw attention in this chapter to the diverse practices and processes through which different groups of residents in Gurgaon access sewage infrastructures. In doing so, I describe how this form of technopolitics not only relies on the familiar tropes of "clean," "dirty," "visible," and "invisible," but also how it is historically situated in relations of caste, class, place, and property. I argue that the absence of sewage networks in the urbanizing hinterlands surrounding the city does the technopolitical work of (re)producing spaces of rurality and social categories associated with these spaces – villagers, small landholders, landless pastoralists, factory workers, and large numbers of migrants living in these villages – as "inferior" and "dirty," thus "disavowing" their right to inclusion in circuits of urbanity and infrastructures.

Questions of place and property are critical for both rural and urban areas, but they become particularly vexed in the "transitional areas"[4] that surround large urban metropolises. Since the mid-1980s – and far more aggressively in the 1990s – cities have started to urbanize their hinterlands in India, but these transitional areas were left in a policy vacuum. Even though urban peripheries have grown exponentially[5] and hundreds of thousands of new residents have come to live and work in these transitional areas, policy has not kept pace with the changes. There is a great deal of uncertainty over the jurisdiction of different state- and municipal-level administrations. Such vagueness is not unusual for new and changing urban spaces, as the loci of authority can often be initially unclear and contested. However, uncertainty has been pervasive for the last thirty or more years in the city of Gurgaon, and I argue that this uncertainty has had far-reaching consequences. Uncertainty has cultivated an uneven geography of access, provisions, claims, and rights that in part describes how, why, and where the "system is designed to fail" in urbanizing frontiers. This chapter seeks to elucidate what undergirds such uncertainty: how it enables some systems to work, while others are overlooked; what constitutes the spatiality of uncertainty; and what it reveals about emergent urban forms, subjectivities, and social-material practices.

In the following discussion, I situate the competing regimes of sewage in the broader political and economic shifts that began to take root in India in the mid-1980s. I begin by arguing that the so-called infrastructural deficit or failure, which is often explained in the language of technological incapacity or dysfunction and "rendered technical," is in part due to the shift I describe further on as a shift from state-funded public works projects to privately funded projects of infrastructure (see Li, 2007). As India opened its economy, urbanization was seen as the mantra for attracting foreign capital and propelling economic growth. During this time, agrarian hinterlands were feverishly incorporated into the metropolitan areas, but *not* as urban spaces. The urban peripheries served as rural residuum for factory workers and migrant labour from the rest of India and housed small landholders, landless pastoralists, and lower castes. They played the role of dormitories for the labour force that sustained and serviced the everyday needs of urban residents living in high-end gated enclaves. While the propertied classes were able to privately access commoditized infrastructure, these peripheries were largely left to make their own arrangements. The ambiguity around how different spaces were administratively designated as urban, rural, and transitional areas, the overlap and lack of clarity between different levels of government, and the uncertainty around the responsibilities of different governance bodies were crucial in shaping this chequered infrastructural landscape. Such convoluted multiscalar institutions, practices, and processes led to a system that AK in the opening statement aptly describes as "designed to fail." For instance, in Gurgaon, there was confusion about the administrative responsibilities and territories of the state-level body HUDA, the role of the new Municipal Corporation of Gurgaon, and the extent of village councils' (*panchayats'*) authority over rural areas that lay outside the official boundary (*lal dora*) area of the village. Such uncertainty is clearly indicative of extreme governmental ineptitude, but I suggest that uncertainty reveals a technopolitics of disavowal and irresponsibility that is constitutive of the new geographies of the urban.

Technopolitics of Disavowal

The anthropologist Brian Larkin has argued that the concept of technopolitics is the most dynamic approach to the study of infrastructure from an anthropological perspective, as technopolitics "reveal forms of political rationality that underlie technological projects and which give rise to an apparatus of governmentality" (Larkin, 2013: 328). In bringing discourses on power into a conversation with science and technology studies that are attentive to materiality and non-human forms of agency, recent writings on infrastructure have drawn attention to the ways in which technical devices such as pipes, pumps, wires, and roads are constitutive of diverse ethical-political terrains and how they shape politics of places and subjectivities. It is this technopolitical perspective of infrastructure that I bring to this analysis of sewage in Gurgaon.

In 1985, the National Capital Regional Act was passed in response to New Delhi's growing population, which redrew the boundaries of the capital and extended the limits of National Capital Territory (NCT) to a National Capital Region (NCR). In order to make room for a growing city and accommodate the mounting pressures of housing, work, and mobility, the boundaries of the NCR incorporated the hinterland of Delhi, including the satellite towns of the three neighbouring states, Uttar Pradesh, Rajasthan, and Haryana, which are close to Delhi. The creation of the NCR unleashed simultaneous forces of deagrarianization[6] and urbanization – bringing significant change to neighbouring towns/villages like Gurgaon, Noida, Ghaziabad, and Faridabad.

The Census of India indicates that, of the four towns, Gurgaon grew most rapidly, with a 73 per cent increase in its population between 2001 and 2011. It is now home to multiple multinational corporations, hundreds of private residential colonies and enclaves, over two dozen malls, golf courses, cyber parks, and more. Private developers have played a key role here. Delhi Land and Finance (DLF) was the primary player in the making of Gurgaon, entering the housing market in Gurgaon with the support of state and federal governments and purchasing over 3,500 acres of land directly from farmers. Other large and small land developers followed suit, and, through the 1990s, hundreds of thousands of new residents came to populate expanding Gurgaon. In order to attract the middle classes, private developers promised luxurious amenities and created islands of world-class infrastructures, but residual spaces of urban villages were mired in uncertainty. It is incomprehensible, on the one hand, that a city of two million would lack sewerage; on the other hand, the case of Gurgaon's sewage infrastructure is a classic example of how governmental denial takes shape in rural/pastoral spaces.[7] Before I turn to discuss everyday infrastructural practices, the next section lays out the political-economic context in which the politics of urban infrastructure changed.

Changing Priorities: From Public Works to Infrastructures

Not too long back, roads, pipes, water, sewage, transport, ports, and highways were considered to be *public works*. Public works projects were an integral part of India's post-independence developmentalist goals, and road and highway construction, dams, public buildings, and water and sewage management were, in principle, mandated by the state to promote economic development, encourage growth, reduce poverty, and, most centrally, provide basic services to the *public*. But, starting in the mid-1980s, a categorical shift in these public-oriented projects took place. This shift can be traced back to 1994, when the World Bank published its annual report, *World Development Report 1994: Infrastructure for Development* (World Bank, 1994). The report heavily influenced the direction of infrastructure funding and investment all over the world. In India, an Expert Group on the Commercialization of Infrastructure was set up by the Ministry of Finance, echoing the World Bank's recommendation that "infrastructural facilities and services [be] priced according to the market for greater efficiency, etc." (Ghosh, Sen, & Chandrasekhar, 1997: 803). The enhanced role of private funding in infrastructure was a "major departure from the axiomatic reliance on the government in fulfilling this role that has marked both economic theory and policy practice over this century" (Ghosh, Sen, & Chandrasekhar, 1997: 803). This shift, from public works to infrastructure investment, moved basic infrastructure from the public good realm to that of private finance, speculation, and commoditization. Not only did this policy shift the role of the state, but the World Bank report and the Indian adoption of it meant that the Indian government now acted as a "facilitator" for private investment in the provision and distribution of infrastructure services. Buoyed by the ideological assumptions that public funding is inferior to private, and private funding can ensure transparency, efficiency, and accountability, the government made a concerted effort to prepare an investment-friendly context for private investment. Over time, the government's expenditure in infrastructure has significantly decelerated and even declined in real terms across many sectors since the economic liberalization (Ghosh, Sen, & Chandrasekhar, 1997: 804), producing a highly differentiated and contested terrain of infrastructure in which those who could afford the high cost of privatized infrastructure were able to buy these services, while others were left to "make do" with sub-adequate conditions.[8]

In this milieu of neoliberal urbanization, Haryana was the first state in India to creatively manipulate the Urban Land (Ceiling and Regulation) Act (ULCRA)[9] and, through a regime of "exemptions" and "license applications," permitted private developers/colonizers[10] with "good track records" to acquire extensive tracts of land for private housing and commercial purposes. I have argued elsewhere that as the private developers hastily carved urban sectors out of agricultural land, the trajectory of the urban did not conform to urban planning norms (Gururani, 2013). The city of Gurgaon was planned "flexibly," largely driven by the vision of private developers and politicians who frantically built a city, but in this case, one without sewers.

The uneven landscape of Gurgaon can, however, be traced as far back as the heyday of agrarian development, when the state of Haryana was carved out of Punjab in 1966. Around that time, Sanjay Gandhi, son of former prime minister Indira Gandhi, set up India's first car plant – Maruti Udyog, a subsidiary of the Suzuki Japanese automobile and motorcycle manufacturer – in Gurgaon. Maruti in many ways was a game changer, and while a complete discussion is beyond the scope of this paper, suffice it to say that the automobile plant, manufacturing facilities for its parts, residences for workers, and other related needs required extensive land acquisition, necessitating the passage of the Haryana Urban Development Authority Act (HUDA Act) in 1977 to oversee urban development. In 1977, Gurgaon was a rather featureless village, and was considered to be "handicapped"[11] due to lack of water and irrigation. Its future as an urban destination was non-existent, and it was largely outside the regimes of urban planning. In any case, urban development was not on the minds of central or state governments, and they did not anticipate the upcoming urban boom.

With the setup of the Maruti automobile plant in Gurgaon, planners primarily envisioned the city as a factory town, which would be serviced by factory workers who would commute from Delhi to work in the car factory but return to live in Delhi. Since Gurgaon was not envisioned as a residential destination, the issue of waste and sewage was not given due attention. Because Haryana was primarily agricultural, there was lack of clarity about which department was responsible for infrastructure provision in this changing landscape. Typically, urban planning departments are responsible for land acquisition, land use, and preparing land for development, while municipal bodies are responsible for town planning, engineering, and infrastructure provision. However, until recently, there was no local municipal corporation in Gurgaon. There was a municipal council that was directly under the chief minister's office and worked with HUDA. But, despite some semblance of coherence in their planning and infrastructure activities, in practice there was no coordination between urban development and infrastructure because there was no governmental apparatus to mediate and coordinate the two arms at the city level. For example, in the HUDA Act of 1977, there is only a brief discussion of "drains," and it is tucked under "engineering operations" and kept outside the expert domains of urban planners and engineers (Government of Haryana, 1977).[12] In Gurgaon too, until recently, drains/sewerage remained a technical matter under the aegis of engineers, and town and country planners did not address the issue. It was only in 2002 that the government of Haryana made amendments to the HUDA Act, and the issue of sewerage was addressed directly; even then, a careful distinction was made between "amenities" and "basic amenities." Sewerage, along with "metaled roads, wholesome water, and electrification" was considered as a basic amenity for *urban* areas, but in rural areas, sewerage was not deemed to be a basic amenity. As a result, in this unfolding patchwork of urban and rural infrastructures, while the urban authorities made provisions for basic infrastructure in urban enclaves, large parts of the city that are designated rural and are densely populated are still largely left on their own.

This distinction between rural and urban areas was critical, as the designations of rural, urban, and transitional areas determined how connections for water and sewage would be made, who would be allocated funds, and from which administrative pocket. The complicated categorization is related to the way in which India designates specific territories as urban and rural at central and state levels, and it informs the regime of governmental technopolitics. Interestingly, the Census of India, an organ of the central government, classifies settlements into rural and urban, but it is state governments that grant municipal status to urban centres (Bhagat, 2005). The definition of urban includes areas that have a population not less than 5,000, a population density of 400 persons per square kilometre, and 75 per cent of the workforce employed in the non-agricultural sector; it identifies two types of towns: municipal or statutory towns and census or non-municipal towns. In 1991, 36 per cent of towns in India were classified as census/non-municipal towns compared to over 26.6 per cent in 2001.

In other words, a settlement may be classified as urban in the Census of India but it may or may not be granted an urban status by the state and brought under corresponding municipal authorities. Conversely, states may grant settlements municipal status but that status may be at odds with what is designated as "municipal" under the Census of India definition. In 1992, the 74th Constitution Amendment Act was passed, recognizing for the first time transitional areas and granting them civic status by creating *nagar panchayats* or town councils. This change was a hopeful move for urban peripheries, but since municipal governance is a state matter in a federal system, most states did not develop criteria for the identification of different categories of urban areas and establish institutions of municipal governance (Bhagat, 2005: 70). While some states like Tamil Nadu went ahead and classified many such urban areas, making it possible to set up town or municipal councils, many others have still not created this new civic category in spite of the rapidly changing configuration of areas adjoining the large metropolitan cities.

There are other reasons why states have not identified transitional areas at the urban periphery. Towns considered non-municipal qualify for rural development funding from the central government's large reserve (INR 2,000 crores = USD 450 million) for water supply, roads, schools, or healthcare, while the programs supported by the Ministry of Urban Development are only able to draw upon a comparatively small sum. In addition to accessing central government funds, the rural category, despite the population growth, helps avoid higher levels of municipal taxation, gives access to electricity at lower charges and tariffs, and provides almost free primary education and healthcare (Bhagat, 2005: 65). As a result, retaining the rural category is a carefully negotiated decision. The challenge, however, is that areas designated as non-municipal, or census towns, are considered to be urban by the Census of India such that their population is included in the urban census category, but they are governed by rural local bodies (Dupont, 2007; Gururani & Kose, 2013). In other words, large urban populations in transitional areas live in places designated as villages, which are not considered to be urban and hence are not governed by urban bodies, so they cannot access funds allocated for urban amenities, like sewerage.[13]

Gurgaon is a case in point. Based on the 1991 census conducted by the government of India, Gurgaon did not meet the Ministry of Urban Development criteria for an urban local body (ULB);[14] nor did the state government of Haryana identify Gurgaon as an urban area (that is, an area with a population over 50,000) between the 1991 and 2001 censuses. In 1994, a Municipal Corporation Act was passed, but since Gurgaon was not classified as urban, the municipal corporation did not administer it, thus preventing urban planning and development authorities from taking charge of urbanization in the periphery. Instead, the town was governed by a municipal council, which reported directly to the chief minister of Haryana. This arrangement placed Gurgaon in the category of villages that were designated as rural despite their high population between 10,000 and 60,000 people. Meanwhile, infrastructures for sanitation, garbage collection and disposal, water supply, street cleaning, lighting, and so on were considered responsibilities specific to designated urban areas; thus high-population "villages" that were still designated as "rural" lacked the funding and/or institutional structure to take care of these basic amenities. The village councils (*panchayats*) were supposed to undertake village improvements, but their efforts were partial and inadequate. The private housing colonies were able to secure infrastructure in their gated enclaves through privatized services, but there was no government authority with infrastructure development responsibilities to provide sewage facilities in the villages that housed half of Gurgaon's population.

As a result, even though the villages and their outskirts are central to the urban fabric and constitute some of the city's densest areas, they are largely abandoned by the state. In other words, the prevailing confusion, overlapping and arbitrary categorization of rural and urban spaces in the context of chequered urbanism, not only contours an uneven landscape, it also unravels a deep technopolitics of disavowal through which working classes and subaltern groups are virtually neglected by the state and left to simply make their own arrangements.

Message in a Tank

During my field visits in the months of April and May 2011, I met the retired army officer *Subedar* DC. He wanted to talk about the pipe he and his villagers had installed twenty years ago. He went into great length describing how he convinced Delhi Land and Finance to install the pipe, how he mobilized the villagers, and how he himself dug the ground for the pipe.

> I worked in the army for thirty years. I travelled all over the country. I lived in Bombay, Bangalore, Madras, Nagaland, Amritsar, everywhere. I joined when I was seventeen, I was *ganwar* [can be translated as rural, uncouth], I had no idea, but the army teaches you discipline, manners, style, the proper way to live. When I retired from the army, I wanted to have a pipe built in the village. People here are uneducated, they are backward, they did not understand why I was asking DLF to put a pipe in the village, but I persisted. The pipe runs from here and there. In the whole of Gurgaon, ours is the only village which has a pipe. It has not been maintained and it is not adequate for the village now, but it serves some families. (DC, retired army officer, personal communication, 12 May 2011)

A few days later, in another village, Jhankot, as I sat down to talk to the former village headman (*sarpanch*), he turned around and pointed to the large septic tank that lay buried right at the entrance of his house in front of the open courtyard. He said proudly, "This is a new one. I just got it installed two summers back. The small one was not working" (Village headman, personal communication, 1 May 2012). In yet another meeting, Rishi Kumar, who was from the scheduled caste,[15] first showed me his new motorcycle, then introduced me to his two sons, Varun and Arjun, who studied in a "private" school, and finally to his wife, Sheela, who ran a small beauty shop in the village selling women's make-up and fake jewellery. He showed me the septic tank that he had recently installed at the back of his house and pointed to the bathroom that stood close to the entrance of his house. He said proudly, "Bathroom is a must if your duty is in the office. Without one, it was getting very difficult. Children have to go to school and I have my duty, and there is no time in the morning to wait for our turn in the shared toilet" (R. Kumar, personal communication, 20 February 2012).

Having grown up in Delhi, I was aware of the economy of water tanks, as big black Sintex water tanks were ubiquitous on every rooftop in Delhi and beyond. I had run up and down the stairs turning the pump meter on or off many times; in contrast, the everyday mechanics of septic tanks initially were invisible to me, and only upon reflection over the months did I come to see what septic tanks in the villages of Gurgaon represented.

Matthew Gandy, in his discussion of the "bacteriological city" in nineteenth century Europe, points out that, with plumbing technology, bathrooms moved inside the homes, which altered moral social behaviour:

> The spread of the private bathroom marked a new bashfulness towards the body as emerging fashions for washing, hygiene and bodily privacy fostered increasing aversion to human excrement. The modern home became subject to a new moral geography of social behaviour that enabled the development of modern technologies to be incorporated into an "invented tradition" of domesticity. (Gandy, 2004: 366)

In contemporary India too, bathrooms and hygiene – public or private bathrooms, flushable toilets, European style toilets, number of bathrooms, showers, sewage, access to networked pipes, lack of pipes, septic tanks, latrines, and open defecation – are central to the moral geography of class, caste, and gender. The existence and style of bathrooms operate as important indexes of power and status, and the absence of a "modern" bathroom is a source of shame and embarrassment. More importantly, in the context of entrenched caste politics – where discrimination and violence against lower castes, and especially against those castes associated with the dirty work of waste and sewage, is prevalent – waste, dirt, and filth work as critical social-material axes spatially organize residential patterns in urban villages. Waste and pollutants, unlike any other matter, as Mary Douglas (1966) has pointed out, are constitutive of social relations and categories, and in contemporary India the violent politics of caste, along with religion and gender, continues to powerfully configure the ethico-political terrain of everyday life in urbanizing frontiers.

In urban peripheries where sewers are absent and one finds piles of garbage and open drains, and where choked gutters mark rural spaces, networked sewage facilities, septic tanks, private bathrooms, and latrines become metonyms for upward mobility, middle class-ness, upper caste-ness, and, above all, urbanity. In describing the sewer line, DC was drawing my attention to how urban and urbane his life was, how his travels and army training disposed him to cleanliness, style, and order. Similarly, Rishi Kumar was insistent that if you have to go to work in an "office," you need a motorcycle *and* a bathroom.

In the village of Banai, which was one of the first villages to be urbanized, two traditionally pastoralist castes of Gujjars and Yadav constitute the majority. Landholding families who had sold small plots of land in the eighties and nineties and then invested the small sums they received by building rental housing, purchasing land elsewhere, or starting small businesses are better off. For instance, most Yadav families have installed their own septic tanks and made sewer connections. Yadavs are also active in the local electoral and village politics, and are able to mobilize their political status to influence decisions about water tankers and location of septic tanks. Some Gujjar families, those who had modest land holdings, have also succeeded in securing access to borewells, septic tanks, and private bathrooms. But the ones who were landless, like some landless Gujjars and lower caste members from Balmiki, Prajapati, Nai, Koli, and Kumhars, and others from scheduled castes and other backward classes, did not gain from selling land and are still unable to pull together enough funds to invest in borewells or septic tanks. It is their spatially demarcated settlements that are typically dilapidated and without water or sewer connections.

In contemporary India, in the context of the volatile and violent politics of caste where discrimination and violence against lower castes, and especially against those castes associated with the dirty work of waste and sewage, is prevalent, the ethico-political terrain of quotidian life, among other material and symbolic networks, is also articulated through the circuits of waste. Waste, dirt, and filth work are critical social-material axes that spatially organize residential patterns, especially in the urbanizing frontiers where public service and public sector structures are non-existent. The upper classes and upper castes, which are typically land and property owners, somehow secure access to sewage facilities, but others without private property are left on their own. In other words, the social-material practices of infrastructure align with caste, property ownership, and place-based dynamics of differentiation and exclusion.

When I asked Mr SK, a planner in the Town and Country Planning Office (TCPO), about the lack of basic amenities in the villages, he responded:

> Villages are under their *panchayats* [councils] – they get funds to provide these facilities, they also get funds from the state. They have a lot of money but they mishandle it. The city's governance does not reach into the *lal dora* [village boundary] – they are responsible for their upkeep ... The villagers in Gurgaon are very well off. You may get the impression that they are not, but they have made a lot of money, they just don't know how to spend it. They don't mind the filth; they are used to it. They are uneducated and uncouth. They will buy a car, buy a television, air conditioner, but they will not keep the village clean. They are well off, probably better so than you and me. (SK, TCPO planner, personal communication, 23 February 2012)

SK's remarks echoed the sentiments of many planners, architects, property workers, and even environmentalists I met during field visits in 2009, 2011, 2012, 2013, and 2015. Mr SK's statement was telling, as he made it clear that it is not lack of funds or lack of knowledge about the state of affairs in villages that inform the uneven terrain of sewage infrastructures. Instead, rural spaces and people who live in them are considered backward and deemed to be less worthy of infrastructural support. They can be overlooked because "they are used to it." Such islands of rurality in a sea of urban residential towers and malls draw attention to the highly differentiated practices through which people manage their everyday sewage needs and highlight the technopolitics of disavowal. Due to pervasive societal faith in urban transformation, even though rural spaces sustain the urban fabric by housing labour and service needs of the urban, they are considered as the abject other, overlooked and anomalous.

It is important to acknowledge that in the absence of public provision of sewerage, there are a wide range of arrangements that service *both* the propertied classes who live in state-of-the-art gated enclaves *as well as* the villages that house migrant workers, small landholders, pastoralists, and members of lower castes. While the propertied residents are able to secure these services through private means and do not have to worry about them on a day-to-day basis, those living in the villages are largely abandoned and rely on ad hoc infrastructures. In the absence of public provision of basic infrastructures, all residents make their own arrangements of access, networks, distribution, and disposal of sewage, all of which are mediated by relations of caste, class, and place.

Ethnography of the Unspeakable

In this paper, I have tried to shed light on the administrative ambivalence around the categorization of rural and urban at different levels of governance and to suggest that the prevailing uncertainty around different spatial settlements is productive, as it allows political elites, propertied classes, private land developers, land mafia, as well as rural elites and upper castes to benefit from this seemingly chaotic landscape. Even though the absence of sewers is depoliticized and interpreted as a technical issue, I have tried to show that a governmental technopolitics of disavowal works in the context of entrenched politics of caste, class, and place to overlook undesired places and people who live in them. The everyday infrastructure of sewage has a specific geography, working in some places and not in others, and is illustrative of a system described by the Gurgaon private housing colony resident as "designed to fail."

In conclusion, I wish to briefly return to the theme of pervasive uncertainty. Not only is there uncertainty about the jurisdictional layers of governance and spatial categories of urban, rural, and transitional areas, but there is also uncertainty about the infrastructure of sewage itself. How can we document and analyse the invisible, the unspeakable, and the unaccountable? For instance, in Gurgaon, it is not easy to evaluate the question of sewage through standard technical or governance categories. There is no accounting of how much sewage is generated, and there are no figures or facts that can be tabulated, compared, and analysed.[16] During the course of field work, even talking about sewage or drains (*nala*) was not easy, due to awkwardness and reticence on the part of interviewees.[17] The questions around water generated a lot of energetic exchange focusing on the number of ponds (*jhors*), how they have dried up, their exact location, where the animals used to drink water, the digging of borewells, their ever-deepening depths, the ban on digging borewells, and so on. Talking about garbage (*kurha karkat*) was also possible, but there was awkwardness around the subject of sewage. This uneasiness in part was the result of a linguistic challenge, as there is only one term in Hindi for drains[18] – clean or dirty – and hence the specificity of sewage talk necessitated using awkward terms such as "*tatti aur peshab*" (shit and urine) or "*ganda nala*" (dirty drain). In India, like rest of the world, there is unease when discussing fecal matter.[19] However, in the context of the charged undercurrents that arise from social differentiation based on caste, and where indicators of dirt and filth are central to configuring the social-material-political terrain of places, awkwardness and uncertainty point to a more complex societal issue than simple social aversion to talking about human waste. If there is so much uncertainty and lack of knowledge regarding hidden infrastructures, their study not only poses a methodological challenge, but, more importantly, urges us to revisit our knowledge base that rests on evidence.

The contribution of chapter 6 to the volume lies in the light it casts on aspects of infrastructures that are often left in the dark. The chapter has addressed two issues characterizing hidden infrastructures: first, the unwillingness to talk about these infrastructures, especially sewerage, and second, the need for organizations with the coordination and financial capacity to assure their existence. Without such agencies, infrastructure networks cannot exist. The chapter has revealed the unease people feel talking about sewerage and its purpose. This reticence is the case in different parts of the world, hence the lack of political enthusiasm for underground infrastructures, especially sewers. Everywhere, political discourse about highways, public transit, and power grids is abundant, but there is much less interest in sewage and sewage treatment plants. The chapter presents an extreme case: a large urban area deprived of an integrated sewerage system. It discusses the historical events leading to this outcome, and, above all, the impossibility of developing a sewerage system without the existence of an institutional structure capable of coordinating and funding such a system. In this sense, the chapter has explored the connection between institutions and infrastructures, first introduced by chapter 3. The main lesson emanating from chapter 6 is that, without a coordinated approach to the provision of sewerage, access to sanitation becomes yet another source of social distinction. In such circumstances, there is pride in having toilets in the house, a mark of middle-class status. Another consequence of the absence of citywide sewerage is the need to rely on complicated alternative arrangements to have access to sewers, which make sewerage expensive and a constant object of concern for households.

ACKNOWLEDGMENTS

The author is grateful to the organizers and participants of "The Global Suburban Infrastructure" workshop held at the School of Planning, University of Waterloo, Ontario, in June 2015, where an earlier version of this paper was presented. She wants to thank the SSHRC–MCRI project on "Global Suburbanisms: Governance, Land, and Infrastructure in the 21st Century" for partially funding this research. She is also grateful to Kajri Jain and Sukrit Nagpal for their comments and suggestions, and to Prasad Khanolkar for research assistance. An early version of the paper, "'Designed to Fail': Techno-politics of Disavowal and Disdain in an Urbanizing Frontier," was published in *Economic and Political Weekly*, *52*(34) on 26 August 2017.

NOTES

1 GDL is a pseudonym. In order to maintain anonymity, names of all individuals and private developers have been changed.
2 "If all the private colonies are connected to the HUDA main line, it will choke" (Kumar, 2012: para. 2).
3 R.S. Rathee, quoted in Behl, 2016: para 7.
4 The Census of India uses the term "transitional area" to refer to urban peripheries in India. See Bhagat, 2005.
5 Between the 2001 and 2011 censuses, "the populations of four adjacent districts of Delhi, Gurgaon and Faridabad in Haryana and Ghaziabad and Noida in Uttar Pradesh, have exploded." Gurgaon has seen the most dramatic increase of 73 per cent in a decade, while Noida grew about 50 per cent and Ghaziabad 41 per cent (Varma, 2014).
6 Hall, Hirsch, & Li (2012) describe "deagrarianization" as "the process by which agriculture becomes progressively less central to national economies and the livelihoods of people in even rural areas."
7 Unfortunately, Gurgaon is not alone. In a recent report, "Urban Shit: Where Does It All Go?," Rohilla et al. (2016) note that "only a third of urban houses in India are connected to sewer systems. The majority of the houses – 38.2 per cent, as per Census 2011 – use toilets connected

to septic tanks." In the National Capital Territory alone, there are 350 to 400 vacuum tankers that are run by individuals or unions who, for a small sum (Rs. 1,000), suck fecal waste in a drain behind one of the private hospitals, which then is emptied into the river Yamuna.

8 According to Ghosh, Sen, and Chandrasekhar, "the share of capital expenditure in total expenditure of the government has declined sharply from around 30% at the beginning of the Eighth Plan (1992) to only around 24 per cent now." They go on to add that "every single indicator of physical infrastructure output shows a *deceleration* since 1990, with the single exception of some expansion in telecom facilities in terms of the number of telephones available" (Gosh, Sen, & Chandrasekhar, 1997: 804).

9 The Urban Land (Ceiling and Regulation) Act was passed in 1976 to prevent concentration of urban land in the hands of a few persons and to undermine speculation and profiteering. It was intended to "bring about an equitable distribution of land in urban agglomerations to subserve the common good" (Government of India, 1976: 1). The legislation fixed a ceiling on the vacant urban land that a "person" in urban agglomerations could acquire and hold. This ceiling limit ranges from 500 to 2,000 square metres, and excess vacant land is either to be surrendered to the competent authority appointed under the act for a small compensation or to be developed by its holder only for specified purposes. The government acquired any land owned in excess of the prescribed limit by following a specific method of calculation, which was based on the income the acquired land was able to generate.

10 Colonizers are land developers who prepare land for building housing enclaves that are referred to as housing colonies.

11 In Delhi's first master plan, which was drafted in 1962, two new neighbouring towns of Ghaziabad and Faridabad were proposed as future "industrial towns," with projections of relatively higher proportions of workers engaged in manufacturing – close to 14 per cent of the total population in 1981 – but Gurgaon was considered to be a "handicapped" district town "for want of water sources," and hence proposed to have modest growth (Delhi Development Authority, 1962).

12 Throughout the 1990s in many Indian cities, infrastructure governance was pulled out of town planning and urban development, and placed under different boards, such as sanitation or water boards, which allowed them to coordinate with each other.

13 There is also an electoral calculus at work here, whereby competing pressures and calculations guide how territories are classified as rural or urban.

14 To be declared a "census town," (that is, urban), a settlement has to fulfil the following three conditions: (1) the population must be 5,000 or more;

(2) the density must be at least 400 persons per square kilometre; and (3) 75 per cent of the male workforce should be employed in the non-agricultural sector. The population of all human settlements that are not classified as urban by the Census of India is included in the total rural population (Bhagat, 2005).

15 Scheduled castes refer to a group of the lowest castes, formerly considered untouchables, which were placed in a schedule when the Indian constitution was drafted after independence in 1950. The scheduled caste status gives access to some affirmative action provisions.

16 According to the Central Pollution Board's report, Gurgaon generated 80 million litres daily (MLD) in 2005, but other calculations suggest the figures to be 130 MLD, 160 MLD, or even 260 MLD – more than three times the initial figure (see Narain & Srinivasa, 2012). Even international consultants like the CH2M Hill, who are involved in waste management all over the world, declined to estimate the amount of waste generated in Gurgaon. According to them, since many private colonies do not have sewers and empty out their sewage into privately managed septic tanks, the amount of sewage generated and/or collected cannot be accounted for. There is, similarly, an ambiguity surrounding collection. Again, it is estimated that 50 to 60 per cent of Gurgaon has a sewerage system, and 95 per cent of this sewerage consists of closed drains, but these figures are contested (Narain & Srinivasa, 2012).

17 Some middle-class residents living in private colonies were well aware of the challenges of sewerage, but many were unsure about the underlying networks of liquid and solid wastes. They paid maintenance charges to their resident associations but did not know how their garbage or sewage was disposed. They were more vocal about garbage, and many, because of its unsightly nature, smell, and health hazard, had organized various marches and filed countless petitions. Since we spoke mostly in English, it was not hard to have a conversation about sewage, sewerage, gutters, and wastewater.

18 There is only one term in Hindi and the local dialect of Haryanavi for drains, sewers, gutter, and sewerage: *nala* (big) or *nali* (small), which encompasses the general category conduit or channel for any kind of flow. *Nalas* refer to drains both above and below the ground, visible and invisible, big and small, clean and dirty, water and sewage. There were many *nalas* above the ground that carried clean and dirty waste, and one could trace their tracks. But talking about the hidden *nalas* – those that carried the unspeakable – came up against the taboo and was initially challenging.

19 Typically, the terms "bathroom" or "toilet" spoken in English are used as code words for shit, sewage, and sewerage.

REFERENCES

Anand, N. (2012). Municipal disconnect: On abject water and its infrastructures. *Ethnography, 13*(4), 487–509. https://doi.org/10.1177/1466138111435743

Behl, A. (2016, 16 July). Gurgaon sewage lines choke on increased load. *Hindustan Times*. Retrieved from http://www.hindustantimes.com/gurgaon/gurgaon-sewage-lines-choke-on-increased-load/story-Oz6dhOGBVTWtf5knw0tiPJ.html

Bhagat, R.P. (2005). Rural-urban classification and municipal governance in India. *Singapore Journal of Tropical Geography, 26*(1), 61–73. https://doi.org/10.1111/j.0129-7619.2005.00204.x

Delhi Development Authority. (1962). *Delhi Master Plan, 1962*. Retrieved from https//dda.org.in/planning/mpd-1962.htm

Douglas, M. (1966). *Purity and danger: An analysis of the concepts of pollution and taboo*. London: Routledge.

Dupont, V. (2007). Conflicting stakes and governance in the peripheries of large Indian metropolises: An introduction. *Cities, 24*(2), 89–94. https://doi.org/10.1016/j.cities.2006.11.002

Elyachar, J. (2012). Before (and after) neoliberalism: Tacit knowledge, secrets of the trade, and the public sector in Egypt. *Cultural Anthropology, 27*(1), 76–96. https://doi.org/10.1111/j.1548-1360.2012.01127.x

Gandy, M. (2004). Rethinking urban metabolism: Water, space and the modern city. *City, 8*(3), 363–79. https://doi.org/10.1080/1360481042000313509

Ghosh, J., Sen, A., & Chandrasekhar, C.P. (1997). All dressed up and nowhere to go: India infrastructure report. *Economic and Political Weekly, 32*(16), 803–8.

Government of Haryana. (1977). Haryana Urban Development Authority (HUDA) Act. Retrieved from http://www.lawsofindia.org/pdf/haryana/1977/1977HR13.pdf

Government of India. (1976). Urban Land (Ceiling and Regulation) Act, 1976. Retrieved from https://indiankanoon.org/doc/1005850/

Gururani, S. (2013). Flexible planning: The making of India's "Millennium City," Gurgaon. In A. Rademacher & K. Sivaramakrishnan (Eds.), *Ecologies of urbanism in India: Metropolitan civility and sustainability* (pp. 119–44). Hong Kong: Hong Kong University Press.

Gururani, S., & Kose, B. (2013). Shifting terrain: Questions of governance in India's cities and their peripheries. In P. Hamel & R. Keil (Eds.), *Suburban governance: A global view* (pp. 278–302). Toronto, ON: University of Toronto Press.

Hall, D., Hirsch, P., & Li, T.M. (2012). *Powers of exclusion: Land dilemmas in Southeast Asia*. Singapore: National University of Singapore.

Kumar, Y. (2012, 5 March). City's sewage system stinks. *The Times of India.* Retrieved from http://timesofindia.indiatimes.com/city/gurgaon/Citys-sewage-system-stinks/articleshow/12139689.cms

Larkin, B. (2013). The politics and poetics of infrastructure. *Annual Review of Anthropology, 42*(1), 327–43. https://doi.org/10.1146/annurev-anthro-092412-155522

Li, T.M. (2007). *The will to improve.* Durham, NC: Duke University Press.

Narain, S., & Srinivasa, R.K. (2012). *Excreta matters: How urban India is soaking up the water, polluting rivers and drowning in its own waste.* Delhi: Centre for Science and Environment.

Rohilla, S.K., Luthra, B., Padhi, S.K., Yadav, A., Watwani, J., & Varma, R.S. (2016, 12 April). Urban shit: Where does it all go? India is staring at a big sanitation crisis and needs to reinvent its excreta disposal mechanism. *Down to Earth.* Retrieved from http://www.downtoearth.org.in/coverage/urban-shit-53422

Simone, A.M. (2004). People as infrastructure: Intersecting fragments in Johannesburg. *Public Culture, 16*(3), 407–29. https://doi.org/10.1215/08992363-16-3-407

Varma, S. (2014, 22 June). Delhi downsizes as NCR population booms, census data show. *The Times of India.* Retrieved from https://timesofindia.indiatimes.com/india/Delhi-downsizes-as-NCR-population-booms-census-data-shows/articleshow/36983385.cms

World Bank. (1994). *World development report 1994: Infrastructure for development.* New York: Oxford University Press

7 Governance by Crises and Failing Infrastructure in Michigan: The 21st-Century Republican Strategy

IGOR VOJNOVIC, ZEENAT KOTVAL-K, JEANETTE ECKERT, AND XIAOMENG LI

Introduction

In April 2015, considerable controversy surrounded Michigan Sales Tax Increase for Transportation Amendment, Proposal 1 to fund repairs to Michigan's crumbling roads and bridges. The proposal involved a complex reform package to amend the state constitution and various statutes to make changes to the sales tax, use tax, and motor fuel tax, in addition to restructuring other state finances (Senate Fiscal Agency, 2015b). Driving these complex proposed changes is a model of "governance by crisis," which emerged during the state's post–Great Recession economic upswing and following the inauguration of Republican Governor Rick Snyder in 2011.

Michigan's economic recovery was evident over the last year of Democratic Governor Jennifer Granholm's term, in 2010, and has continued under Governor Snyder's two terms in office. Yet, by 2015, the state of Michigan entered a fiscal crisis considered more severe than anything faced in the midst of the Great Recession (Fulton & Hymans, 2015). Despite economic expansion, the state has experienced steep revenue shortfalls and deteriorating infrastructure.

As noted by Michigan Department of Transportation (MDOT) officials, with roads and bridges continuing to "fall further into disrepair, dragging down our economy and quality of life" Michigan finds itself in a "transportation funding crisis" (Michigan Department of Transportation [MDOT], 2015). According to the Michigan Transportation Asset Management Council (2014; 2015), Michigan roads are rapidly deteriorating, with only 17 per cent considered "good," 45 per cent considered "fair," and 38 per cent considered in "poor" condition. In his 2012 "State of the State Address," Governor Rick Snyder (2012) noted that the lack of investment in infrastructure was a "problem" in Michigan, particularly the funding of roads. Snyder would go on to acknowledge in the address that, on an annual basis, "[w]e are underinvesting in roads by $1.4 billion" (Snyder, 2012: 2).

Pat Shellenbarger's (2014) article in *Bridge Magazine*, titled "Michigan Roads Now among Nation's Worst," reports on the alarming state of Michigan's crumbling infrastructure. Across Michigan, freeway overpasses are being temporarily supported by steel posts. There are highway overpasses in such poor condition that they have sheets of plywood installed between beams to protect motorists from falling concrete. There is an overpass completely closed due to safety concerns. Even under these conditions, "the state's bridges are in better shape than its roads" (Shellenbarger, 2014: para. 11). As Kirk Steudle, director of MDOT, noted, "It's easy to see why people say we have the worst roads in the country. We put the least amount of emphasis on roads" (quoted in Shellenbarger, 2014: para. 12).

While this chapter focuses on neoliberal governance and its impact on deteriorating Michigan transportation infrastructure under the Snyder administration (2011–2018), it explores more broadly the strategy of "governance by crises." Promoted by US Republican governors, reducing taxes and moving to flat tax structures have been stripping states of public revenues needed to fund critical infrastructure, including roads, bridges, schools, policing, and waterworks. In the midst of the fiscal chaos and the breakdown in infrastructure and service provision – caused by the tax cuts – the states partially or fully withdraw from basic governing responsibilities, claiming lack of funding, a condition that they themselves have created. In the process, they use strategically planned and executed mismanagement to showcase the "ineffectiveness" of the public sector, an intentional attack on government by the Republican Party (GOP).

Reflecting on broader global political processes, Naomi Klein (2007) introduced the "shock doctrine," whereby public disorientation after massive collective shocks – wars, natural disasters, or economic recessions – is used to wipe out resistance and introduce neoliberal policies. The public disorientation allows legislation to be implemented that would never have been accepted otherwise, producing unfavourable outcomes for the vast majority, including the disadvantaged. In the context of provoking "shocks" and "public disorientation," the Republican

Party's central platform has focused on reducing taxes, recognizing fully that the basic infrastructure of US society would collapse under the pressures of the debt. The growing debt would force government to withdraw from basic public services – including social security, healthcare, and education – and strip the population of benefits that have been achieved over years of hard-fought battles.

In the 2016 national election, with a primary concern supposedly being the debt, the GOP candidates' major economic proposals centred on moving towards flat taxes and promising trillions of dollars in tax reductions. Donald Trump's proposed tax cuts amount to $9.5 trillion over just one decade (Gleckman, 2016). The claim is that the reduced taxes will "grow the economy," increasing tax revenue through job growth. However, as seen with the Ronald Reagan and George W. Bush administrations, the reduced taxes only produced record-setting deficits, debts, and collapsing economies, which have resulted in calls by Republican leaders for government to withdraw from public service and infrastructure provision, including entitlements, claiming that the state can no longer afford the programs.

This chapter will explore the governance by crises strategy in Michigan, and in the context of rapidly crumbling state infrastructure. There is an added sociospatial dimension to this analysis, which focuses on unique urban/suburban segregation patterns evident across the state, in part shaped by decades-long racial and class conflicts. The flight of whites and the wealthy from Michigan cities – and from populations they consider a threat, the minority and the poor – became a widespread imprint on urban regions across the state in the post–Second World War period. This excessive decentralization has not only facilitated disinvestment and decline within Michigan cities, but also extensive costs in infrastructure outlay and maintenance to service the dispersing low-density suburbs. In addition, the inefficiencies of these development patterns are sustained by shifting their costs to the broader population, including the urban poor. This analysis illustrates the intricacies involved in implementing such policies.

Under the Republican governance by crises strategy, the state has not only withdrawn from funding critical public infrastructure, but considerable resources are devoted to ensuring that the costs of the remaining state infrastructure fall disproportionately on lower income groups. The governance by crises strategy is not only economically costly, significantly reducing the business capacity and potential of the state, but it is excessively burdening the urban poor with the costs of public services.

With collapsing infrastructure – whether lead poisoning in water, crumbling schools, or inadequate policing – particularly evident in cities, this strategy is accentuating the urban/suburban socio-demographic divide across Michigan. The policy is reinforcing inequities and inequalities between cities and suburbs, and actively suppressing the opportunities of urban populations, ensuring the ongoing decline of their communities.

The chapter begins with an introduction into US suburbanization, urban decline, and the Detroit region, exploring regional segregation and travel behaviour. This analysis will provide insight into automobile reliance by population subgroups and the socioeconomics of transportation infrastructure use, critical to understanding inefficiencies and inequities associated with the current manoeuvring in transportation infrastructure financing. The study then shifts to statewide funding of transportation infrastructure and public revenues in Michigan, and analysis into the current fiscal and infrastructure crises.

Suburbanization and Urban Decline

The 1960 U.S. Census revealed a turning point in American city building. Between 1950 and 1960, the urban population grew by almost twenty-three million people, an unprecedented increase in absolute numbers, as the country became over 63 per cent urban (U.S. Census Bureau, 1993). While urban population increased by 38 per cent, urban land use increased twofold during the same period (Boyce, 1963). The average urban density decreased from 5,410 persons per square mile in 1950, to 3,759 in 1960. There were a number of variables – changing demographics, transportation technology, and public policy – responsible for not only the urban decentralization, but also the resulting urban decline.

The increasing post-war birth rate and growing population were important influences on US suburbanization. So were changes in transportation technology, including the ongoing diffusion of the affordable automobile and transport truck. Public policy would also exert its own impact on urban form and the resulting spatial segregation. The extensive network of roads and highways, built under the 1956 Interstate Highway Act, is considered consequential in reshaping American cities. Other policies facilitating the coupling of suburbanization and urban decline included preferential subsidization of suburban infrastructure, Federal Housing Administration (FHA) mortgage guarantees favouring suburban homes, racially restrictive covenants, exclusionary zoning, and redlining (Vojnovic, 2000; 2006; 2009; Vojnovic & Darden, 2013).

Throughout the second-half of the twentieth century, a new spatial structure was imprinted across US metropolitan centres, characterized by urban decentralization and dispersion, as an increased use of land became evident in the production of built space. Between the years 1970 and 2000, the average lot and house size doubled across the United States, with lots averaging 14,000 square feet and houses averaging 2,400 square feet by the year 2000 (Burchell et al., 2002). A new socio-demographic imprint also emerged across American cities, and along with it came a distinct imprint of spatial segregation that generated a landscape of inequity and inequality, evident with the blackening of cities as whites suburbanized. It was not only that the whites were suburbanizing, but so was the urban tax base, facilitating a rapid decline of central cities.

Michigan cities were at the forefront of these suburbanization trends. Between 1982 and 1992, Michigan lost 10 acres of farmland every hour to urban development, 854,000 acres over the decade (Burchell, Downs, McCann, & Mukherji, 2005). By 2000, more than 60 per cent of the Michigan population lived in suburbs (U.S. Census Bureau, 2000). And while many urban centres in the state are examples of excessive suburbanization and urban decline – including Saginaw, Flint, and Bay City – it is the decentralization of Metro Detroit, along with the decline of the city, that has captured international attention.

In addition, by the 2010 Census, new patterns of urban (re)development became evident across the United States, with urban reinvestment and growing ethnic/racial diversity in many inner cities. It was also apparent, however, that, as in Detroit, Midwest cities – whether Buffalo, Rochester, or Cleveland – continue to be vulnerable to excessive suburbanization and urban decline.

Socio-Demographic Patterns of Segregation across the Detroit Region

The City of Detroit's population peaked at 1.85 million people in 1950 and declined to a population of 677,116 people by 2015. The exodus from the city between the years 2000 and 2010 averaged sixty-three people every day, or about three people every hour, and it continues (U.S. Census Bureau, 2015). In addition, a clear racial and class imprint is evident across the region. By 2010, 83 per cent of City of Detroit residents were black, while over 97 per cent of all whites living in broader Metro Detroit (3.86 million people) resided in the suburbs.

Between 2007 and 2011, the national poverty rate across the United States averaged 14.3 per cent, but the poverty rate in the City of Detroit was 36.2 per cent. In contrast, the poverty rate in the Detroit suburbs of Oakland County was 9.5 per cent. Between 2007 and 2011, national per capita income averaged $27,915, but it was only $15,261 in the City of Detroit. In the Detroit suburbs of Oakland County, however, per capita income averaged $36,314 (U.S. Census Bureau, 2013). It should also be recognized that while Metro Detroit is an extreme example, these racial and class imprints are generally evident across urban Michigan (Vojnovic, 2009).

As the City of Detroit experienced rapid disinvestment and decline – including the complete breakdown of infrastructure and public services as the city's tax base deteriorated and the city filed for Chapter 9 bankruptcy in 2013 – many of its suburbs retained status as some of the most exclusive communities in America. These suburbs maintain a healthy tax base, some of the best public services in the country, and fashionable high-end neighbourhoods.

This contrast in municipal fiscal capacity between the city and suburbs is perhaps most evident in the provision of public education. In 2003, while average instructional spending per pupil was $3,100 in the City of Detroit, it was $6,148 in the wealthy suburb of Bloomfield Hills (Heath, 2003). This differential in expenditures played an important role in shaping educational outcomes. In 2003–2004, the Detroit School District maintained a graduation rate of 24.9 per cent, while the Bloomfield Hills School District was among the top nationally ranked public school systems (Swanson, 2008).

Such unequal access to services perpetuates the disparities along lines of class and race in Metro Detroit. It should be reinforced, once again, that this outcome is partly a result of discriminatory public policies that have been responsible for shaping the opportunities and burdens of populations across the region (Vojnovic & Darden, 2013). The assessment of transportation infrastructure financing will show how these urban/suburban divides continue to be perpetuated though tax policies, with considerable effort devoted to disproportionately burdening the urban poor.

Automobile Ownership and Travel: The Detroit Region

The dispersed development patterns not only altered travel, but also imprinted specific socio-demographic and spatial dimensions to automobile dependence. There are considerable differences in automobile reliance across the Detroit region, shaping distinct sociospatial patterns of transportation infrastructure use. A 2008 mail survey was carried out by the authors of this chapter. It assessed the relationship between urban form, socio-demographic composition, and travel in selected urban and suburban Detroit neighbourhoods. The neighbourhoods were chosen to represent variability in urban form and socio-demographic characteristics. A total of 1,191 responses were received, a 20 per cent return rate. The survey also collected car ownership data – verified through the Michigan Secretary of State vehicle registration records – enabling vehicle characteristics to be obtained by household and owner.

The data show that as income increases, vehicle ownership increases, overall travel distance increases, as do driving distances (table 7.1). The highest income group (above $100,000 annual income) drives about three and a half times the distances compared to the lowest income group (lower than $25,000 annual income). In addition, as incomes increase so does vehicle weight and engine size. Drivers in the top 20 per cent of household incomes are twice as likely to own SUVs, minivans, and trucks compared to drivers in the bottom 20 per cent of household incomes. Vehicle type and weight have important impacts on road wear and tear, since even small increases in vehicle weight result in large increases in road damage (Freeman & Clark, 2002).

Average annual household income in urban Detroit was just over $30,000, while for the low-density suburbs it was about $108,000. These income patterns not only influence automobile access, but also the spatial dimension of vehicle ownership. While 45 per cent of all households earning less than $25,000 annually had no access to a vehicle, over 99 per cent of households earning more than $100,000 annually did. In addition, while 23 per cent of Detroit residents had no access to a vehicle, close to 100 per cent of households in low-density suburbs did, with 42 per cent of these vehicles being SUVs, trucks, or minivans.

Table 7.1 Households and vehicle characteristics by income, the Detroit region

Household Income (000s)	Households without Access to a Vehicle (%)	Vehicles that Are SUVs, Minivans, and Trucks (%)	Average Annual Distances Driven (miles)	Average Annual Distances Travelled (miles)
Under $25	45	20	2,113	3,126
$25–$50	13	25	4,843	5,371
$50–$100	<1	33	7,064	7,829
Over $100	<1	45	7,486	7,891

In assessing transportation infrastructure financing, it is worth recognizing that these vehicle ownership and travel patterns across the Detroit region – and also revealed within Lansing – are likely replicated across Michigan, a result of the prevailing urban/suburban socio-demographic imprint (Kotval-K & Vojnovic, 2015; 2016; Vojnovic et al., 2013; 2014). As income increases, one is more likely to live in the suburbs, with greater likelihood of owning a vehicle – one that is heavier and likely driven over longer distances. Because of the nature of the built environment, sociospatial segregation, and resulting travel, it is the wealthy in the suburbs who are most reliant on the state's roads, highways, and bridges.

Urban residents, in contrast, consist disproportionately of lower income and minority populations, and they travel far less than the wealthier, maintaining lower reliance on the state's transportation infrastructure. The urban poor also live in communities with lower levels and quality of public service provision. Yet, as this analysis will show, they face higher relative economic burdens in supporting the state infrastructure. In addition, under the Republican administration, extensive resources have been devoted to shifting an even greater monetary burden of state transportation expenditures onto the poor, even as urban infrastructure within their communities deteriorates to unprecedented lows.

Equity and Efficiency Implications of Transportation Charges and Infrastructure Financing

There are policy dimensions associated with Detroit regional, and wider Michigan, travel patterns, with the focus here placed on the financial aspects of transportation levies and infrastructure funding. This analysis helps frame the discussion into efficient and equitable public charges associated with private vehicle use, transportation infrastructure financing, and more generally, personal travel.

At the federal level, Detroit's regional travel profile reinforces the inefficient and inequitable structure of the "gas guzzler tax." The gas guzzler tax was implemented in the Energy Tax Act (1978) in an effort to discourage the manufacture and purchase of fuel-inefficient vehicles. The tax is levied on manufacturers for sales of vehicles that do not meet the minimum 22.5 miles per gallon (mpg) fuel efficiency standard. The tax ranges from $1,000 per vehicle for vehicles that have a combined fuel economy of at least 21.5 mpg but less than 22.5 mpg to $7,700 per vehicle if the combined fuel economy is less than 12.5 mpg.

This tax, however, is not levied on SUVs, minivans, and pickup trucks, which make up some 46 per cent of the registered vehicles.

These are large vehicles with high levels of fuel consumption and pollution emission. In addition, as travel within the Detroit region illustrates, these vehicles also tend to be driven by wealthier populations, who drive long distances and are responsible for the most significant environmental harms; yet, they avoid paying the gas guzzler tax.[1]

At the state level, Michigan Republican policies demonstrate a similar disinterest, not only in efficiency and equity but in transportation infrastructure financing more broadly. In part, this situation is associated with Michigan's shift to a flat – and regressive – tax structure, a revenue strategy pursued by Republican administrations (Democratic Governors Association [DGA], 2014; Slemrod, 2006). These initiatives are generally accompanied with a governance by crises approach to state management.

The move to flat taxes reduces state revenues, which places pressure on the financing of public services, generating a breakdown, deterioration, and even collapse of local infrastructure and, more generally, government. The resulting chaos is the intention of the Republican leadership, forcing the public sector to withdraw from the funding of services while providing evidence of dysfunctional government. In Michigan, in this fiscal disarray, not only has the state withdrawn from funding public services, but it has also used the public disorientation to attempt to shift costs of the remaining state infrastructure disproportionately onto lower income groups, including the urban poor.

Governance by Crisis: Michigan's Dwindling State Finances and Deteriorating Infrastructure

In early 2015, the crisis over Michigan's crumbling infrastructure occurred as the state faced a $456 million general fund shortfall, and hence the push for the Michigan Sales Tax Increase for Transportation Amendment, Proposal 1 (House Fiscal Agency, 2015b). Money was eventually appropriated from the School Aid Fund to cover this deficit, but the fix came with the further gutting of education, and uncertainty over state revenues remains (House Fiscal Agency, 2015b; Senate Fiscal Agency, 2014). Despite a Michigan economic recovery, which was into its sixth year, the state fiscal crisis appeared as intense as during the Great Recession.

Many variables have contributed to Michigan's current fiscal crisis. As noted by the non-partisan legislative Michigan Senate Fiscal Agency (2015b), one aspect of the decreased state revenues for road funding can be traced back to the gasoline tax rate, which was set in 1997 and has not been changed since. In addition to the impacts of inflation, the increasing average fuel economy of vehicles further contributed to the decrease in fuel tax revenue.

The current road funding crisis is just one of many state financial pressures evolving since the 1990s. In 1995, the Michigan Economic Growth Authority Board was created under Republican Governor John Engler to authorize tax credits, including business tax credits (State of Michigan, 2015). The tax credits were used by Democratic and Republican administrations, including Republican Governor John Engler (1991–2002), Democratic Governor Jennifer Granholm (2003–2010), and Republican Governor Rick Snyder (2011–2018). Outstanding business tax credits are awarded into 2032 and are valued at $6.5 billion by the Senate Fiscal Agency (2015d), although the agency indicates that a more realistic cost of the credits is about $9.4 billion.

Added fiscal stress emerged when newly elected Republican Governor Rick Snyder instituted a major restructuring of Michigan's business and personal income taxes. In 2012, the Michigan business tax was repealed and replaced with a corporate income tax, with an estimated reduction of some $1.8 billion annually in business taxes. The losses in business taxes were partly covered by increased personal taxes, which placed disproportionate burdens on retirees and low-income workers (through reduced earned income tax credits), and by cuts to education (Senate Fiscal Agency, 2014). While the tax code restructuring reduced tax burdens on Michigan's businesses, Snyder's reforms retained the business tax credits. Thus, not only were corporate responsibilities to state taxes almost eliminated – at the expense of the most vulnerable taxpayers – but the tax overhaul locked in Michigan's commitment to billions of dollars in corporate tax credits.

In two of the first three Senate Fiscal Agency's monthly revenue reports for 2015, the business tax collections in Michigan were negative, with the largest in March totalling a negative $51.7 million for the month, which is central to the state's financial crisis (Senate Fiscal Agency, 2015c). While the ongoing fiscal crisis will leave an impact on Michigan for decades, and the state's infrastructure is collapsing under an even more regressive tax structure, Governor Rick Snyder considers his tax overhaul a major accomplishment, arguing that he "made the tax system simple, fair and efficient" (Snyder, 2012: 2).

Snyder came into office in 2011 calling for "shared sacrifice," but within months of being elected, the business tax was cut by about two billion dollars while individual taxes were raised – by over a billion dollars – on retirees, the poor, and also by cuts to education (Egan, 2011; Senate Fiscal Agency, 2014). The full impact of the restructuring was also not revenue neutral. The increases in individual taxes did not offset the business tax cuts, with a public revenue shortfall after the "reforms" of over $300 million in FY 2012–2013 (Senate Fiscal Agency, 2014).

Snyder argued that reduced business taxes would generate jobs, eventually leading to higher revenue generated from income taxes. Michigan union leaders fought against the tax code restructuring, arguing that Snyder's assumption about businesses hiring made little sense, since in the state and nationally, businesses had unprecedented cash reserves during the economic recovery but were not hiring (Hoffman, 2011).

Since the tax code restructuring, despite the new cash infusion for business, hiring in the state has been steady, that is, no dramatic increase. In fact, according to the Senate Fiscal Agency (2014: 13), as of April 2014, "payroll employment has remained relatively flat for 12 months." Hiring in the state began under the Granholm administration, before Snyder's inauguration on 1 January 2011. After over 400,000 jobs lost in 2008 and 2009, there were 65,000 jobs added in 2010 and 97,000 jobs added in 2011 (House Fiscal Agency, 2015a: 6). Snyder's repeal of the Michigan business tax became effective 1 January 2012, and yet, for the next three years, employment growth was modest. In 2012, employment slowed to 37,000 jobs added and then increased to 71,200 in 2013 (House Fiscal Agency, 2015a). The first eleven months of 2014 witnessed modest job growth of roughly 37,700 new jobs.

It was in the context of these infrastructure and fiscal pressures that Proposal 1 was advanced to address Michigan's rapidly deteriorating roads, the costs of which were considered significant, with increased vehicle operating costs and serious traffic accidents. In the TRIP (2015) review of Michigan's transportation system, the added total vehicle operating costs to Michigan motorists associated with driving over poor quality roads was about $4.8 billion annually, an average of $686 in extra vehicle operating costs per motorist. The poor quality roads and highways also contributed to serious traffic crashes, injuries, and fatalities, with roadway features responsible for about one-third of Michigan's serious and fatal crashes.

Proposal 1 was a tax code restructuring intended to generate revenue, and, according to the Senate Fiscal Agency (2015b), the most significant impact on taxpayers was to come from a sales tax increase. The key aspect of Proposal 1 involved a sales and use tax increase from 6 per cent to 7 per cent. The legislation would exempt motor fuel from sales and use taxes, but would increase the state motor fuel tax (Senate Fiscal Agency, 2015b; 2015e). An impact assessment by the Senate Fiscal Agency (2015b: 11) stated:

> From the consumer and taxpayer standpoint, the provision of the reform package that would have the most impact is the sales tax increase. The sales tax increase on nonfuel goods from 6.0% to 7.0% would increase the cost of purchasing these goods. On the other hand, the fuel tax increase combined with the repeal of the sales tax on fuels would result in a nominal difference for those buying fuel at the pump.

The overall impact of Proposal 1 was to be a net revenue increase to the state of $1.2 billion (Senate Fiscal Agency, 2015e).

The state of Michigan removed the corporate/business tax, which is considered by economists as the most progressive tax. In addition to increasing personal taxes on vulnerable populations, the Snyder administration tried to offset revenue shortfalls through state funding that is considered among the most regressive: the sales tax (Davies, 1959; Davies, St-Hilaire, & Whalley, 1984; Slemrod & Bakija, 2008; Suits, 1977). The regressive nature of the sales tax – which disproportionately burdens moderate-income families – is a result of lower- and middle-income earners consuming a greater percentage of their income, and hence paying sales tax on a greater share of income.

Michigan's Regressive Tax Structure

As of the 2015 March Monthly Revenue Report (Senate Fiscal Agency, 2015c) – a pre–Proposal 1 finance profile – the estimated revenue from the combined corporate income and Michigan business taxes is expected to be 1.2 per cent of FY 2014–2015 state revenue. Revenue generated from sales and use taxes is thirty-five times more, making up an estimated 42 per cent of state revenue. Tobacco taxes – separate from the sales and use taxes – are expected to generate 4.2 per cent of revenue, a figure three times higher than state corporate taxes. The Michigan personal income tax is estimated to generate some 39 per cent of state revenue.

In the midst of the 2015 financial crisis, Republican lawmakers were actually pushing for an additional decrease in the state income tax, which would further erode the state's revenue capacity. Within this context, it is important to consider the burden of state taxes on Michigan families. In general, based on Internal Revenue Service (IRS) and census data, it is

Figure 7.1 Michigan total state and local taxes, 2015

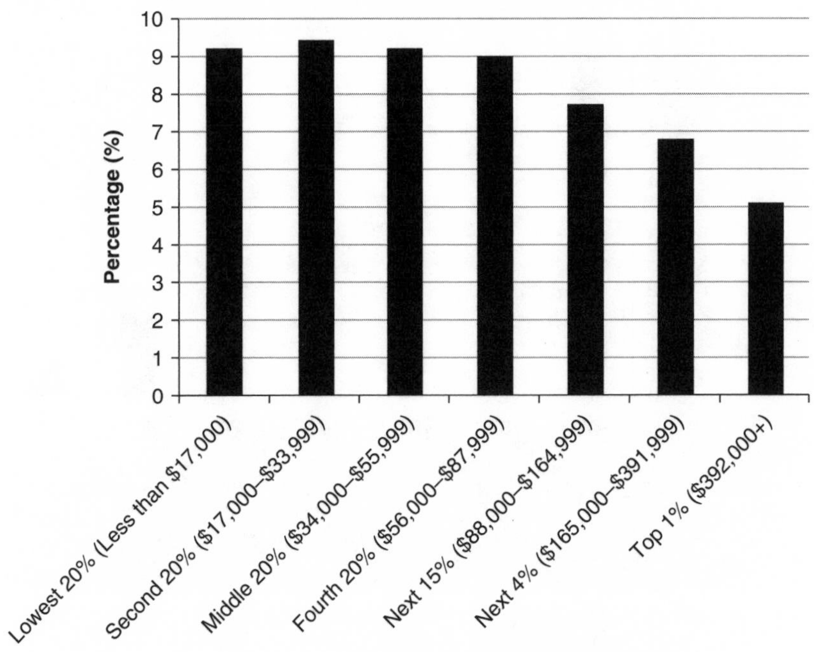

Source: Data from Institute on Taxation & Economic Policy. (2015). *Who pays?* Washington DC: ITEP.

recognized that flat taxes have regressive impacts in the states that have adopted this formula (Vojnovic, 2003a; 2003b; 2007).

The Institute on Taxation and Economic Policy's annual reports examine tax policies across the United States. In 2015 in Michigan, the lowest effective total local and state tax, as a share of family income (for non-elderly tax payers), is expected on families with average annual family incomes of $1,164,700 (Institute on Taxation and Economic Policy, 2015). The effective tax rate on this group of 1 per cent of the highest income earners is almost half of that placed on the poorest 20 per cent of Michigan families, which have an average annual family income of $9,500 (figure 7.1). This rate is Michigan's revenue burden based on the pre–Proposal 1 budget, but it does incorporate the impact

Figure 7.2 Sales and excise tax share of family income, 2015

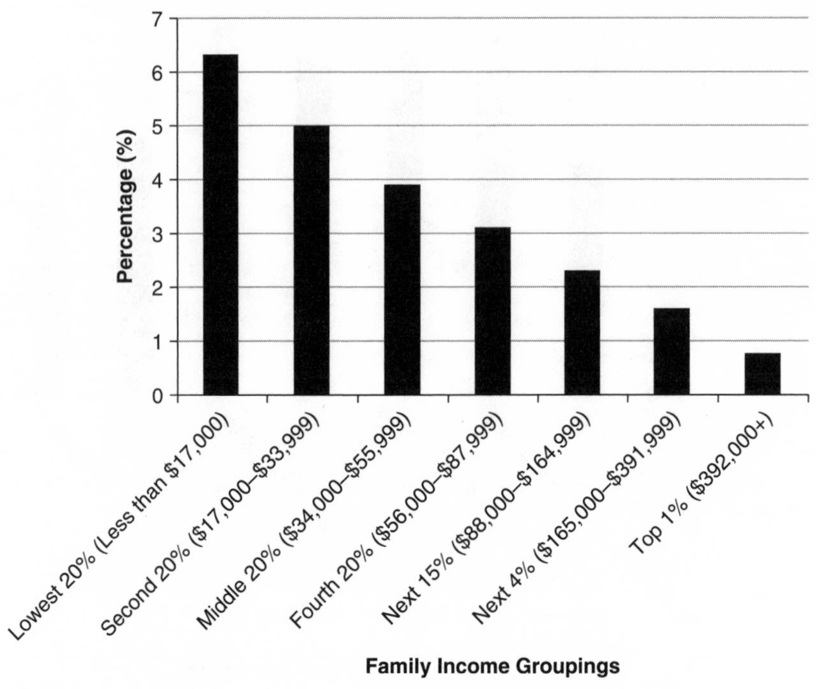

Source: Data from Institute on Taxation & Economic Policy. (2015). *Who pays?* Washington DC: ITEP.

of Snyder's removal of the Michigan business tax and the increases in taxes on individuals.

The largest portion of state finances comes from the highly regressive sales and excise taxes (figure 7.2). Their rate on the family incomes of the poorest 20 per cent of Michigan families, with incomes below $17,000 annually, is about eight times greater than it is on the richest 1 per cent, with annual incomes of $392,000 or higher. In contrast, the most progressive Michigan tax is the personal income tax (figure 7.3). It is relevant to consider changes to the state income tax, since they too have contributed to Michigan's current fiscal stresses. The Michigan Individual Income Tax was enacted in 1967. It was introduced as, and still remains, a flat tax. Over its forty-eight years, the median tax rate has

Figure 7.3 Personal income tax share of family income, 2015

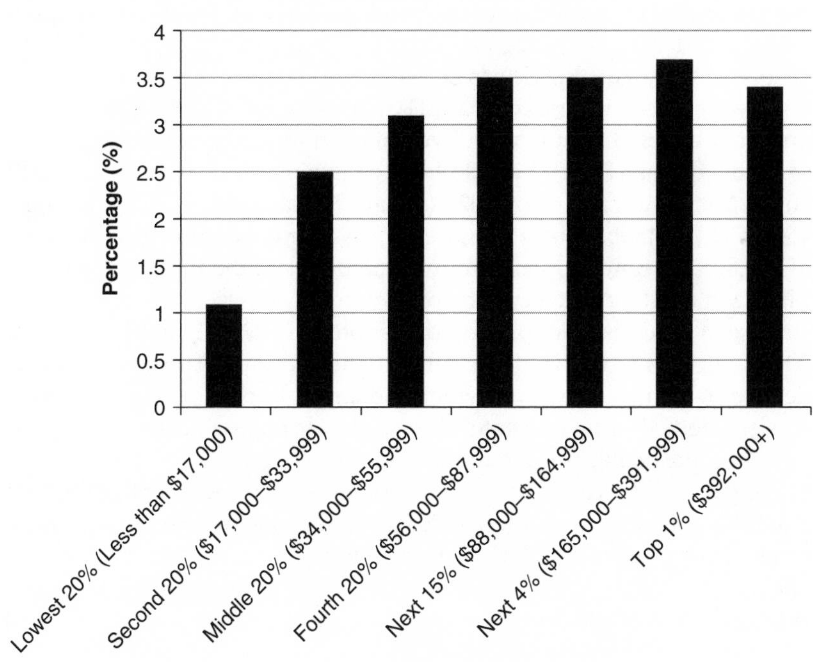

Source: Data from Institute on Taxation & Economic Policy. (2015). *Who pays?* Washington DC: ITEP.

been 4.4 per cent (Senate Fiscal Agency, 2015a). The current income tax rate is 4.25 per cent, reduced by Snyder from 4.35 per cent in 2011. The state income tax rate had been increased from 3.9 per cent to 4.35 per cent by Granholm in FY 2007–2008, responding to fiscal difficulties of the Great Recession. At 4.35 per cent, the 2008 state income tax rate was below its historical median of 4.4 per cent. In addition, due to various exemptions and credits, the effective tax rate (the actual tax collected) in Michigan is far less. Over the last decade, on average, the effective tax rate has been less than half the nominal rate (Senate Fiscal Agency, 2015a). It is also due to these exemptions that the flat state income tax is progressive, but this aspect of Michigan's income tax is relatively recent.

In 2006, Jennifer Granholm introduced Michigan's earned income tax credit (EITC) for low-income earners. If Michigan taxpayers qualify for a federal EITC, they are also eligible for the Michigan EITC. In 2006, the Michigan EITC was set at 20 per cent of the amount that a taxpayer can claim for a federal EITC. For example, in FY 2010–2011, a qualified married couple with two children maintaining an annual income of less than $46,044 would have received a federal credit of $5,112 and a Michigan credit of $1,022 (Senate Fiscal Agency, 2011). In 2008, some 711,000 taxpayers – maintaining an average annual gross income of $15,805 – claimed $145.2 million in credits. It is the EITC that enables Michigan's flat income tax to be progressive.

Once in office, Snyder proposed to abolish the Michigan EITC, and in 2011 a bill was introduced to eliminate the program. The proposal was tied up in political disputes, when the larger Republican tax overhaul was enacted, shifting the state's tax burden from businesses to individuals, and particularly the more vulnerable. As part of the tax reforms, Michigan's EITC was reduced from 20 per cent to 6 per cent of the federal EITC (Senate Fiscal Agency, 2015a). Thus, the income tax was more progressive prior to Snyder's inauguration.

Where Do We Stand Now?

On 5 May 2015, Michigan voters showed their contempt for the Michigan Sales Tax Increase for Transportation Amendment, Proposal 1, rejecting it 80 per cent to 20 per cent. The ballot was the object of even more disdain once it was discovered that Michigan corporations and political supporters of Proposal 1 had raised $9.3 million to promote the plan across the state, spending seventeen times more in their failed lobbying efforts than the opponents of Proposal 1 (Wall Street Journal, 2015).

Leonard Fleming and Gary Heinlein (2015) of the *Detroit News* recognized this result as "one of the worst ballot measure defeats in more than a generation." Republican voters were unhappy that higher taxes were being proposed. Democratic voters were outraged that an even greater tax rate was being placed on individuals, while corporations were almost completely exempt from state taxes. House Democratic Leader Tim Greimel, who backed Proposal 1, declared: "Our families have given Lansing $1.6 billion in new taxes with nothing to show for it – not better schools or better roads – because all that money went to give tax cuts for corporations" (Fleming and Heinlein, 2015). John Yob, the head of Citizens Against Middle Class Tax Increases, normally considered an ally of Snyder, exclaimed: "The legislature failed to do their job in funding basic infrastructure and instead tried to convince voters to raise taxes on themselves because they didn't have the political strength to cut the budget or raise taxes" (Fleming and Heinlein, 2015).

Following the failed ballot, there were six months of bipartisan talks for a resolution that would address Michigan's rapidly deteriorating infrastructure and ongoing fiscal crisis. The negotiations failed due to Republicans' insistence that any fuel tax increase needed to be coupled with income tax reductions. Finally, on 3 November 2015, the Republican-led legislature passed a plan, almost exclusively along party lines, to appropriate $1.2 billion of annual funding for the repair of Michigan roads. It was signed into law by Snyder a week later.

As with Proposal 1, the plan is a complex package involving multiple bills, which also include state revenue reductions, including provisions for a roll-back in state income taxes. Overall, the measure provides an estimated increase in state revenues of only $407 million by FY 2018–2019 for a $1.2-plus billion expenditure plan (House Fiscal Agency, 2015c). The full $1.2-plus billion is also not planned until FY 2020–2021.

In January 2017, there were increases in Michigan's 15 cents per gallon diesel tax and 19 cents per gallon gasoline tax, bringing both up to 26.3 cents per gallon. These tax increases have generated approximately $400 million annually since their implementation in FY 2017–2018. In January 2017, vehicle registration fees were also raised for vehicles and trucks, with the average registration of $100 increasing to an average fee of $120. The registration fee increase has generated about $200 million annually since its FY 2017–2018 implementation. According to the House Fiscal Agency (2015c), these increases produce an extra $617 million annually in state revenue. However, in FY 2018–2019, the expansion of the homestead property tax credit, also part of the restructuring, will reduce the net increase in state revenues to just $416 million.

The remaining $806 million required for the $1.2-plus billion FY 2020–2021 plan will come from Michigan's general fund, which means that funding to education, public safety, child protection, and healthcare for the state's working poor – all supported by general funds – will be cut. This two-thirds sum for the proposed infrastructure investment is also not fleshed out, requiring considerable reductions to these programs by the next governor and legislature. In addition, the full $1.2-plus billion funding package does not start until 2021, a decade after Snyder acknowledged that annually roads are being underinvested by $1.4 billion.

The plan also involves an "income tax reduction trigger," Senate Bill 414, which reduces general fund revenues for years in which prior-year general fund revenue growth exceeds the rate of inflation. These revenue reductions would then continue in subsequent years. In reviewing Senate Bill 414, the House Fiscal Agency (2015c: 5) estimated the impact of the income tax rate reductions using Snyder's current budget, indicating that "the income tax rate for tax year 2016 would drop from the current level of 4.25% to approximately 3.96%, resulting in a revenue reduction of $593 million."

The Cost of Inaction: The Incompetence Gets Worse

Among the many controversial aspects of the transportation infrastructure financing plan is that nothing has been resolved on the funding of roads, highways, and bridges, pushing the ultimate resolution of the crises to a new governor and legislature. This fact is an important recognition, since the Republican administration has placed the state under even greater financial stress than what has been revealed, simply because of policy inaction.

In his 2012 "State of the State Address," Snyder recognized that the underinvestment in roads was $1.4 billion annually. This figure was based on a 2011 estimate – *Michigan's Road Crisis: How Much Will It Cost to Maintain Our Roads and Bridges?* – which calculated the lowest long-term costs of "reasonably" maintaining Michigan's transportation infrastructure (Olson and Schmidt, 2011). These projected lowest long-term costs were $1.4 billion per year between 2012 and 2015, with costs then increasing to about $2.6 billion per year by 2023, all part of what was considered to be the minimum maintenance costs on roads. The report states:

> Bottom line: if the investments projected by these models are not done, either the deferred costs of maintaining our roads will be much higher OR we choose to accept lower quality roads. From a business perspective, the set of investments recommended is the lowest long-term costs of maintaining our roads. (Olson and Schmidt, 2011: 13)

The 2011 assessment recognized that if road maintenance and improvements were deferred, the future costs of repair would be higher, since having to replace poor quality roads is costlier than maintaining good roads. For instance, based on 2015 costs, the costs would be $85,000 per lane mile for a road considered in "good" condition to ensure adequate preventive maintenance. However, for a road considered "fair," resurfacing costs increase to $575,000 per lane mile. For a road in "poor" condition, greater reconstruction would be required, increasing costs to $1,625,000 per lane mile, nineteen times greater than the cost of preventive maintenance (TRIP, 2015). It was thus recognized in 2012 by the Republican-led legislature that there was a cost associated with legislative inaction.

In 2014, MDOT analysts and Rick Olson – former Republican state representative, House Republican policy adviser, and co-author of the 2011 Road Crisis Report – updated the costs of repairing Michigan's roads. The 2014 Update Report concludes that the total in first-year costs have increased from $1.4 billion in 2011 to $2.183 billion in 2014, with the infrastructure repair costs increasing to over $3.2 billion by 2025, up from the $2.6 billion per year that was expected by 2023 (Olson, 2014). Thus, the current costs of addressing Michigan's deteriorating roads, due simply to inaction, are considerably above the full amount of the road funding of $1.2 billion expected in 2021, and the costs continue to increase with every year of inaction.

The Republican road funding program not only appropriated inadequate funds for the repair of Michigan transportation infrastructure, but it also left the burden of what will need to be cut to allow for the funding of roads to a new governor and legislature. Their inaction has also ensured that the costs of fixing roads will continue to increase. The next administration will confront total costs in transportation infrastructure repairs that will be billions of dollars more than what the Snyder government faced in 2011. Thus, it's not just that Snyder and the Republican legislators "kicked the can down the road" to subsequent administrations, but their inaction is costing Michigan taxpayers billions of dollars to solve the state's road crisis. In addition, Snyder (2012) himself recognized that Michigan's inadequate infrastructure is also costing the state in potential economic investment, which is a whole other layer of costs.

Across the United States, the political spin on Michigan's fiscal chaos was immediate and familiar. In the *Wall Street Journal*, for instance, after reviewing Michigan's road and financial crises, the authors stated: "Voters can elect Democrats if they want a tax increase. They elect Republicans to make hard spending choices and grow the economy" (Wall Street Journal, 2015).

Conclusion

Michigan represents yet another case of Republican experiments in minimal government, and perhaps the clearest impact is evident with the deterioration of its cities to new lows of distress and devastation. The collapse of Detroit – where residents are living among over 70,000 abandoned properties – is an urban condition spreading across the state. This collapse is all taking place during an economic upswing, the impacts of which are highly selective, shaped by Republican policies giving extensive preference to the privileged, generally suburbanites.

The Detroit region not only illustrates the segregation and stark socio-demographic distinctions between cities and suburbs, but, with regard to transportation, shows the significant spatial differences in automobile ownership and travel. Wealthy suburbanites are more likely to own vehicles, and larger and heavier vehicles, and travel longer distances, especially by driving. So not only do the suburbanites generally travel more, but their larger vehicles disproportionately damage road surfaces.

Despite these differences in who is responsible for the wear and tear of Michigan's roads, extensive effort was placed by the Snyder administration and corporate interests to ensure that transportation infrastructure – through sales tax increases – are disproportionately covered by moderate-income earners, including the urban poor. This initiative was advanced despite an already regressive state tax structure, where the poor are financially burdened at much higher levels, in terms of proportion of income, for supporting state expenditures. Such schemes also reveal the importance of public policy in shaping economic opportunities and burdens, with clear urban/suburban distinctions, which have been evident throughout Michigan's history.

The Snyder administration had over half a decade in office to come up with a solution to Michigan's infrastructure crisis. Upon being elected, they reduced the income tax and virtually eliminated taxes on corporations. At the same time, they proceeded to raise the income tax on the poor and retirees and gut funding for education. In their second term, they attempted to transfer the disproportionate cost of upgrading the state's transportation infrastructure on low- and moderate-income families. With the failure of this initiative, the Snyder administration simply shifted the crises to the next governor and legislature, but now with significant increases in repair costs. Snyder's administration has ensured that transportation infrastructure maintenance costs will be billions of dollars more than what they needed to spend to repair state roads; and this increase does not include losses in potential state investment due to Michigan's lack of infrastructure capacity.

The push by the Snyder administration towards a flat tax structure – with increased reliance on the sales tax as a means of supporting state finances – despite the statewide collapse in infrastructure, not only illustrates the complete disregard for basic equity and efficiency but also, broadly, for general governance. The ongoing manoeuvring by the Republican administration in the middle of these crises to reduce the income tax only reinforced their complete disinterest in responsible government.

This situation is part of the Republican governance by crises strategy. They cripple the state and leave it completely dysfunctional. The next administration is required to devote extensive energy and resources simply responding to the various crises created by the Republican leadership, whether the deteriorating conditions of urban schools, poisoned water, or the crumbling roads and bridges. The Snyder government has shown, as has the Republican Party more broadly, that their mantra of flat, or reduced, taxes is capable of "tearing-down" with widespread damage, but has little capacity in "building" or being "innovative."

The twenty-first century Republican strategy is based on ongoing financial and political crises because it supports the party's platform that government is ineffective and that it cannot afford basic public services, whether education, roads, or safe water.[2] With an efficient and well-financed public sector, it becomes difficult to argue that government is dysfunctional or that it cannot finance public services. With one of the principal objectives of the Republican Party in the twenty-first century being governance by crises, one can assume that anytime elected, their term will be characterized by fiscal chaos and public sector breakdown.

For Michigan's richest, these policies have cut a few percentage points from their taxes, but for Michigan's urban poor these programs are fundamentally transformative. They have not only burdened them with higher tax rates, but facilitated a collapse in urban public infrastructure, including education, policing, and water provision.

Republican-driven chaos in the Michigan context also has a spatial dimension, since their strategically planned and executed mismanagement targets cities and the urban poor. The governance by crises strategy is accentuating the urban/suburban socio-demographic divide, reinforcing inequities and inequalities between cities and suburbs, while actively suppressing the opportunities of urban populations. And while "lower taxes" are promoted in the interest of "economic growth," the reduced taxes – as evident since the late 1990s – have only encouraged Michigan's ongoing economic decline and allowed inadequate state infrastructure to diminish the state's economic capacity and performance (Vojnovic, 2006; 2007; 2009).

The chapter has brought to light the critical importance of the financial dimension of infrastructures. It has done this from different angles by focusing on the relation between the conditions of infrastructures and their funding; on the social equity consequences of funding formulas; and on infrastructure funding as a political strategy. Insufficient finances as an obstacle to the provision of adequate infrastructure to meet needs are an object of concern present in virtually all chapters, reflecting the close ties between the condition and scale of infrastructure systems and the financial capacity of agencies providing them. Relative to other chapters in this volume, however, chapter 7 goes further in exploring the link between infrastructures and public sector financial strategies. It presents a case where the availability of funding for infrastructures was dictated by political strategies inspired by neoliberalism and had far-reaching economic and social equity repercussions. The political ideology guiding highway funding in Michigan was probably expressed more blatantly here than in the cases reported in other chapters. The Michigan case also stands out for the extreme inequity of the fiscal load required to fund highway maintenance. But we can assume that everywhere funding policies determining the form and scale of infrastructures bear the mark of dominant political ideologies.

NOTES

1 The Corporate Average Fuel Economy (CAFE) standards have been a more effective tool in reducing energy consumption by increasing the fuel economy of cars and light trucks. CAFE, enacted in 1975, is responsible for almost doubling passenger car miles per gallon by 1985 (National Research Council, 2002).

2 This strategy is evident across Republican-led states, Kansas providing another classic example, and also federally, as with federal Republican administrations.

REFERENCES

Boyce, R.R. (1963). Myth versus reality in urban planning. *Land Economics, 39*(3), 241–51. https://doi.org/10.2307/3144362

Burchell, R., Downs, A., McCann, B., & Mukherji, S. (2005). *Sprawl costs: Economic impacts of unchecked development*. Washington, DC Island Press.

Burchell, R., Lowenstein, G., Dolphin, W., Galley, C., Downs, A., Seskin, S., Still, K., & Moore, T. (2002). *Costs of sprawl 2000*. Washington, DC: National Academy Press.

Davies, D. (1959). An empirical test of sales-tax regressivity. *Journal of Political Economy*, *67*(1), 72–8. https://doi.org/10.1086/258131

Davies, J., St-Hilaire, F., & Whalley, J. (1984). Some calculations of lifetime tax incidence. *The American Economic Review*, *74*(4), 633–49.

Democratic Governors Association (DGA). (2014, 15 April). On tax day, the super-rich thank their lucky stars for GOP govs. Washington, DC: Author. Retrieved from https://democraticgovernors.org/news/on-tax-day-the-super-rich-thank-their-lucky-stars-for-gop-govs/

Egan, P. (2011, 18 February). Snyder's budget calls for "shared sacrifice." *Detroit News*, pp. 1A and 7A.

Fleming, L., & Heinlein, G. (2015, 6 May). Michigan voters reject Proposal 1 tax hike. *Detroit News*. Retrieved from http://www.detroitnews.com/story/news/politics/2015/05/05/proposal-one/26952783/

Freeman, T.E., & Clark, T.M. (2002). Performance of pavements subject to higher truck weight limits in Virginia. *Transportation Research Record: Journal of the Transportation Research Board*, *1806*, 95–100. https://doi.org/10.3141/1806-11

Fulton, G., & Hymans, H. (2015). *Michigan outlook a summary for 2016–2017*. Ann Arbor, MI: Center for Labor Market Research, University of Michigan.

Gleckman, H. (2016, 17 February). *How the GOP candidate's tax plans stack up against one another?* Washington, DC: Tax Policy Center Retrieved from http://www.taxpolicycenter.org/taxvox/how-gop-candidates-tax-plans-stack-against-one-another

Heath, B. (2003, 25 May). Michigan still shortchanges poor schools. *The Detroit News*, p. 1A.

Hoffman, K. (2011, 25 May). Mich. Gov. Snyder signs sweeping tax legislation. *Associated Press*. Retrieved from http://archive.boston.com/business/taxes/articles/2011/05/25/mich_gov_snyder_signs_sweeping_tax_legislation/

House Fiscal Agency. (2015a). *Economic outlook and revenue estimates for Michigan: FY 2014–15 through FY 2016–17*. Lansing, MI: Author.

House Fiscal Agency. (2015b). *FY 2014–2015 Supplemental Appropriations, Public Act 5 of 2015, House Bill 4110 (H-1) as enacted*. Lansing, MI: Author.

House Fiscal Agency. (2015c). *Legislative analysis: Road funding package – Enacted analysis*. Lansing, MI: Author.

Institute on Taxation & Economic Policy. (2015). *Who pays?* Washington, DC: Author.

Klein, N. 2007. *The shock doctrine*. New York: Henry Holt and Company.

Kotval-K, Z., & Vojnovic, I. (2015). The socio-economics of travel behavior and environmental burdens: A Detroit, Michigan, regional context. *Transportation Research Part D: Transport and Environment, 41*, 477–91. https://doi.org/10.1016/j.trd.2015.10.017

Kotval-K, Z., & Vojnovic, I. (2016). A socio-ecological exploration into urban form: The environmental costs of travel. *Ecological Economics, 128*, 87–98. https://doi.org/10.1016/j.ecolecon.2016.04.009

Michigan Department of Transportation (MDOT). (2015). *Transportation funding continues its decline*. Lansing, MI: Author. Retrieved from http://www.michigan.gov/mdot/0,4616,7-151--292908--,00.html

Michigan Transportation Asset Management Council. (2014). *Michigan's roads and bridges, 2014 annual report*. Lansing, MI: Author.

Michigan Transportation Asset Management Council. (2015). *Michigan Transportation Asset Management Council update*. Lansing, MI: Author. Retrieved from https://www.ctt.mtu.edu/sites/ctt/files/resources/bridge/6safford.pdf

National Research Council (NRC). (2002). *Effectiveness and impact of corporate average fuel economy (CAFE) standards*. Washington, DC: National Academy Press.

Olson, R. (2014). *Michigan's road crisis: How much will it cost to maintain our roads and bridges? Update 2014*. Lansing, MI: Work Group on Transportation Funding, House of Representatives Transportation Committee. Retrieved from http://media.mlive.com/lansing-news/other/2014-olsen-road-report.pdf

Olson, R., & Schmidt, R. (2011). *Michigan's road crisis: How much will it cost to maintain our roads and bridges?* Lansing, MI: Work Group on Transportation Funding, House of Representatives Transportation Committee. Retrieved from http://www.annarbor.com/Michigan'sRoadsCrisis.pdf

Senate Fiscal Agency. (2011). *Memorandum: The impact of eliminating Michigan's EITC on the TANF MOE*. Lansing, MI: Author.

Senate Fiscal Agency. (2014, August). *State budget overview.* Lansing, MI: Author.

Senate Fiscal Agency. (2015a). *A history of the Michigan individual income tax rate*. Lansing, MI: Author.

Senate Fiscal Agency. (2015b). *The long and winding road: Proposal 1 and road funding reforms.* (State Notes). Lansing, MI: Author.

Senate Fiscal Agency. (2015c). *Monthly revenue report, March 2015.* Lansing, MI: Author.

Senate Fiscal Agency. (2015d). *A primer on certified credit under the Michigan business tax*. (State Notes). Lansing, MI: Author.

Senate Fiscal Agency. (2015e). *Transportation funding: Bill analysis.* Lansing, MI: Author.

Shellenbarger, P. (2014, 8 April). Michigan roads now among nation's worst. *Bridge Magazine.* Retrieved from https://www.bridgemi.com/economy/michigan-roads-now-among-nations-worst

Slemrod, J. (2006). The role of misconceptions in support for regressive tax reform. *National Tax Journal, 59*(1), 57–75. https://doi.org/10.17310/ntj.2006.1.03

Slemrod, J., & Bakija, J. (2008). *Taxing ourselves: A citizen's guide to the debate over taxes.* Cambridge, MA: MIT Press.

Snyder, R. (2012, 18 January). *Governor Rick Snyder state of the state address.* Lansing, MI: State of Michigan.

State of Michigan. (2015). *Michigan Economic Growth Authority (MEGA) Board.* Lansing, MI: Author. Retrieved from https://www.michigan.gov/snyder/0,4668,7-277-57738_57679_57726-255865-,00.html

Suits, D. (1977). Measurement of tax progressivity. *American Economic Review, 67*(4), 747–52.

Swanson. C.B. (2008). *Cities in crisis: A special analytical report on high school graduation.* Bethesda, MD: Education Research Center.

TRIP, A National Transportation Research Group. (2015). *Michigan's top transportation challenges: Providing a transportation system to support and sustain Michigan's economic revival.* Washington, DC: Author.

U.S. Census Bureau. (1993). *Population and housing unit counts: United States summary.* Washington, DC: U.S. Government Printing Office.

U.S. Census Bureau. (2000). *Population estimates of metropolitan areas, metropolitan areas inside central cities, metropolitan areas outside central cities, and nonmetropolitan areas by state.* Washington, DC: U.S. Government Printing Office.

U.S. Census Bureau. (2013). *State and county quickfacts.* Washington, DC: U.S. Government Printing Office.

U.S. Census Bureau. (2015). *State and county quickfacts.* Washington, DC: U.S. Government Printing Office.

Vojnovic, I. (2000). Shaping metropolitan Toronto: A study of linear infrastructure subsidies, 1954–66. *Environment and Planning B: Planning and Design, 27*(2), 197–230. https://doi.org/10.1068/b2620

Vojnovic, I. (2003a). Governance in Houston: Growth theories and urban pressures. *Journal of Urban Affairs, 25*(5), 589–624. https://doi.org/10.1111/j.1467-9906.2003.00004.x

Vojnovic, I. (2003b). Laissez-faire governance and the archetype laissez-faire city in the USA: Exploring Houston. *Geografiska Annaler: Series B, Human Georgraphy, 85*(1), 19–38. https://doi.org/10.1111/1468-0467.00128

Vojnovic, I. (2006). Urban infrastructures. In P. Filion & T. Bunting (Eds.), *Canadian cities in transition* (3rd ed.) (pp. 123–37). Oxford: Oxford University Press.

Vojnovic, I. (2007). Government and urban management in the 20th century: Policies, contradictions, and weaknesses of the new right. *GeoJournal*, *69*(4), 271–300. https://doi.org/10.1007/s10708-007-9110-z

Vojnovic, I. (2009). Urban settlements in Michigan: Suburbanization and the future. In R. Schaetzl, J. Darden, & D. Brandt (Eds.), *Michigan: A geography and geology* (pp. 487–507). Cambridge, MA: Pearson Publishing Group.

Vojnovic, I., & Darden, J. (2013). Class/racial conflict, intolerance, and distortions in urban form: Lessons for sustainability from the Detroit region. *Ecological Economics*, *96*, 88–98. https://doi.org/10.1016/j.ecolecon.2013.10.007

Vojnovic, I., Kotval-K, Z., Lee, J., Ye, M., LeDoux, T., Varnakovida, P., & Messina, J. (2014). Urban built environments, accessibility, and travel behavior in a declining urban core: An exploration of the extreme conditions of disinvestment and suburbanization in the Detroit region. *Journal of Urban Affairs*, *36*(2), 225–55. https://doi.org/10.1111/juaf.12031

Vojnovic, I., Lee, J., Kotval-K, Z., Podagrosi, A., Varnakovida, P., LeDoux, T., & Messina, J. (2013). The burdens of place: A socio-economic and ethnic/racial exploration into urban form, accessibility and travel behaviour in the Lansing capital region, Michigan. *Journal of Urban Design*, *18*(1), 1–35. https://doi.org/10.1080/13574809.2012.683403

Wall Street Journal. (2015, 6 May). Michigan's road kill: Voters demolish a $2 billion sales tax increase. *Wall Street Journal*. Retrieved from http://www.wsj.com/articles/michigans-road-kill-1430953555

8 Infrastructure Interludes: Sociotechnical Disposition and Planning for Water and Wastewater Systems in the Stockholm Archipelago

JONATHAN RUTHERFORD

Introduction

In two of Ingmar Bergman's films of the early 1950s (*Summer Interlude* and *Summer with Monika*), there is a striking contrast between the depiction of light, space, shimmering water, freedom, and happiness as his characters pass their summers enveloped in the beauty of nature in the Stockholm archipelago, and the dour reality and confinement of urban life as they return to jobs and everyday routine in the city when summer games (and love affairs) are over (see Koskinen, 1995).[1] This portrayal is not just a kind of metaphysical distinction between the innocence and playful terrain of nature and the grim solemnity of the urban, but it serves as a useful representation of the importance of escaping temporarily into "suburban" recess in Swedish culture and lifestyle, where immersion in landscape and water offers regeneration before resurgence into the rigours of modern life. Infrastructure configurations are enmeshed in the making and remaking of these places, accompanying (and increasingly blurring) these distinctions and nuances. Water systems in the Stockholm archipelago are not the dense, interconnected, ubiquitous industrial networks of the city. They have a temporary, hybrid, sparse, "natural" quality to them, marking out these areas as, in a Bergmanian sense, an "interlude"[2] full of freedom and possibility, albeit also with a foreboding of the coming encroachment of the urban, which can be measured by tensions and conflicts over matters of development, resource distribution, and (state) regulation of lives/living spaces that are otherwise a little "freer" than in the city.

In practical terms, in the peripheral and less dense areas of some European cities – areas that are not (yet) served by centralized infrastructure systems – local planners, technicians, and residents are indeed questioning the relevance of water and energy network extensions. In these contexts of low population density and distinctive modes of residence, actors and authorities are weighing between the possible return on investment from network deployment and use, and the technical difficulties and additional costs of laying and maintaining the necessary cables and pipes. These areas that lie beyond the network may be included in future extension plans or may remain more dependent on alternative forms of technology, which already offer arrangements adapted to suburban living.

Infrastructure systems are crucially important in extending and reworking socio-natural (sub)urban spaces. They have always either accompanied and supported or preceded and stimulated urban growth. Yet, paradoxically, the actual forms and processes of making and remaking suburban infrastructure configurations are largely ignored (although see Keil, 2013). Much work on infrastructure in suburban contexts has usually restricted itself to the study of the relative financial cost of sprawl or extended urban development compared to that of densification (for an overview, see Jaglin & May, 2010), reducing both suburbanization to a narrow (and preset) technoeconomic issue, which can objectively inform more efficient planning practice, and infrastructure to generic and passive equipment, the features and specificities of which have little influence on development. Eschewing this limited perspective, this chapter makes a contribution to the study of "massively unknown" suburban infrastructures (see chapter 1, the introduction to this volume) by focusing on sociotechnical and sociopolitical issues as they emerge locally in the Stockholm archipelago area, where particular hybrid configurations of water infrastructure are at the heart of changing forms, modalities, and outcomes of local development and modes of residence.

At the margins of the urban and the grid are important transformations of sociotechnical infrastructure space, which both reflect the status of "in-between" areas as a major zone of recent residential, functional, and infrastructure change (Young, Burke Wood, & Keil, 2011) and offer a vision and some wider lessons for urban futures. Exploring infrastructure interludes here provides analytical purchase on constantly shifting frontiers between collective and individual, public and private, industrial and vernacular technology, untamed "nature" and societal reproduction (see Desfor & Keil, 2004). The dynamics of infrastructure, where it is actually being deployed or changed, thus reshuffle a host of given interactions and arrangements in a new politics of infrastructure beyond networked urbanism.

Understanding Suburban Infrastructure Disposition

Urban infrastructure has become a key site or arena for analysis of forms, meanings, and outcomes of sociotechnical change, for "heterogeneous engineering" work, and for an emerging material politics (see, for example, Graham & Marvin, 2001; McFarlane & Rutherford, 2008; Monstadt, 2009). Sustained and constantly reworked through a lively discussion between science and technology studies and urban studies, it has been productively analysed as a loose, interdependent arrangement of different systems and networks of provision, service configurations, rules and norms, finance, material and resource flows, and individual and social practices and experiences. This analysis translates what appear to be, on the surface, quite mundane and passive urban objects into captivating mediators of all kinds of relations, flows, symbolisms, powers, and politics. It thus demands engagement with its material forms, technical workings, and reconfigured territorialities through which it becomes a lively urban ingredient in a dynamic, contingent, and contested arrangement of things, processes, and relations.

This emergent and processual notion of infrastructure can be relevant across different urbanizing contexts and help to unpack the specificities of infrastructural change outside of dense, central cities in areas where economies of scale, modes of residence and living, and the technical processes of deploying, maintaining, reconfiguring, and using infrastructure are quite distinct. Some recent work on urban infrastructure has dispensed with the traditional narrative of growth and extension of centralized technical networks as an epistemological, practical, and aspirational norm (see, for example, Coutard & Rutherford, 2015), and instead focused on the significant and consequential "outsides" of networked urbanism (which are only outsides within a networked urbanism frame of reference) and on how infrastructure is actually lived and experienced in diverse ways (see, for example, Graham & McFarlane, 2015). It is a fair assumption that suburban areas constitute, and are constituted by, a hybrid and persistently evolving set of configurations and practices of dealing with, planning for, and experiencing infrastructure that do not necessarily relate or respond to a singular, settled networked urbanism ideal or model. Studying, knowing, and accounting for suburban infrastructure thus requires dispensing with any assumption that the suburbs are marginal areas (within larger infrastructure territories) whose residents lie passively "in wait" for the usual urban connections, flows, and accessibilities. There are likely to be distinctive systems, relations, and mixes of components that fit or work together very differently. Indeed, the coexistence of differentiated and functioning juxtaposed configurations of infrastructure in cities in the South is a crucial lesson of wider importance (see, for example, Furlong, 2015; Jaglin, 2012; Kooy & Bakker, 2008). By looking at these kinds of embedded hybridities and resistant interludes, we might expand and advance our understanding of wider infrastructure transitions and of new, more heterogeneous landscapes and territories of sociotechnical change through which the (sub)urban political is being made and remade.

Infrastructure involves very technical systems that mediate and translate some crucial political debates around urban futures: questions of finance and investment, scales of governance, issues of organization and ownership, resource flows, and sociospatial solidarities, among others. These crucial questions concern not so much the categorization of spaces or degrees of urbanity (for example, infrastructural differentiation of urban and suburban), but a developing critical understanding of processes of urbanization in terms of where, how, and for whom materiality becomes political. In this respect, the politics of sociotechnical change are no less material and salient in the outer suburbs than in the central city.

I argue for a concern for the activity and materiality of infrastructure and the tracking of "a spatio-temporal *process* in which the more tangible, physical stuff of the city is a lively participant" (Latham, McCormack, McNamara, & McNeill, 2009). This kind of view focuses on the different ways in which objects, points of contention, and policy orientations are made visible, tangible, and/or durable, that is, material, by or for specific groups through, for example, practices of "ordering, circulation and manipulation" (Latham et al., 2009). The production of infrastructure configurations is always a material-political process shaped by the interests, values, and resources of particular actors and how these are translated into contested practices that put "things" to work for or to particular purposes and ends. Gandy (2014: 24) argues that "the story of urban environmental change, and the politics of infrastructure networks, are marked to a significant degree by the largely unseen and unrecorded dimensions to human experience." In this view, the question becomes how infrastructure comes to matter in a given situation, moment, or locality, and for whom, and thus how particular infrastructural situations are made or rendered visible and thus political (see Barry, 2013).

Easterling's (2014) recent excavation of infrastructure space is a useful intervention into this process, tracing a series of "active forms" that always give infrastructure a disposition or a potential, that is, the underlying markers in which we can detect the organization and outcomes of sociotechnical activity. She shows the pertinence of an immanent approach that tracks always emerging configurations in situ rather than against pregiven ideas of networks and their components. Infrastructure is always doing something (Latour, 2005), reflecting how it is put to work through "ordering, circulation and manipulation" of "active forms" such as multipliers, switches, topologies, and stories. But it is, at the same time, in constant flux, where/when things and technical bits overflow or exceed their immediate constitution and become live or "vibrant" (Bennett, 2010) or gain momentum. The respective influence of human practices putting things to work for their interests and the inherent activity of things, and the articulation between the two, in the make-up of sociotechnical configurations and change is subject to some debate. Here, the point is that infrastructure becomes a site or arena of material politics as particular social interests engage with and appropriate in contested ways a variety of systems, objects, and components that are not necessarily passive in their make-up, functioning, and activity (see, for example, Braun & Whatmore, 2010). Sociotechnical disposition "is the character of an organisation that results from the circulation of these active forms within it" (Easterling, 2014: 73), where circulation and activity are material-political processes combining intentional agency and lively participation of non-human components.

In order to flesh out this view of suburban sociotechnical disposition and infrastructure change, I use recent work on water and wastewater systems in the Stockholm archipelago and how these systems are bound up in shifting planning practices, residential patterns, and sociopolitics of this low-density living space. Infrastructure is tremendously active here in shaping living spaces, and its material components and their constantly shifting relations to each other within a systemic arrangement are constitutive of residential and sociopolitical forms and possibilities.

Planning for Water and Wastewater Systems in the Stockholm Archipelago

For a number of years, the Swedish capital region of Stockholm has seen increasing tensions between demographic and residential change and infrastructure dynamics, especially in relation to water and wastewater systems. The population of the region (or county) has been steadily rising every year; modes of residence are transforming; and, in some areas, previously second homes are becoming permanent residences. At the same time, water demands are increasing, policies are promoting connection to centralized networks, and the costs and financing of infrastructure extension are evolving, all while national and European regulation becomes stricter on environmental and health grounds. In a large part of the county, and especially in the archipelago area, water and wastewater are now at the centre of local planning preoccupations. These matters are, at once, economic, environmental, and sociospatial concerns. First, the question of financing local services takes on new importance in a context of residential change (extension of permanently inhabited areas within municipalities), local budgetary constraints, and an absence of available information about and calculations of costs of supplying services. Second, water systems have an impact on resources and, more generally, on an increasingly fragile suburban environment. Third, local planning and infrastructure decisions and projects (for example, whether to extend local networks) influence residential and social equilibriums in municipalities, notably through the redistribution among households of (new) costs associated with water infrastructure and systems.

Stockholm County already has very contrasting forms of urban fabric and urbanization processes. Its increasing population of more than two million is heavily concentrated in a central "belt" in and around the main city (one of twenty-six municipalities). Almost half the surface area of the county is forest, while agriculture and water still occupy more land than built-up urban areas (only 14 per cent of land use) (Statistiska centralbyrån [SCB], 2008). This situation translates into strong variations in density between the 4,000 inhabitants per square kilometre of central Stockholm and archipelago municipalities, where there are sometimes fewer than 100 inhabitants per square kilometre. Regional planners have strongly promoted densification ("building the city inwards") and tried to protect countryside and coastal areas from unwarranted sprawl, but local municipalities are responsible for planning policy and have interpreted these guidelines in different ways (Pemer, 2006; Regionplane-och trafikkontoret [RTK], 2007). However, suburban transformation in Stockholm takes on a particular form constituted by heterogeneous and non-contiguous residential dispersion (rather than any standard linear peri-urban diffusion or sprawl). Sparsely populated areas of mostly summerhouses are transformed from secondary summer residences into permanent all-year residences requiring constant access to water and wastewater services. The drivers of this shift are multiple: urban regional demographic growth, retirees moving out of the city, and the search for cheaper alternatives to increasingly expensive and scarce homes in central Stockholm.

One of the main issues for local and regional planners – "the primary concern" according to one regional planner (Regional planner, personal communication, October 2010) – is the presence of around 100,000 second homes or summerhouses owned mainly by Stockholm residents and located primarily in the coastal archipelago area. Swedish anthropologists have captured quite neatly the importance of these second homes in Swedish culture:

> To many Swedes the summerhouse is of much higher symbolic value than the permanent home. The summerhouse is the happy place where you spend the high value time of vacation, while the home is a necessary requirement, tied to work, the daily rat race and sheer survival. Many urban Swedes dream of moving to the summerhouse, simply because this means moving to a happier place. The opportunity comes when commuting time is shortened, when you work part-time or when you don't have to go to work each day. (Arnstberg and Bergström, 2002: 6–7)

The issue is the increasing practice of households moving permanently into these summerhouses. Over the last twenty years, the regional planning office (RTK, now SLL) estimates that up to 1,000 residences have been transformed every year (Regional planning official, personal communication, October 2010). This change can lead to problems because the homes were not built for permanent all-year living (they sometimes only have access to summer water supply),[3] while municipalities must deal with a population increase and new requirements in the management of local services:

> Municipalities say that when more than 30 per cent of the people live there permanently, then the area begins to change, because they have greater requirements for services for which the area was not originally intended.

> And the people only there part-time want to keep it in its half-organized state. And for the municipalities to be able, for example, to draw out municipal water, generally they have to change the whole area and put in twice as many houses, to sell land areas to be able to pay for the whole changing growth, building out the sewerage systems. So it's a whole change of area that needs to be planned very carefully if everybody's going to get organized. (Regional planning official, personal communication, October 2010)

Provision and management of water and wastewater systems are mandated in Sweden to local municipalities, which either take care of these systems themselves or group together in inter-municipal organizations (Sveriges Riksdag, 2006). Local decisions are therefore made about which zones to connect to networks and how to regulate or control the use of water installations. Costs are financed by tariffs such that what is paid by households is reinvested into local systems (Svenskt Vatten, 2000). On a regional level, this arrangement inevitably means that there is great variation[4] between tariffs paid by a household in an apartment in central Stockholm and a household living in an isolated house on an island in the archipelago:

> There is a direct correlation between density or pipe length and the costs. As long as you have a wider spread network, you not only have to invest in longer pipes, but also in pump stations and so on. (Inter-municipal organization official, personal communication, October 2010)

Within municipalities, tariffs are standardized with no differentiation operated between households, depending on where they live (in relation to the pipes, for example). This standardization simplifies billing processes and allows a reasonable level of solidarity between households in the same local area, but it also produces a substantial degree of redistribution across sometimes very large and diverse municipal territories, as illustrated later on.

Crucially, not all households are connected to local centralized water infrastructure. It is estimated that around 90,000 households in Stockholm County use alternative systems to municipal networks. The most recent regional plans have indeed promoted "small-scale solutions" in sparsely populated areas of the region, which "may be significantly better than today's systems" (Office of Regional Planning [SLL], 2010: 151). In these areas, what is infrastructure beyond the networked city?

How do local officials plan (for) infrastructure? How do residents (and other actors) contribute to configurations/dispositions? The central question at hand is to understand how infrastructure matters here as it becomes a nexus for disputes around modes of residence, environment, finance, and technology, and thus articulates a wider politics of "suburban" sociotechnical arrangements. In order to investigate how changing modes of residence and population dynamics are accounted for in local planning decisions and infrastructure provision, I examine the forms, stakes, and wide-ranging implications of water and wastewater systems in one municipality of the Stockholm archipelago.

Infrastructures of Permanence?

Norrtälje is the largest municipality in Stockholm County, taking up fully a third of its surface area in the north. Although it is around seventy kilometres from central Stockholm, roads and public transport connections make it an increasingly attractive residential area. Its population of around 60,000 is widely dispersed in a patchwork of small urban centres, hamlets, and quite isolated settlements inland, along the coast, and on numerous small islands. During weekends and over the summer, its population expands two or threefold as Stockholmers and others escape to their summerhouses (of which there are 25,000 in Norrtälje).

Recent local planning policy has taken on a liberal slant under the Moderate (conservative) majority,[5] as local politicians have pushed to attract incoming residents for increased tax revenue (Local planning official, personal communication, October 2010) and become permissive of new constructions and extensions, including along the formerly protected coastal area. They have also supported permanent residence in summerhouses (Local planning official, personal communication, October 2010; figure 8.1), with more than 200 having been transformed per year on average over the last twenty years (Norrtälje kommun, 2013: 22). For the mayor of Norrtälje, "it's a question of local democracy, which does not concern the departments of the state" (quoted in *Dagens Nyheter*, 26 February 2008). As a core part of this urbanizing policy, the municipality has explicitly recognized the need to expand and adjust water infrastructure and service provision for a growing and increasingly permanent population (Norrtälje kommun, 2008; 2013; 2015b).

This policy is an attempt at reconfiguring a multiplicity of systems for a changing residential and suburban fabric. As we can see from table 8.1, 44 per cent of the population is not connected to local centralized water

Table 8.1 Percentage of population connected to centralized water and wastewater systems in Stockholm County municipalities

Municipality	Population Connected to Water Network (%)	Population Connected to Wastewater Network (%)
Norrtälje	56.07	56.07
Värmdö	58.64	58.64
Nykvarn	72.60	72.01
Österåker	76.47	76.71
Nynäshamn	79.15	79.15
Vallentuna	79.61	79.83
Vaxholm	81.53	81.53
Södertälje	84.00	84.00
Upplands-Bro	86.61	86.51
Haninge	87.11	85.71

Source: Data from Svenskt Vatten, 2007

and wastewater grids,[6] with particularly important water access and wastewater removal issues on the many islands, some of which are not accessible by bridge (Länsstyrelsen i Stockholms län, 2008). Consequently, there are, in effect, at least four existing and juxtaposed configurations, all with distinctive divisions of ownership, organization, responsibility, and financing:

- Direct connection of a house or apartment to the centralized network: households pay water tariffs to the municipality, which owns and manages the network
- Community connection to the centralized network: a collective cluster of houses is interconnected by a mini-network installation with a single connection point to the municipal grid; households pay tariffs as a community
- Autonomous community systems: a collective cluster of houses installs their own mini-network with a small treatment plant and wastewater tank, authorized and regulated by the municipality
- Individual home systems: wells and septic tanks for one property are financed and maintained by households, and authorized and regulated by the municipality

8.1 Summerhouse residence in Norrtälje. Photo: Courtesy of the author

The response of Norrtälje municipality was to develop an ambitious, long-term water and wastewater strategy and investment program in 2007 (Norrtälje kommun, 2008). The foundation of this strategy and program was a categorization of local areas into three zone types: (1) zones of development where homes are being constructed or expanded; (2) zones in transformation with primarily summerhouses that are being transformed into permanent homes; and (3) zones for treatment where there are environmental problems or risks including poor water quality, infiltration, eutrophication, leaching, and seepage of wastewater into either potable water systems or the Baltic Sea.[7]

These zones are then dealt with through a three-pronged sustainability strategy for water and wastewater systems. This strategy involved, first, major long-term investment (1.6 billion kronor between 2008 and 2030, the biggest-ever single municipal investment project)[8] in new and extended infrastructure in order to support urban growth and extend/upgrade pipes and installations to developing residential areas.[9] Second, a series of measures were introduced for mini-network

8.2 Off-grid home in Norrtälje with well and septic tank. Photo: Courtesy of the author

collectives, where a small number of clustered homes group together and maintain a community water system with (or sometimes without) a connection point to the municipal network. Third, there has been reinforced regulation of individual systems (off-grid homes with wells and septic tanks: see figure 8.2), which have been the source of many of the environmental problems but are expected to increase in number by 2030. This strategic intervention around, and planned/lived experience of, infrastructure can be read as working through particular configurations of technology, materiality, and responsibility.

Heterogeneity of Technology

What is interesting here, first, is the acknowledgment that connection to local municipal networks is neither feasible nor desirable in many sparsely populated areas. Rather than targeting a reduction in the number of households with alternative water systems, the local authority actually envisages this number to increase by 2030 from around 23,000 to 36,000

(Norrtälje kommun, 2008). There is, then, an explicit hierarchy of local areas and households in terms of available and possible water systems in which local planners, residents, and a variety of technologies, resources, waters, finance arrangements, and regulations become embroiled to create and maintain a working solution for sustainable water provision and wastewater removal. In between a direct connection to a centralized network at one end of the scale, and off-grid individual solutions such as wells and septic tanks at the other, are several options that are utilized for small groups of dwellings, which are often located some distance from urban centres. Either the dwellings cluster to connect to a centralized network, or they implement collective autonomous solutions, such as a small treatment plant, for example, which are used only by the local residents. This stratification of technical solutions according to several local factors (density, distance from network, geographical conditions, technical possibilities, costs, and so on) therefore allows the adaptation of services to vary to some extent, depending on specific contexts and living conditions. This "fluid" approach to infrastructure (de Laet and Mol, 2000), operating on the principle that a number of sparsely populated areas will never be supplied by centralized network systems, opposes the largely dominant view that network expansion tends to follow, accompany, or even anticipate the urbanization of new suburban areas.

Matters of Engagement

The municipality's technicians and planners came up with a total of 122 zones that required analysis and some form of intervention based on a variety of criteria: residential pressure envisaged by 2030, distance to existing network installations, physical conditions for eventual alternative system use, local environmental concerns, and economic conditions for extending the municipal network. After analysis using a "decision tree method" (Norrtälje kommun, 2008: 10), sixty-four of these zones were classed as suitable for connection to the municipal network system, and the other fifty-eight zones were classed as requiring alternative solutions (either off-grid homes or autonomous community systems). In the former case, the connection solution may take the form of a collective group of around twenty homes creating a mini-network system, which can then be connected to the central network. This form necessitates a land survey first, but is encouraged by the municipality as it creates a sustainable solution for homes with wells and/or wastewater systems that cannot be feasibly maintained in the long run due to pollution problems, and it allows technicians to connect these homes through a single point of connection with the household collective taking on the responsibility for installing and maintaining the mini-network (Norrtälje kommun, n.d.)

While planners are engaging in the usual deployment of maps, zones, pipes, and money to extend big infrastructure and future-proof for growth, local residents are themselves constantly defining and redefining how their modes of residence can adapt and be adapted to materialities of a particular system configuration. Adaptation may involve the creation of a sociotechnical collective with shared community space, routines, and division of labour to organize the deployment of a small-scale system installation; the associated forms and paperwork for regulatory conformity; or the negotiation and collection of costs and tariffs. Off-grid households are responsible themselves for doing the work for and paying the costs of ensuring quality of water drawn from wells, organizing lorries to come for regular emptying of septic tanks, maintaining non-permeability of their installations, and so on. Thus, sociotechnical arrangements in the Stockholm archipelago suggest different meanings of and negotiations with infrastructure. Infrastructure is "inhabited" in a variety of contingent ways, and through entanglement and engagement with multiple "fluid" objects and processes, which cannot be taken for granted but have to be produced and reproduced on a regular basis (see also de Laet and Mol, 2000).

Responsibility and Learning

For the local authority, the differentiated sociotechnical infrastructure arrangements privilege "shared responsibility" (Norrtälje kommun, n.d.), with their own planning and investment being supplemented, or even replaced, by some degree of resident self-organization, such that residents are required to take on economic and environmental costs and risks. As an example, in order to encourage the creation of self-organized community systems as well as conformance to environmental norms and rules, the municipality has offered extra building rights to households as a "carrot." It is clear, then, that the production of suburban infrastructural lives here is always an emerging, incremental, collective arrangement. This production may be bounded by certain rules and regulations and overseen by a local authority, but it depends on a series of interventions by households, technicians, and specialists, which contribute to reproducing (and improving) a system, albeit one that is differently lived or experienced.

Systemic reproduction is thus multifaceted, involving repetitions at once of practices of use, components/zones for intervention, norms and rules, and so on. It implies not only repetition in the sense of a series of regular patterns and components, but also as rehearsal or trial performance, through which authorities and other actors are constantly seeking "configurations that work." Whether it is improvements in relation to levels of pollution, redistribution of costs, or performance of pipe or tank materials, this systemic reproduction can be viewed as a situated place-based sociotechnical learning process on and across scales of households, communities, and the municipality as a whole. Sustainability is thus a process involving a particular, contingent, situated, and contested layering of repetitions through which actors learn and finesse their "performances" in relation to one another and a host of active materials. But the particular goal, disposition, and outcome of this system of repetitions and responsibilities – status quo, spirals of growth for some but not others, new social and ecological organizations of/through infrastructure – can be subject to dispute.

Disputes in Suburban Infrastructure Space: The Politics of Sociotechnical Disposition

Suburban sociotechnical change in Stockholm is constituted by (and brought about through) persistent friction between residential dynamics, local growth strategies, and the diversity of actually existing water systems. While the hierarchy of sociotechnical arrangements adapts diverse local contexts and modes of residence to particular system configurations to some extent, there are tensions and points of dispute over various aspects of these differentiated arrangements, underscoring the contested nature of infrastructure disposition. This contestation can be seen, for example, in the interplay between three "active forms" within intervention wherein infrastructure comes to matter here: the classification and zoning of residential areas used to manage infrastructure; the repetitive components or practice routines and regulatory paperwork required to carry out infrastructure management; and the narrative constructs required for future planning and visioning.

First, classifications, zones, and hierarchies of residential areas, technical systems, and modes of access to services are used as a tool of infrastructure management. The way in which zones are defined and bounded, tariffs decided, and intervention accomplished is not neutral and has political basis and consequences. This process of infrastructural emergence reflects spatial and temporal priorities (where and when) such that certain zones can expect development, transformation, and treatment earlier or later, or at differing rhythms, in a twenty-year program of intervention. Priority is given to the big pipe for the port (and therefore the local economy) and doubling the capacity of the main treatment plant in support of further overall population growth (and local taxes). A "zone of development" clearly has different economic, ecological, and infrastructural prospects than a "zone for treatment," so being close to new big water pipes offers potential connection and inclusion in the municipal project. This prioritization adds another layer of hierarchy onto that which distinguishes in practice four technical forms of water system.

This series of classifications allows infrastructure to actively constitute and immanently bring into being (in different ways) living space. Infrastructure is not just deployed passively and then taken up and used, and it is not just another sector of local policy. Infrastructure interventions are direct reconfigurations of residential possibilities. When a local authority decides to make its biggest public investment ever in extending and improving local water systems, it knows that this decision is about much more than service provision and access. Hierarchy and classification of infrastructure zones actually orient urban policy more broadly, so quality of access to water and wastewater services partially determines decisions about construction and urban extension. It enacts ongoing and future development of suburban areas as a whole. It brings the suburbs into being.[10]

The second active form of suburban infrastructure intervention is the circulation of repetitive components such as paperwork, norms, tariffs, routines, and system materials. These elements are necessary to produce particular infrastructure configurations, but always work in specific ways. Zones and classifications of particular infrastructure configuration become "multipliers" that allow the deployment of repetitive components and forms. These repetitive components make heterogeneous territory manageable and governable for the authority; they facilitate knowing and intervening in an otherwise vast and diverse set of contexts. Recurring elements include norms and routines governing individual wastewater systems, year-on-year water tariff rises permitting the program of work and intervention, and the circulation of the workers and companies that intervene in laying pipes, maintaining systems, emptying tanks, and so on. Autonomous or quasi-autonomous households are not excused from regular regulatory incursions into domestic or community space. They are obliged to work for their autonomy through routines of ensuring that forms are filled in correctly, environmental norms and standards are met, and checks and controls undertaken and paid for. Individual arrangements must accord with (reproduce), or at least not unduly hamper, the collective good. As Easterling (2014: 74, 81) argues, these kinds of "multipliers of activities" that "spread spatial changes throughout a field" are the "levers of disposition in infrastructure space" taking sociotechnical change in a particular direction.

Most notably, repeated increases in water tariffs used to fund major investment programs are associated with shifting distributions of costs and benefits between households depending on where they live (Former mayor, personal communication, October 2010). Households living in the "dense" central towns of Norrtälje already connected to networks are paying rising prices to extend networks to those Stockholmers arriving to live in newly permanent and/or more distant homes. In the context of a municipality with the lowest average income per household in the county,[11] this practice is steadily amounting to a contested subsidization of new, usually well-off residents and their lifestyles by existing residents who have already paid for the water systems they use.

A third active form of infrastructure intervention is the construction and interlinking of narratives that serve as productive techniques for envisioning desirable futures, creating collective trajectories, and enrolling people into public concerns. The water program is tied into a story of municipal growth and expansion, whereby more permanent residents bring more tax revenue, thus feeding into an attractive image of the municipality as a dynamic place to live. This approach fits coherently into a wider regional strategy oriented around "making our environments available for even more residents and businesses" (Stockholms läns landsting, 2013: 2). Concerned publics are constructed through provision of information, attribution of responsibility, and system regulation. Consequently, infrastructure narratives matter, such as demonstrated here. They might lead to or become a tool that local politicians and practitioners can use as a local growth strategy, prioritizing some areas over others, regulating modes of residence, and making residents responsible for the infrastructure systems they inhabit. Yet, as Easterling (2014) showed, sociotechnical disposition can be seized on, diverted, or subverted by other actors, so that here it enables residents to create diverse adapted infrastructure solutions, sustain their lifestyles and routines, and engage with or contest official suburban story-making for the years to come. Equally, while nature is being transformed by liberal urban growth and the deployment of bigger infrastructure for more supply, it also resists change and speaks through infiltration, seepage, and a number of physical transformation processes, forcing further regulation of systems and a sanctioning of offenders. Infrastructure acts thus in multiple, contingent ways: "For if the pump must act, what is it to do: provide water or provide health? Build communities or make a nation?" (de Laet and Mol, 2000: 247).

Norrtälje's story is a story of the suburbs producing an evolving "infrastructure interlude," actively adapted and adjusted to their needs, situations, and specificities, and not being subjected to or passively enrolled in the expansion of a generic network model imported from elsewhere. There is the mobilization of a circular metabolism centred on suburban spaces and uses rather than subsumed into a linear system organized elsewhere. But this interlude is persistently adjusted, and actually comes to matter, through a series of unresolved tensions between residents, development practitioners, and techno-natures as to the sociotechnical disposition of suburban infrastructure. These heterogeneous groups order, circulate, and manipulate configurations in different ways, reflecting competing views of, and articulations between, lifestyle modes (suburbanisms), local planning and politics, and materialities, flows, and resistances of water systems (see White and Wilbert, 2009). This contested disposition and mobilization of infrastructure reworks a view of the politics of sociotechnical systems as unfolding immanently through processes and practices of making and remaking configurations.

Infrastructure always has the potential to overflow policy aims and actions. It is thus a key site of material struggle over the direction, means, and outcome of sub/urban change. The articulation here between collective and individual, public and private, redistribution of costs and benefits, and the degree of agency attributed to the environment all emerge as charged issues in unsettling infrastructure arrangements.

Infrastructure here is always a relational process, and never just an isolated object. It evolves continually through the shifting entanglements and interactions of its components and arrangements, and it is deployed and used processually (to develop, to transform, to treat). There are no pre-existing areas with static forms of water system and installation. In all its configurations, infrastructure is immanent and at once is brought into being by, and itself brings into being, a living space, which is configured or reconfigured through systemic interaction and intervention.

Through this reading, the politics of (suburban) infrastructure is shown to be less about presence/absence and inequalities in absolute access to a stable configuration than about the contested forms and outcomes of specific dispositions of infrastructure (and notably the sometimes quite local uneven distribution of costs and benefits of those dispositions). The embedding of choices, routines, and forms of regulation do produce difference and have sociospatial consequences. There are shifting frontiers between the collective and the individual, and between public and private, that are fundamentally reworking suburban living and planning. There are different meanings of and engagements with infrastructure, which is thus inhabited and not taken for granted. Infrastructure and sociotechnical systems are a primary means through which wider decisions about and planning of the future development of suburbs are taking place.

So while this narrative appears to be quite a prosaic story on the surface (although the particular does matter locally), it quickly and continuously overflows into issues and questions of environmental pollution, local and regional futures in an age of climate change, collective addressing of suburban futures through cross-subsidization between infrastructure users and resource transfers, and short-term versus long-term planning. In short, there is a meaningful and substantive politics of sociotechnical change in/to the suburban living space.

Conclusion

Infrastructure is at the heart of changing forms, modalities, and outcomes of local suburban development and modes of residence. However, this evolution does not necessarily signify the usual singular network extensions of the city. In the Stockholm archipelago here, but also in a variety of other contexts, exploring modes and meanings of infrastructure means going beyond networked urbanism and its narrative, paradigm, and model of connection to centralized, industrial technical systems. Building on an emergent and processual understanding of infrastructure, the idea developed in this chapter has been that infrastructure interludes are useful junctures to study transitional spaces and/or times of sociotechnical change that point to new or other possibilities not encountered in usual or mainstream infrastructure compositions in the central city. An interlude becomes a hybridized, flexible (and reflexive) sociotechnical disposition that is constantly being reconfigured for shifting and multilayered infrastructure spaces. It suggests freedom and possibility and has adaptive potential, but at the same time reflects and enacts an encroaching urban moment and its constitutive tensions. Infrastructure interludes are thus constantly shifting, fluid frontier spaces mediating between contested notions, practices, and implications of collective and individual, public and private, industrial and vernacular technology, "vestiges of nature" and societal reproduction (see Gandy, 2014: 18).

Mobilizing Easterling's (2014) notion of sociotechnical disposition, which captures the unfolding "agency," "capacity," or "character" of infrastructure arrangements, I sought out contested "orderings, circulations and manipulations" of a variety of systems, objects, and components that are not always passive in their make-up, functioning, and activity. In Stockholm, infrastructure intervention worked through heterogeneity of technology, contingent engagements with materiality, and divisions of responsibility and learning. The sociotechnical disposition of this intervention could be detected through (the interplay of active forms such as) classifications, zones, and hierarchies of residential areas, technical systems, and modes of access to services; circulation of repetitive components such as paperwork, norms, tariffs, routines, and system materials; and narratives of growth and expansion, and of enrolling concerned groups through provision of information, attribution of responsibility, and system regulation.

The overall capacity, meaning, and concern of infrastructure only emerge immanently in the shifting relative positions of its components. But relations between components are inherently disputed, overflowing any singular and immediate attention, and here the social, economic, and environmental constitution and consequences of the various classifications, repetitive elements, and narratives were differently interpreted, reflecting competing views of, and articulations between, modes of residence (suburbanisms), ways and whys of local planning and politics, and materialities, flows, and resistances of water systems.

Just as Bergman's ballerina was able to use memories of her "summer interlude" to embrace a hopeful future, so excavating suburban infrastructure interludes maps out new possibilities and pathways of sociotechnical change. It becomes a meaningful and potentially transformative site or arena of material politics, which is excessive of technical systems per se and works through specific forms of disposition made visible in the relations between infrastructure components. Infrastructure interludes offer reflexivity and a degree of reversibility in contexts of a forthcoming/encroaching urban where and when transition pathways and dynamics remain to be forged and negotiated. Sustainability of interludes appears to rely on the capacity of actors and components to continue to promulgate or cultivate the diversity and freedom of their living spaces and adaptive sociotechnical systems. In Bergman's films, interludes come to an end, brought back to an inevitable urban reality. But suburban configurations and spaces are perhaps not always so transient.

Chapter 8 has shared chapter 7's concern with the unequitable distribution of the financial burden of infrastructures. In the case described by Chapter 8, however, this inequity is not so much the outcome, described in chapter 7, of a financial strategy driven by the neoliberal ideology; rather, it is the result of multiple local decisions attempting to address water issues caused by a growing population and a tendency for year-long rather than seasonal living. Neoliberalism does play a role in the chapter 8 case study, but its impact has been more limited. Its influence has been felt in a relaxation of development regulations, which has resulted in houses being built in sectors unserved by water systems. Still, social equity consequences are the same – in both cases lower income households shoulder a disproportional financial load. Beyond the financial impact of infrastructures, the chapter brings to light the multiple ramifications they can have on human settlements. The chapter shows the role new water infrastructures can play in the transition of a sector from a summertime recreational area to a year-long suburban sector. The presence of these new infrastructures changes the way the area is perceived and how people behave therein. The chapter has documented the clash of values that comes with such a transition. Perhaps the main observation the chapter brings to the volume is that infrastructure decisions have reverberations that far exceed the scope of infrastructures themselves.

ACKNOWLEDGMENTS

I acknowledge the research assistance of Frédérique Boucher-Hedenström during the fieldwork in Stockholm.

NOTES

1 Koskinen (1995) also suggests other dualisms: youth versus ageing, innocence versus experience, play versus seriousness, paradise lost versus gloomy urban present.

2 *Merriam-Webster*'s definition of "interlude" (http://www.merriam-webster.com/dictionary/interlude) is as follows:

- a usually short simple play or dramatic entertainment
- an intervening or interruptive period, space, or event: interval
- a musical composition inserted between the parts of a longer composition, a drama, or a religious service

3 Summer water supply is available six months of the year between May and October (Local planning official, personal communication, October 2010).
4 It is up to three times more expensive in outer municipalities than in central ones according to figures from Svenskt Vatten.
5 Norrtälje had a Moderate-led majority in the city council from 1998 to 2014.
6 There are almost 40,000 properties without connection to the municipal wastewater system (Norrtälje kommun, 2015b: 7).
7 Pollution of the Baltic Sea from alternative systems of wastewater management has been a major problem along the Swedish coast for many years, leading to high-level political discussions among national governments of the Baltic region (Stockholms läns landsting, 2013).
8 The project was financed primarily by increasing the tariffs paid by households (in accordance with Swedish law), with regular 7 per cent annual increases in recent years (Norrtälje kommun, 2015a, and other annual reports), leading to it being the municipality with the highest tariffs in the region (Fastighetsägarna, 2015; Holgersson, 2015). The local press note the "sharply increased" tariffs (Sverke, 2014), representing a more than 50 per cent rise in the price of a cubic metre of water in the last ten years. The original program plan also mentioned the possibility of resorting to "tax-funded liquidity support" in case of overall losses over three consecutive years (Norrtälje kommun, 2008: 25–6).
9 The infrastructure extension included deploying a big water pipe to the port of Kapellskär and doubling the capacity of the main treatment plant.
10 In a similar way, de Laet and Mol (2000: 237) argue that a water pump technology in Zimbabwe has a number of boundaries that help to constitute, for example, a community or a nation.
11 Norrtälje is not a low-income municipality per se, but has average income levels a little below the other municipalities in Stockholms län (Stockholm County).

REFERENCES

Arnstberg, K.-O., & Bergström, I. (2002). *URBS PANDENS project case study report on Stockholm: An introduction*. Stockholm: Stockholm University.

Barry, A. (2013). *Material politics: Disputes along the pipeline*. Chichester, UK: John Wiley & Sons.

Bennett, J. (2010). *Vibrant matter: A political ecology of things*. Durham, NC: Duke University Press.

Braun, B., & Whatmore, S. (Eds.). (2010). *Political matter: Technoscience, democracy, and public life*. Minneapolis, MN: University of Minnesota Press.

Coutard, O., & Rutherford, J. (Eds.). (2015). *Beyond the networked city: Infrastructure reconfigurations and urban change in the north and south*. Abingdon, UK: Routledge.

Dagens Nyheter. (2008, 26 February). Skyddet av littoralen mer och mer vattnas ner. *Dagens Nyheter*.

de Laet, M., & Mol, A. (2000). The Zimbabwe bush pump: Mechanics of a fluid technology. *Social Studies of Science, 30*(2), 225–63. https://doi.org/10.1177/030631200030002002

Desfor, G., & Keil, R. (2004). *Nature and the city: Making environmental policy in Toronto and Los Angeles*. Tucson, AZ: University of Arizona Press.

Easterling, K. (2014). *Extrastatecraft: The power of infrastructure space*. London: Verso.

Fastighetsägarna. (2015, 26 August). Norrtälje har högsta taxan för Vatten och Avlopp i Stockholms län. Stockholm: Author.

Furlong, K. (2015). Rethinking universality and disrepair: Seeking infrastructure coexistence in Quibdo, Colombia. In O. Coutard & J. Rutherford (Eds.), *Beyond the networked city: Infrastructure reconfigurations and urban change in the north and south* (pp. 94–113). Abingdon, UK: Routledge.

Gandy, M. (2014). *The fabric of space: Water, modernity, and the urban imagination*. Cambridge, MA: MIT Press.

Graham, S., & Marvin, S. (2001). *Splintering urbanism: Networked infrastructures, technological mobilities and the urban condition*. London: Routledge.

Graham, S., & McFarlane, C. (Eds.). (2015). *Infrastructural lives: Urban infrastructure in context*. Abingdon, UK: Routledge.

Holgersson, N. (2015). *Fastigheten Nils Holgerssons underbara resa genom Sverige-en avgiftsstudie för 2015*. Stockholm: Author.

Jaglin, S. (2012). Services en réseaux et villes africaines: L'universalité par d'autres voies? *L'Espace géographique, 41*(1), 51–67. https://doi.org/10.3917/eg.411.0051

Jaglin, S., & May, N. (2010) Étalement urbain, faibles densités et "coûts" de développement: Introduction. *Flux, 79/80*(1), 6–15. https://doi.org/10.3917/flux.079.0006

Keil, R. (Ed.). (2013). *Suburban constellations: Governance, land and infrastructure in the 21st century*. Berlin: Jovis Verlag.

Kooy, M., & Bakker, K. (2008). Technologies of government: Constituting subjectivities, spaces, and infrastructures in colonial and contemporary Jakarta. *International Journal of Urban and Regional Research*, *32*(2), 375–91. https://doi.org/10.1111/j.1468-2427.2008.00791.x

Koskinen, M. (1995). The typically Swedish in Ingmar Bergman. In R.W. Oliver (Ed.), *Ingmar Bergman: An Artist's Journey* (pp. 126–35). New York: Arcade Publishing.

Länsstyrelsen i Stockholms län. (2008). *Förbättring av enskilda avlopp inom Norrtälje kommuns skärgårdsområde*. Stockholm: Author.

Latham, A., McCormack, D., McNamara, K., & McNeill, D. (2009). Constructions. In A. Latham, D. McCormack, K. McNamara, & D. McNeill (Eds.), *Key concepts in urban geography* (pp. 53–87). London: Sage.

Latour, B. (2005). *Reassembling the social: An introduction to actor-network theory*. Oxford: Oxford University Press.

McFarlane, C., & Rutherford, J. (2008). Political infrastructures: Governing and experiencing the fabric of the city. *International Journal of Urban and Regional Research*, *32*(2), 363–74. https://doi.org/10.1111/j.1468-2427.2008.00792.x

Monstadt, J. (2009). Conceptualizing the political ecology of urban infrastructures: Insights from technology and urban studies. *Environment and Planning A*, *41*(8), 1924–42. https://doi.org/10.1068/a4145

Norrtälje kommun. (2008). *Program för utveckling av kommunalt vatten och avlopp 2008-2030*. Norrtälje: Author.

Norrtälje kommun. (2013). *Oversiktsplan 2040*. Norrtälje: Author.

Norrtälje kommun. (2015a). *Taxa för kommunalt vatten och avlopp*. Norrtälje: Author.

Norrtälje kommun. (2015b, 28 September). *VA-policy*. Norrtälje: Author.

Norrtälje kommun. (n.d.). *Att ansluta till kommunalt vatten-och avloppsnät*. Norrtälje: Author.

Office of Regional Planning (SLL). (2010). *RUFS 2010 Stockholm regional plan*. Stockholm: Author.

Pemer, M. (2006). *Creating a sustainable city in a polycentric region: The Stockholm example*. Stockholm: City of Stockholm Planning Department.

Regionplane-och trafikkontoret (RTK). (2007). *Vision, mål och strategier för regionens utveckling. Program för ny regional utvecklingsplan (RUFS 2010)*. Stockholm: Author.

Statistiska centralbyrån (SCB). (2008). *Statistical yearbook of Sweden*. Stockholm: Author.

Stockholms läns landsting. (2013). *Regional miljöstrategi för vatten*. Stockholm: Author.

Svenskt Vatten. (2000). *Facts on water supply and sanitation in Sweden*. Stockholm: Author.

Svenskt Vatten. (2007). *Taxestatistik 2007: Sammanställning över kommunala vatten- och avloppstaxor gällande den 1 januari 2007. Rapporten är under utveckling*. Stockholm: Author.

Sveriges Riksdag. (2006). *Lag (2006:412) om allmänna vattentjänster*. Stockholm: Author.

Sverke, J. (2014, 20 November). Kraftigt höjda taxor för vatten och sopor. Norrtelje Tidning. Norrtälje.

White, D., & Wilbert, C. (Eds.). (2009). *Technonatures: Environments, technologies, spaces, and places in the twenty-first century*. Waterloo, ON: Wilfrid Laurier University Press.

Young, D., Burke Wood, P., & Keil, R. (Eds.). (2011). *In-between infrastructure: Urban connectivity in an age of vulnerability*. Kelowna, BC: Praxis (e)Press.

9 Suburban Constellations of Water Supply and Sanitation in Hanoi

SOPHIE SCHRAMM AND LUCÍA WRIGHT-CONTRERAS

Introduction

Vietnam's transition from a socialist state towards a socialist-oriented market economy roughly thirty years ago, known as *doi moi*, has triggered massive urban growth in the country. National urbanization rates were 33 per cent in 2014 (United Nations [UN], 2014) and are now projected to reach 58.8 per cent by 2049 (General Statistics Office of Vietnam [GSO], 2011: 27). In particular, the capital city of Hanoi has been facing massive urban growth since the beginning of the 1990s (Quang & Kammeier, 2002). This growth has drastically transformed the city's urban edges, where residents, real estate development agencies, international organizations, and urban and national administrations shape the process of urbanization in creative as well as contested ways. The diverse suburban landscapes of Hanoi are particularly representative of the increased social stratification in Vietnam since the introduction of a market economy in the 1990s (Labbé & Musil, 2011; Leaf, 2002). Water and sanitation infrastructures reflect and reinforce these broader sociospatial dynamics.

Urban planning, as well as sectoral plans and projects, continue to promote the expansion of large technological networks in order to centralize and standardize service provision for the entire metropolitan region of Hanoi. However, in the wake of Vietnam's insertion into the global economy after *doi moi*, urbanization dynamics and the organization and governance of infrastructure service provision in water supply and sanitation have resisted the project of unification and centralization. Central and local governments have actively supported massive real estate development at the edges of Hanoi, promising residents access to modern networked services, while urban and peri-urban villages have been incrementally growing and densifying at a rate that is too rapid for formal planning.

Public sanitation utilities have undergone few institutional changes, severely restricting their capacity to serve citizens, despite formal water supply organization and governance reflecting current tendencies towards commercialization and liberalization of service provision. Particularly in the suburban areas of Hanoi, large-scale, path-dependent infrastructure networks are relatively weak and lead to sociospatial diversifications and redundancies of water and sanitation provision, contradicting the current government's plans to roll out uniform water supply and sanitation networks integrating the Hanoi metropolitan region. Multiple technologies to access water and sanitation, such as privately owned wells, decentralized septic tanks, and open sewage drains, as well as central and local water and sewerage networks, are shaping Hanoi's suburban spaces. New actors are emerging and providing services exclusively within residential estates, so-called new urban areas, creating spatially restricted "satellite network systems," which often ignore and at times disturb surrounding topologies of water and sanitation provision. In peri-urban villages, a hybrid mix of self-organized and community-based forms of service provision from wells or through septic tanks and self-organized connections to central water networks or local drainage systems has developed (Schramm, 2014). Thus, the suburban areas of Hanoi are experiencing drastic sociospatial diversifications and decoupling of service provision.

Starting from the perspective of urban infrastructure studies that view flows of water and sewage to be closely intertwined with urbanization patterns and reflective of broader dynamics of urban resource distribution and access (Kooy & Bakker, 2008), we consider Hanoi's "splintered" suburban landscapes of water and sanitation provision as a crucial lens to explain the (sub)urbanization of Hanoi and respective dynamics between sociospatial fragmentation and cohesion. Scholars of urban infrastructure studies have articulated the need to analyse such dynamics beyond simplistic conceptualizations of "splintering" networks and "infrastructural bypassing" of poorer urban areas in the wake of neoliberal projects (Coutard, 2008; Graham & Marvin, 2001), particularly with regards to cities of the global south (Kooy & Bakker, 2008). Such analyses have to take into account specific urbanization dynamics and urban morphologies that shape the local ways to access infrastructure services through and beyond centralized networks (Zérah, 2008).

Furthermore, the diversity of practices in the provision of and access to services in different sectors deserves scholarly attention (Coutard, 2008).

Our study of suburban water supply and sanitation in Hanoi contributes to this endeavour by comparing water and sanitation provision in relation to particular urbanization patterns between centralized urban expansion and incremental growth of peri-urban villages. Our analysis of urban planning, central strategies of infrastructure provision, and users' practices provides a more nuanced picture of place-specific dynamics of infrastructure access throughout and beyond centralized networks. As opposed to notions of "infrastructural bypassing" (Graham & Marvin, 2001) of poorer areas, which imply a certain victimization of urban residents who are left to passively accept their exclusion from modern service provision, our study emphasizes the fact that suburban residents of Hanoi have to access services beyond centralized networks in water supply as well as sanitation. However, this fact does not mean that current dynamics of infrastructural access in Hanoi's suburban areas are unproblematic and free of contestation. It is particularly the focus on both water and sanitation that reveals how the ongoing decoupling of the sectors, with sanitation and wastewater management largely lacking funds and the water supply sector becoming an important arena for investments of national as well as global companies and financial corporations, contributes to very place-specific problems for residents in terms of access to water services, sanitation, and drainage. As we elaborate further on, it is thus the (sub)urban flow of water and sewage that unveils the contradictions and challenges that come along with current spatial planning and housing provision as well as infrastructural management in Hanoi.

The broader societal change in Vietnam in the past thirty years has become apparent in suburban infrastructure provision, with tendencies towards sociospatial stratification being closely interrelated with a hybrid mix of ways to access sanitation and water supply infrastructures. We argue that suburban areas are therefore particularly revealing of the contradictions inherent in centralized infrastructure policies insisting on the modern ideal of unitary centralized network provision. At the same time, the relative weakness of centralized infrastructure networks in suburban areas leaves space for creativity and innovative or alternative ways to access basic services for residents (see Monstadt & Schramm, 2013). In the following sections, we analyse the water cycle of Hanoi and how it has been shaped by technologies for potable water sourcing and distribution and wastewater collection and disposal from the early stages of the city's development onward. We then explicate the ways in which central and local governments currently aim to reshape this water cycle through interventions into infrastructural artefacts and networks as well as organizational structures. We also focus on suburban dynamics of water supply and sanitation, and particularly the place-specific perspectives and practices of suburban residents. Thus, we reveal contradictions between formal policies and actual patterns of service provision shaped by broader suburbanization dynamics and sociopolitical constellations.

Hanoi's Water Cycle: Technologies of Water Supply and Sanitation

Since the founding of Hanoi more than two thousand years ago, the question of the fragile "balance between land and water" has been central to Hanoi's urbanization. This challenge is due to the geographic and climate conditions of the city situated within the delta of the Red River at a maximum of twenty metres above sea level (World Health Organization [WHO], 2010). Hanoi is an amphibious city with several distributaries of the Red River flowing through the city and its region from north to south, feeding numerous lakes and ponds. Hanoi has a high rainfall, approximately 1,600 millimetres per annum, which replenishes the aquifer below the city (Nguyen & Helm, 1995). Hanoi's flat terrain and low position, its high groundwater levels, its natural occurrences of arsenic in the groundwater, and its increasing pollution with heavy metals are major challenges for the management of the water cycle (Berg et al., 2008; Winkel et al., 2010). Furthermore, high evaporation rates, the sealing of land by construction, and growing consumption of groundwater contribute to a depletion of the aquifer. As groundwater is the major water source for Hanoi's citizens, supply becomes more and more precarious despite the ubiquity of water in the rapidly growing city. This situation affects the health of those residents who access untreated water.

Still, decentralized means of water supply, such as privately owned wells, remain popular. According to a resident of Hanoi, up to the beginning of the twenty-first century nearly every family living in the urban areas of Hanoi owned a tube or dug well (Resident of the inner-city district Hoan Kiem, personal communication, 2015). Today, Hanoi's residents rarely access water through decentralized means exclusively; they mostly use these means to complement water from the city's large technological networks (GHK, 2005). Centralized piped water distribution networks were initially built by French colonists in the late nineteenth century. Since then, Hanoi has continued to increase its capacity for centralized water production (Ngo, 2014).

Hanoi's first water treatment plant, Yen Phu, was built in 1894. After Vietnam's independence from France in 1954, the city expanded the groundwater treatment plant's capacity, and during the period from 1987 to 1997 allowed for funding from the Finnish government to significantly modernize and increase water production to supply the distribution networks (Ngo, 2014). Initially, these networks largely depended on groundwater pumped from centralized treatment plants. The deteriorating groundwater quality presents a challenge to serve the calculated demand of a million cubic metres per day (Hanoi Water Limited Company [HAWACO], 2014a: 17). Despite moderate amounts of non-revenue water, which account for approximately 20 to 30 per cent (Owen, 2012: 221; HAWACO, 2014a: 17), the situation is tense, as water demand is projected to increase to 2.7 million cubic metres per day in 2030 and 3.3 million per day in 2050 (HAWACO, 2014b: 1).

On its way from Hanoi's households into the Red River and the aquifers below the city, water flows through a range of sewage canals, rivers, and drains. Some of these are part of an ancient system that was constructed more than a thousand years ago to protect the city from flooding despite the flat terrain and high levels of rainfall (Ministry of Construction [MOC], 1993). This system was designed to separate household sanitation from urban drainage, because people in Hanoi have traditionally collected human manure, transported it out of the city, and used it as fertilizer in agriculture (Fayet, 1939). French colonial engineers complemented the drainage system with a combined underground sewerage network during the colonial occupation from 1905 to 1945 (Ngo, 2009). This first and to date largest attempt towards the combined sewerage of human wastewaters and storm water has proven to be a failure, as its incomplete conception based on gravitation did not account for the low hydraulic slope. The underground system built by the French covers only part of the inner city, and has gradually been extended so that it currently covers about 60 to 70 per cent of the city's urban districts (Wastewater expert, personal communication, 2009; Ngo, 2009). Today, only roughly 10 per cent of the approximately 760,000 cubic metres of wastewater generated daily in Hanoi's urban core flows through centralized treatment plants, and the share for the metropolitan region is even lower (Chairman, Hanoi Sewerage and Drainage Company, personal communication, 2008; Perkins Eastman, Posco E&C, and Jina [PPJ], 2010a: 115).

Like in the precolonial era, households of Hanoi City and region mostly rely on decentralized sanitation via septic tanks installed under individual buildings. Septic tanks allow for the separation of human manure from other wastewaters – potentially leading to less contaminated wastewaters in the sewerage system and allowing for the use of human manure as fertilizer in agriculture. However, Hanoi's septic tanks tend to be undersized and emptied too rarely, resulting in the overflow of domestic wastewater directly into the sewerage system. Thus, they hardly contribute to an improvement of the city's rivers, lakes, and ponds, which often have a concentration of dissolved oxygen too low for any microorganism to survive.

Reshaping the Water Cycle: Towards Centralization and Unification

Central and city governments are alerted to the problems concerning water supply as well as sanitation and the deteriorating environmental situation in Hanoi. Planning documents reveal the intent to reshape the water cycle of Hanoi towards the centralization and unification of water supply and wastewater treatment in the city region (Socialist Republic of Vietnam & Hanoi People's Committee [SRV & HPC], 2009; PPJ, 2010a; HAWACO, 2011: 116, 117, 122–7; 2014a: 1–3). With its colonial and then socialist past, Hanoi has a tradition of extremely ambitious urban planning by foreign experts. Its plans are globally fashionable visions of modernity, symbolized by large-scale infrastructure networks and housing development, with a sharp distinction between the city and a rural hinterland. However, they only partially reflect the dynamics of Hanoi.

The current master plan is in line with this tradition. In opposition to its predecessors, the plan does not limit itself to those areas of the city that are urbanized or planned for urbanization. Its planning areas stretch far into the hinterland of the city, and it proposes the conservation of spaces that it represents as non-urbanized, "green," or rural spaces (PPJ, 2010a). The expansion of the planning areas accompanies the inclusion of parts of neighbouring provinces into Hanoi in 2008, which has enlarged the city's administrative area from approximately 92 hectares and 3.2 million inhabitants in 2007 to 335 hectares and over 7 million inhabitants in the urban and rural districts of Hanoi's metropolitan region in 2014 (GSO, 2009; 2014). The master plan envisions water supply as well as wastewater disposal to be organized via large-scale networks that stretch over the urban core as well as the suburban and peri-urban areas of Hanoi and are operated by parastatal water and sanitation utilities (PPJ, 2010a). Thus, the planned water and sanitation infrastructures are to facilitate the integration of diverse urban and suburban spaces into a uniform metropolitan region under the control of national and local governments. Sectoral plans are in accord with this overall vision of infrastructure provision towards the centralization and unification of sub/urban space in Hanoi. Concerning water supply, it is particularly the hazardous metals found in the water and groundwater depletion that serve as justifications for the planned expansion of the centralized piped water system and the reduction of decentralized privately owned wells (NCERWASS, 2014). The USD $1.2 million National Target Program 3 (NTP3) of the Ministry of Agriculture and Rural Development (MARD) aims to extend the pipe distribution networks in suburban areas of Vietnam and is 90 per cent funded by Australia, Denmark, and the United Kingdom (Ministry of Foreign Affairs of Denmark, 2014). To aid water supply for fast growing areas on the outskirts of Hanoi and ease the overexploitation of groundwater, the city government also plans to build two new water treatment plants that extract water from Hong River and Duong River (HAWACO, 2014a: 1). The initial capacity of the plants will increase the treatment of surface water by 200 per cent and will match the current production of groundwater (HAWACO, 2014b: 1–3, 17). Towards 2050, the Hanoi Water Limited Company (HAWACO) intends to increase the capacity of each water treatment plant in order for surface water to fulfil 80 per cent of the projected demand for water in Hanoi City, while groundwater production is to decrease (HAWACO, 2014b: 23).

The construction and operation of the existing surface water treatment plant from Da River has been aided by private and foreign investments in joint stock companies, while the distribution of water is carried out through one-member limited liability companies such as the HAWACO, which is the biggest distribution company in Hanoi. In general, private sector participation models in water management in Vietnam are expected to increase from 2 per cent estimated in 2012 to 12 per cent in 2025 (Owen, 2012: 46). It is likely that the planned expansion of the city's centralized water supply system will augment the influence of international joint stock companies and the HAWACO over (sub)urban water supply. In terms of wastewater treatment and sanitation, an ongoing project of the Japanese International Cooperation Agency in conjunction with the national and city governments aims to install "central large-scaled wastewater treatment plants" in the metropolitan region of Hanoi. Despite the problems experienced with combined underground sewerage since French colonization, this project is the proposed technological solution for planned urbanization in the suburban areas of Hanoi (SRV & HPC, 2005). A recent feasibility study of the project qualifies the septic tank system of the city as inefficient and ignores it in the technological design (see SRV & HPC, 2009).

Current interventions towards centralization of water supply and sanitation are an intensely contested issue. The ongoing project towards the installation of centralized wastewater treatment plants meets protests from sanitation experts and public media. A local newspaper deems the project "inappropriate, ridiculously expensive and useless" (24h, 2011; original in Vietnamese), summarizing the criticism by sanitation engineers, planners, and academics. Project costs accrue to USD $800 per person. According to an international water expert from the Asian Development Bank, costs between USD $200 and $600 per person are high already, even for such "heavy engineering" sanitation projects (Urban water supply and sanitation advisor, personal communication, 2011). Despite these costs, the artefacts constructed within the project are largely inoperable or have insufficient capacity to fulfil their designed purpose (Quoc, 2010). The state of sanitation artefacts reflects the situation of the Hanoi Sewerage and Drainage Company (HSDC), the public utility responsible for sanitation. It is owned by the city government and supposed to be financed by public fees. However, today the only fee for public sanitation in Hanoi is a 10 per cent surcharge on the water tariff (SRV & HPC, 2009). Thus, revenues are far too low for the utility to operate efficiently. As opposed to water supply, urban sanitation, sewerage, and drainage are not considered profitable by Vietnam's government, private corporations, or international organizations. While bilateral and international organizations, such as the German Association for International Cooperation (*Gesellschaft für Internationale Zusammenarbeit*, GIZ), promote a "commercialization" of the sector, making the revenues and costs of the utility transparent and raising tariffs in order to achieve "full cost recovery," there are few tendencies towards privatization in the sector. International and national joint stock companies and corporations largely limit their activities to the construction and maintenance of a centralized water supply network.

In the course of Vietnam's integration into the global economy and the drive towards privatization and liberalization since *doi moi*, both the sectors of water supply and sanitation have become increasingly disparate, with the water sector largely dominated by private corporations and firms. In line with the profitability of the water supply, actors in the water sector are pushing for an expansion of networked water supply into the suburban spaces of Hanoi. There is no public controversy concerning the feasibility of respective interventions, and there are very few investments by international organizations or national governments into decentralized technologies of water supply, with the exception of pilot projects in marginalized areas.

These networks are encouraged largely through community-based management models, and to a lesser extent through cooperatives and private entities (NCERWASS, 2014). Based on the initial commitment to service a minimum of 60 per cent of the inhabitants of a suburban/rural community, MARD will facilitate the extension of the piped water scheme (NCERWASS, 2014). Depending on the capacity of the system, the process includes three months of survey and design and nine to twelve months of construction, or up to twenty months for larger systems. The expansion of the water networks is made possible through a financial mechanism of government subsidies for people living in marginalized areas, which covers up to 60 per cent of investment in capital construction costs. In order to cover the total costs, users contribute financially for the connection and intake from the main pipeline to their household. Thereupon, users in the community are committed to cover 100 per cent of the operation and maintenance costs (MARD representative, personal communication, 2014). As we elaborate further on, users' practices indicate scepticism towards supply from centralized networks, particularly in the suburban areas of Hanoi.

Diversification and Fragmentation: Access to Services in Suburban Hanoi

The contradictions and contestations around the centralization of water supply and sanitation in Hanoi become particularly apparent in the city's suburban spaces, which have experienced massive transformations since the beginning of the 1990s as relaxed influx controls and new economic opportunities in the wake of *doi moi* have motivated people to migrate to the city (Quang & Kammeier, 2002). The government has reacted to this influx with a step-by-step expansion of the city's administrative area through the creation of new urban districts from formerly rural districts in the course of the 1990s and 2000s, and in 2008 with the institutionalization of Hanoi's metropolitan region (Government of Vietnam [GOV], 2008). The creation of new urban districts has led to massive urbanization, notably in the western and southern edges of Hanoi. Formal urban planning based on master plans has facilitated the development of "new urban areas," large-scale modern housing estates on former farming land designed with provision of networked underground water and sanitation services (figure 9.1; Labbé & Boudreau, 2011). Hanoi's government has promoted these developments and has benefited from them (Han & Vu, 2008). In fact, within "growth coalitions" that are developing these investor-led projects, the boundaries between the state and international and national real estate agencies blur, as individual actors regularly represent several of these entities. The city government's annexation of Ha Tay province in 2008, and thus the massive enlargement of its administrative area, exemplifies the master planning vision of creating a "greater" Hanoi through the development of new satellite cities and urban areas (Labbé & Musil, 2011: 2; PPJ, 2010b).

As the name indicates, such "new urban areas" appear in Hanoi's masterplans as modern urban spaces inserted into an otherwise empty, non-defined, and unpopulated space, often marked as green on maps (see PPJ, 2010a). However, they shape the suburban landscape of Hanoi together with the peri-urban villages on the outskirts of the city, which have expanded rapidly since *doi moi* and have partly merged with the urban fabric (figure 9.2). The urbanization process of these villages is mediated between residents of Hanoi and local administrations. They are often semi-legal, as individuals subdivide the plots they own without formal permission. Local administrations regularly tolerate this process because they are mostly residents themselves and thus deeply involved in village life and respective negotiations and activities around land access and housing construction (UN-Habitat program manager, personal communication, 2009). The incremental subdivision of plots in peri-urban villages leads to densification. Where space becomes rare, former huts are regularly transformed into multistorey houses connected with an intricate system of narrow alleys branching from roads and often ending in cul-de-sacs. "These multiple dynamics dramatically alter the suburban fabric of Hanoi, where a new type of urbanism has emerged ... with a great diversity of intermixed landscapes, including walled residential estates, ad hoc densification of pre-existing villages, and the tight intermingling of small-scale industries with commercial, residential and agricultural activities" (Leaf, 2002: 29). Specific sociotechnical water supply and sanitation arrangements enable and

9.1 The new urban area My Dinh II. Photo: Courtesy of the authors

9.2 Land and water in the peri-urban village Van Phuc. Photo: Courtesy of the authors

are at the same time shaped by these urbanization dynamics at Hanoi's urban edge.

Elite Decentralization: Water and Sanitation in New Urban Areas

While new urban areas are provided with local networks for water and sanitation services in accord with modern standards, there is a lack of connectivity to large-scale networks outside the compounds. As public networks do not reach the new urban areas, the urban utilities lose responsibility. Service provision is managed directly by real estate development companies or by the Ministry of Construction (MOC). These actors supply water through privately owned groundwater wells and treatment plants within the compound (figure 9.3), while wastewater flows from households through a local underground sewerage network directly into the surrounding spaces (figure 9.4). Only in exceptional cases do local plants pretreat sewage before its discharge from the estates (Schramm, 2014). Thus, new service providers come forth, reshaping the spatial and organizational patterns of infrastructure provision in Hanoi as they provide services exclusively within new urban areas. By creating satellite networks, they limit the scope of public utilities in terms of service provision. Such suburban "archipelagos" of networks (Bakker, 2003) at times interfere with surrounding topologies of water and sanitation provision.

Residents and local administrations of new urban areas display a general lack of interest for infrastructure provision in the new urban areas. In the new urban area Linh Dam, 30 per cent of a cluster sample of people interviewed assumed that the water they receive is provided by the real estate development company Housing Development Corporation (HUD), while the remaining 70 per cent simply did not know. In general, the consulted population shows an indifference regarding the technologies and management of water supply, as most of the interviewees didn't know where the water they consume came from and had no particular complaints about the water coverage or access (Survey conducted by the authors in Linh Dam district of Hanoi City, 2014). The perceived profitability of providing services in new urban areas became apparent in the rumour circulating in the estate My Dinh II, where several residents explained that the water supply utility intends to build a second set of pipes in order to compete for this water market with the real estate company (My Dinh II in Tu Liem district residents, personal communications, 2011). Also, sanitation is not

9.3 Water supply plant in the new urban area Linh Dam. Photo: Courtesy of the authors

9.4 Wastewater treatment plant in the new urban area My Dinh II. Photo: Courtesy of the authors

a topic of interest for the estates' local administration or the residents, as one resident and ward representative explained: "The ward is not interested in sanitation provision. This is the task of the HUD alone." (Linh Dam ward representative, personal communication, 2011). The indifference of residents towards water and sanitation infrastructures reflects the expectation that living in modern high-rise buildings goes along with the status of passive consumers of externally provided services.

Hybrid Mixing: Water and Sanitation in Peri-urban Villages

The passive stance of residents in new urban areas is in stark contrast with the active involvement in the provision of sanitation services by inhabitants of peri-urban villages. Here, local administrations and villagers themselves construct sewerage lines incrementally as the village grows, following the typical road patterns of Hanoi's peri-urban villages, where smaller roads branch off the main roads and ultimately end in cul-de-sacs (see Schramm, 2016). As there is no centralized provision of sanitation services, villagers organize the construction as well as operation of the drainage system together with the local administrations. The maintenance of septic tanks, which remain the predominant means of household sanitation, is an individual and private issue (Schramm, 2014). Where farming is still prevalent, farmers use sludge from the tanks as fertilizer in fields or aquacultures (figure 9.5), despite the government's prohibition of these practices (Schramm, 2014). Residents of peri-urban villages are actively involved in sanitation provision and largely consider it to be an unproblematic issue. This attitude is particularly the case where sewage or night soil is reused as fertilizer in agriculture or for feeding fish in aquacultures, which contributes to the perception of these practices as integral parts of local sociomaterial cycles. However, in light of the increasing development of housing estates replacing formerly agricultural spaces and discharging their sewage into surrounding areas, the question of drainage becomes pertinent for those who live in villages. In the absence of unsealed ground or functional drainage systems, they have to deal not only with their locally produced sewage but also with wastewater from surrounding housing estates. A resident from Cau Giay district stated: "Here, we are most interested in wastewater ... Before, there were many fields and vegetables. Now only a few are left. How can we avoid flooding?" (Cau Giay district resident, personal communication, 2011).

9.5 Fishing pond in the peri-urban village Lai Xa. Photo: Courtesy of the authors

Thus, it is the suburban flow of water and sewage that unveils the close and partly conflict-ridden interrelations between new housing estates and peri-urban villages, because it fails to respect the boundaries between these places imagined by planners and represented in architectural and infrastructural artefacts.

The flow of water and its stagnation in peri-urban villages furthermore reveal the contradictions between sanitation and water supply planning, and the problematic effects of the de-coupling of these sectors since *doi moi*, as increasing amounts of water are directed to suburban spaces in the absence of any functional drainage or percolation system. While there is no central-level involvement in the sanitation of peri-urban villages, the central government does support the extension of the existing water supply networks to urbanizing areas and peri-urban villages through the MARD.

Even in those cases where villagers have access to water from large-scale networks, users complement this source with water from shallow wells and, like in new urban areas, with purchased bottled water. Also, they use supplementary filter systems to assure the quality of the piped water. The persistent practice of using piped water just as one among multiple sources of water displays the distrust residents have towards water from central networks. One person in the peri-urban

Van Phuc commune in Ha Dong district conveyed that, at times, the output of the tap water in his house was "dirty water, a lot of sludge, a lot of detergent [possibly chlorine]; the colour of water is yellow or brown. I need to use a water purifier" (Van Phuc commune resident, personal communication, 2011).

Residents in other suburban areas such as Gia Lam reported inconsistent coverage and unreliable quality of the water provided by the distribution network. Thus, they maintain privately owned wells despite a high connection rate to the distribution network. However, concerns about water quality only partially explain the reluctance to fully rely on networked water provision. For example, our own surveys have shown that piped water tested equally positively in both the peri-urban village of Trieu Khuc and the new urban area of Linh Dam, while the quality of groundwater accessed from privately owned wells was poor in both areas, presenting levels of ammonium, arsenic, and iron that exceed the National Technical Regulation on Drinking Water Quality (SRV, 2009). Still, privately owned wells are an important alternative to cover all daily water-consuming activities such as gardening, washing, or construction, even though residents are aware of the presence of heavy metals in groundwater. Maintaining decentralized water infrastructures enables the residents to both have control over the access to water and to gain financial independence and flexibility. Residents may save up to twenty times the expected charge of the water tariff, which may vary from USD $0.70 per month to USD $16 per month (VND 15,000 to VND 350,000 at VND 22,300 for USD $1) (Survey conducted by the authors: 100 cluster samples in ten districts, Hanoi City, 2014).

Suburban residents in Hanoi are not passively consuming services provided to them via large central networks, as these services prove to be too inflexible and costly for their particular needs. At the same time, they do not outright reject their insertion into a city-wide system of water reticulation; rather, they carefully navigate and choose among the broad spectrum of ways to access water available to them. Suburban dwellers' capacity to do so is closely related to their built environment and the degree of regulation of space. For example, tube or dug wells are more likely to be found in the peri-urban villages due to the greater flexibility of the built environment that allows for reconstructions by residents, while residents of the rather strictly regulated new urban areas largely depend on the supply of water from a development company. Thus, residents of peri-urban villages are able to make active decisions and resist the project of unification and centralization of water supply in Hanoi.

Consequently, peri-urban villages display a hybrid mix of ways to access water and sanitation services, ranging from self-organized access through private wells or septic tanks, to community-based management of local drainage and sewerage networks, to centralized networks of water supply (Schramm, 2014). In sum, the multiple technologies of water supply and sanitation in suburban Hanoi beyond centralized network provision reflect patterns of central-level provision favouring corporatization of infrastructure systems in housing estates that operate exclusive, decentralized water supply and sanitation systems, and, paradoxically, the expansion of centralized water supply networks to peri-urban villages. At the same time, these villages remain largely excluded from large sewerage and drainage systems despite central-level projects towards the expansion of these networks. Hence, it is especially the residents of peri-urban villages who suffer from flooding resulting from the decoupling of water and sanitation services, and from sanitation policies based on a modernist dichotomy between urban and rural spaces. As a result, these villagers retain agency by actively taking part in their own provision of sanitation and water supply services. They do so not only in the case of sanitation, where hardly any central-level engagement exists, but also in the case of water supply, where local and central-state actors, together with international and national corporations, push for a centralization and unification of service provision via an aggressive expansion of centralized networks.

Conclusion

Multiple interrelated processes lead to a diversification of water supply and sanitation provision at the edges of Hanoi reflecting wider societal changes: rapid urbanization supported by the central and city governments through the development of modern real estate together with an incremental expansion of villages tolerated by local administrations; the reluctance of central-level administrations to realize the large-scale sanitation systems envisioned in master plans and to equip sanitation utilities with the means to operate such networks; and the expansion of water supply networks together with the insistence of suburban residents on the use of decentralized water sources. These dynamics, the push towards the expansion of large technological networks in the water sector and the preference of residents to use water from the centralized network for some uses only and combine it with water from other sources beyond formal control, clearly contradict conceptualizations of "infrastructural bypassing" as a process leading to the exclusion of passive populations from centralized network provision (see Graham & Marvin, 2001). The suburban spaces of Hanoi, rather, display very place-specific practices between passive consumption and active involvement in service provision. These practices are shaped by and at the same time shape urbanization patterns and urban morphologies at Hanoi's urban edge, where residents of planned estates have a different capacity to rearrange the built environment compared to peri-urban villagers.

Suburban spaces in Hanoi are thus shaped by and at the same time resist current central-level planning and policies towards the expansion of centralized networks. These policies, projects, and plans are inserted into global flows of knowledge and capital channelled through bilateral and international finance and cooperation agencies. In the case of water as well as sanitation, the concentration on the roll-out of centralized networks displays a lack of respect for place-specific practices, the actual capacities of actors within and outside the state concerning the construction and maintenance of large-scale systems for the transport and treatment of water and the financing of interventions. Contradictions of current attempts to reshape the water cycle of Hanoi play out especially in the suburban areas. Suburban residents' practices, such as the use of night soil in agriculture and the use of water from decentralized wells for certain activities in combination with networked services for other activities, do not conform to any urban–rural dichotomy. This hybridization challenges the urban and infrastructural policies and plans of Hanoi, which continue to be affected by such a divide in their attempt to bring order, via the expansion of large networks, to spaces declared as urban. Suburban areas thus make apparent the need for urban and infrastructure planning in and beyond Hanoi to demonstrate trust in residents and their capability, for example, to differentiate uses of water from different sources in accord with respective water qualities instead of aiming at controlling and oppressing such practices. Suburban infrastructures in Hanoi and beyond thus pose the challenge to imagine ways in which such suburban practices as the use of decentralized water sources and the local reuse of night soil in agriculture may become an integral part of, and no longer a threat to, large infrastructure systems without compromising larger goals of sociospatial cohesion and equity.

The chapter has brought to this volume a consideration of the importance of the scale of infrastructure systems. It has exposed the modernist vision of large integrated water and sanitation systems, conveyed in plans and promoted by international aid agencies. But the Hanoi case has identified multiple obstacles impeding the implementation of this vision and the adverse health and environmental consequences of fragmented water and sanitation systems. The message of the chapter thus appears to resonate with the modernist perspective on infrastructures and, thereby, promote large-scale, integrated infrastructures. According to this perspective, such infrastructures would be a prerequisite to healthy living conditions, respect for the environment, and coordinated urban and suburban development. The message of the chapter is subtler, however; it considers difficulties in achieving the modernist infrastructure model and alternatives to this model. Hence chapter 9 is of interest because it shows how effective smaller scale water and sanitation systems can be delivered within the prevailing institutional and financial context, while building on local public mobilization.

REFERENCES

Bakker, K. (2003). Archipelagos and networks: Urbanization and water privatization in the South. *The Geographical Journal, 169*(4), 328–41. https://doi.org/10.1111/j.0016-7398.2003.00097.x

Berg, M., Trang, P.T.K., Stengel, C., Buschmann, J., Viet, P.H., Van Dan, N., Giger, W., & Stüben, D. (2008). Hydrological and sedimentary controls leading to arsenic contamination of groundwater in the Hanoi area, Vietnam: The impact of iron-arsenic ratios, peat, river bank deposits, and excessive groundwater abstraction. *Chemical Geology, 249*(1–2), 91–112. https://doi.org/10.1016/j.chemgeo.2007.12.007

Coutard, O. (2008). Placing splintering urbanism: Introduction. *Geoforum, 39*(6), 1815–20. https://doi.org/10.1016/j.geoforum.2008.10.008

Fayet, L. (1939). Preliminary draft of the sewers of Hanoi. (Original in French; Avant-projet sur les égouts de Hanoi). *Hanoi: Imprimerie d'extrême Orient, 1939*. National Archive Vietnam, Centre I Hanoi (078692–03).

General Statistics Office of Vietnam (GSO). (2009). Average population by province. *Population and Employment*. Retrieved from http://www.gso.gov.vn/default_en.aspx?tabid=467&idmid=3&ItemID=9880

General Statistics Office of Vietnam (GSO). (2011). Part 2: Projection result tables. *Projected Population for Vietnam 2009–2049*. Retrieved from http://www.gso.gov.vn/default_en.aspx?tabid=617&ItemID=11016

General Statistics Office of Vietnam (GSO). (2014). Area, population and population density by province. *Population and Employment*. Retrieved from http://www.gso.gov.vn/default_en.aspx?tabid=774

GHK International (2005). *Decentralised wastewater management in Vietnam – A Hanoi case study*. London: GHK International, CEEITA.

Government of Vietnam. (2008). *Explanatory report on the administrative expansion of Hà Noi capital*. (82/BC-CP 2008). Hanoi: Author.

Graham, S., & Marvin S. (2001). *Splintering urbanism: Networked infrastructures, technological mobilities and the urban condition*. London: Routledge.

Han, S.S., & Vu, K.T. (2008). Land acquisition in transitional Hanoi, Vietnam. *Urban Studies*, *45*(5–6), 1097–1117. https://doi.org/10.1177/0042098008089855

Hanoi Water Limited Company (HAWACO). (2011). Construction and development of water supply sector Hanoi. (Original in Vietnamese; Xay dung va phat trien nganh cap nuoc ha noi). Hanoi: Author.

Hanoi Water Limited Company (HAWACO). (2014a). *Hanoi water supply system*. PowerPoint presentation viewed as part of the guided visit to HAWACO's Hoa Binh WTP during the 37th WEDC International Conference, "Sustainable Water and Sanitation Services for All in a Fast Changing World," co-hosted by Loughborough University and the National University of Civil Engineering (NUCE), Hanoi, 15–19 September 2014 (Date of guided visit: 18 September 2014).

Hanoi Water Limited Company (HAWACO). (2014b). *Hanoi water supply system*. Summary distributed as part of the guided visit to HAWACO's Hoa Binh WTP during the 37th WEDC International Conference, "Sustainable Water and Sanitation Services for All in a Fast Changing World," co-hosted by Loughborough University and the National University of Civil Engineering (NUCE), Hanoi, 15–19, September 2014 (Date of guided visit: 18 September 2014).

Kooy, M., & Bakker K. (2008). Technologies of government: Constituting subjectivities, spaces, and infrastructures in colonial and contemporary Jakarta. *International Journal of Urban and Regional Research*, *32*(2), 375–91. https://doi.org/10.1111/j.1468-2427.2008.00791.x

Labbé, D., & Boudreau, J.A. (2011). Understanding the causes of urban fragmentation in Hanoi: The case of new urban areas. *International Development Planning Review*, *33*(3), 273–91. https://doi.org/10.3828/idpr.2011.15

Labbé, D., & Musil, C. (2011). The extension of administrative boundaries of Hanoi: An exercise of recomposition of territories in tension. (Original in French; L'extension des limites administratives de Hanoi: Un exercice de recomposition territoriale en tension). *Cybergeo: European Journal of Geography*. Article 546. Retrieved from https://journals.openedition.org/cybergeo/24179

Leaf, M. (2002). A tale of two villages: Globalization and peri-urban change in China and Vietnam. *Cities*, *19*(1), 23–31. https://doi.org/10.1016/S0264-2751(01)00043-9

Ministry of Construction (MOC). (1993). *Preserving Hanoi's architectural and landscape heritage*. Hanoi: Construction Publishing House.

Ministry of Foreign Affairs of Denmark. (2014). Sector program support to water, sanitation and hygiene promotion (2012–2014). In *Water and Sanitation – Denmark in Vietnam*. Retrieved 17 March 2015 from http://vietnam.um.dk/en/danida-en/water-and-sanitation/

Monstadt, J., & Schramm, S. (2013). Beyond the networked city? Suburban constellations in water and sanitation systems. In R. Keil (Ed.), *Suburban constellations: Governance, land and infrastructure in the 21st century* (pp. 85–94). Berlin: Jovis.

National Centre for Rural Water Supply and Sanitation (NCERWASS). (2014). *Rural water supply in Vietnam: Evolution and opportunities*. PowerPoint presentation viewed during the 37th WEDC International Conference, "Sustainable Water and Sanitation Services for All in a Fast Changing World," co-hosted by Loughborough University and the National University of Civil Engineering (NUCE), Hanoi, 15–19 September 2014. Presented by Le Thieu Son.

Ngo, H. (2014). *Internship report: Yen Phu water treatment plant*. (Original in Vietnamese; Noi dung bao cao thuc tap cong nhan nha may nuoc yen phu). Unpublished work. Hanoi: Institute of Environmental Sciences and Technology, National University of Civil Engineering.

Ngo, T.T.V. (2009). The existing sewerage and drainage system in Hanoi. *TU International*, *63*, 18–19. Retrieved 22 June 2009 from https://www.alumni.tu-Berlin.de/fileadmin/Redaktion/ABZ/PDF/TUI/63/van_ngo_TUI63.pdf

Nguyen, T.Q., & Helm, D.C. (1995). Land subsidence due to groundwater withdrawal in Hanoi, Vietnam. In F.B.J. Barends, F.J.J. Brouwer, & F.H. Schröder (Eds.), *Proceedings of the Fifth International Symposium on Land Subsidence, held at The Hague, The Netherlands, 16–20 October 1995* (pp. 55–60). The Hague: International Association of Hydrological Sciences. Retrieved from http://hydrologie.org/redbooks/a234/iahs_234_0055.pdf

Owen, D. (2012). *Pinsent Masons water yearbook 2012–2013: The essential guide to the water industry from leading infrastructure law firm Pinsent Masons* (14th ed.). London: Pinsent Masons.

Perkins Eastman, Posco E&C, and Jina (PPJ). (2010a). *Construction planning of Hanoi capital city to 2030 and vision to 2050 conference ceport crban development planning Vietnam*. (Original in Vietnamese; Quy Hoạch Chung Xây Dựng Thủ Đô Hà Nội đến năm 2030 và tầm nhìn đến năm 2050). Retrieved from http://hanoi.org.vn/planning/data/ppj_20100402.pdf

Perkins Eastman, Posco E&C, and Jina (PPJ). (2010b). Greater Hanoi: General construction plan deals until 2030 and vision to 2050. (Original in Vietnamese; Uy hoạch chung xây dựng Thủ đô Hà Nội đến năm 2030 và tầm nhìn đến năm 2050). In *Hanoi's Channel, YouTube*. Retrieved from https://www.youtube.com/watch?v=Q9-RErNFhrU

Quang, N., & Kammeier, H.D. (2002). Changes in the political economy of Vietnam and their impacts on the built environment of Hanoi. *Cities*, *19*(6), 373–88. https://doi.org/10.1016/s0264-2751(02)00068-9

Quoc, D. (2010). Re-evaluation of Hanoi drainage project (Original in Vietnamese; Dự án thoát nước Hà Nội: Phải đánh giá lại). *Tiền Phong*. Retrieved from http://www.tienphong.vn/Thoi-Su/189367/Du-an-thoat-nuoc-Ha-Noi-Phai-danh-gia-lai.html

Schramm, S. (2014). *City in flux: Sanitation in Hanoi in the light of social and spatial transformations* (Original in German; *Stadt im Fluss: Die Abwasserentsorgung Hanois im Lichte sozialer und räumlicher Transformationen*). Stuttgart, DE: Franz Steiner Verlag.

Schramm, S. (2016). Flooding the sanitary city: Planning discourse and materiality of urban sanitation in Hanoi. *City*, *20*(1), 32–51. https://doi.org/10.1080/13604813.2015.1125717

Socialist Republic of Vietnam (SRV). (2009). *National Technical regulation on drinking water quality, Hanoi*. (QCVN 01: 2009/BYT). Compiled by the Department of Preventive Medicine & Environment and promulgated by MOH's Minister at the Circular No.04/2009/TT-BYT, Hanoi. Retrieved from http://www.wpro.who.int/vietnam/topics/water_sanitation/wmq_water_standards_technical_regulation_on_clean_drinking_water_quality.pdf

Socialist Republic of Vietnam, & Hanoi People's Committee (SRV & HPC). (2005). *Feasibility study report and program implementation for Hanoi sewerage and drainage environmental renovation project – Phase II Volume I Main report* (Original in Vietnamese; Báo cáo nghiên cứu khả thi (F/S) Và Chương trình thục hiện (I/P) Cho Dự án thoát nước Nhằm cải tạo môi trướng Hà nội – Giai đoạn II). Hanoi: NIPPON KOEI; SRV; HPC.

Socialist Republic of Vietnam, & Hanoi People's Committee (SRV & HPC). (2009). *Feasibility study for the construction project of central large-scaled wastewater treatment plants for Hanoi environmental improvement–Final report summary*. Hanoi: NIPPON KOEI; VIWASE.

24h (2011). Any solution to the problem of pollution of Tô Lịch? (Original in Vietnamese; Lới giải nào cho bài toán ô nhiễm sông Tô Lịch?). In *24h*. Retrieved from http://us.24h.com.vn/tin-tuc-trong-ngay/loi-giai-nao-cho-bai-toan-o-nhiem-song-to-c46a327224.html

United Nations. (2014). *World urbanization prospects: Highlights*. New York: Department of Economic and Social Affairs, United Nations.

Winkel, L.H., Trang, P.T.K., Lan, V.M., Stengel, C., Amini, M., Ha, N.T., & Berg, M. (2010). Arsenic pollution of groundwater in Vietnam exacerbated by deep aquifer exploitation for more than a century. *Proceedings of the National Academy of Sciences, 108*(4), 1246–51. https://doi.org/10.1073/pnas.1011915108

World Health Organization (WHO). (2010). *International travel and health: Situation as on 1 January 2010*. Geneva: Author.

Zérah, M.-H. (2008). Splintering urbanism in Mumbai: Contrasting trends in a multilayered society. *Geoforum, 39*(6), 1922–32. https://doi.org/10.1016/j.geoforum.2008.02.001

SECTION 3

Reshaping Suburban Infrastructures

10 The "In-Between Territories" of Suburban Infrastructure Politics

DAVID WACHSMUTH

Introduction

There is a contradiction in the contemporary relationship between infrastructure development and local growth politics in the United States. On the one hand, infrastructure development is commonly understood to be a central policy focus of local growth coalitions, since infrastructure provides the material preconditions for new capital investment in localities. The emergence in recent decades of new infrastructure-led development strategies based around trade and logistics has strengthened this focus and supplied new tangible targets for growth politics. But, on the other hand, infrastructure development also challenges the coherence of local growth coalitions, since infrastructure projects frequently exceed the local and metropolitan scales at which such coalitions are preferentially organized (via municipal governments, chambers of commerce, economic development corporations, and regional partnerships).

In this chapter, I argue that this contradiction has been driving a new form of multi-city growth politics in the United States – one that has tended to privilege certain specific suburban development interests. The paper advances two major claims. First, building on recent research (Wachsmuth, 2017b), I argue that supply-chain expansion – the extension of effective supply chains and the intensification of circulatory possibilities within a supply chain – offers a structuring principle for new, emerging multi-city growth coalitions analogous to the role of land-use intensification underlying traditional local growth machines. Second, I argue that in-between territories – suburban jurisdictions located between major growth poles along urban corridors – have become key strategic points in the territorial politics of supply-chain expansion and infrastructure politics more broadly.

The plan for the chapter is as follows: I begin by discussing the relationship between the development of the built environment and the local growth coalitions widely understood to be central to contemporary entrepreneurial urban governance. On the basis of that discussion, I introduce the concept of supply-chain expansion and outline the basic strategic contours of the multi-city growth coalitions that assemble to pursue geographical market-making strategies. I then proceed to argue that supply-chain expansion generates a distinctive suburbanized spatial politics. Where local economic development agendas in polycentric urban regions are driven by infrastructure priorities of the growth poles, specific suburban spaces can emerge as strategically important sites for territorial growth politics, despite their apparently under-institutionalized local governance. This situation is due to (1) the imperative to unite political-economic interests across the entire regions; and (2) strategic opportunities for siting new infrastructure development along established corridors but outside the major growth poles. I call these sites "in-between territories," in the sense that their strategic significance arises from their spatial location in between growth poles along urban corridors. But describing these sites in terms of their relationship to nearby major cities is not meant to imply that they are peripheral. Indeed, while in-between territories are spatially "in-between," they turn out to be strategically central.

In the chapter I develop this argument through a comparison of suburban infrastructure politics in two sites in the United States: Pinal County, Arizona, and Polk County, Florida. Pinal County is a rapidly urbanizing county between the major Arizona growth poles of Phoenix and Tucson. In the last twenty-five years its population has increased four-fold, from 100,000 residents in 1990 to over 400,000 by the end of 2014. Polk County lies along the I-4 corridor in Central Florida, equidistant between Tampa and Orlando. Like Pinal County in Arizona, Polk County has been one of the fastest growing areas of the state thanks to expansion from the two major cities between which it lies. In the last ten years both the Phoenix–Tucson and the Tampa–Orlando corridors have been the sites of new economic development strategies focused on expanding and intensifying the built environment for trade and logistics. And, in both cases, the in-between territories of Pinal County and Polk County respectively have become (surprisingly?, disproportionately?) central to these strategies. The chapter systematically compares these two cases, and draws out the implications for the future of urban growth politics in an increasingly polycentric and suburbanized urban landscape.

Infrastructure Alliances and Supply-Chain Expansion

Recent scholarship has suggested that infrastructure development is fragmenting local urban politics, but I have recently argued that it has had the opposite impact at the multi-city regional scale (Wachsmuth, 2017b). The last several decades have seen new multi-city growth coalitions emerge across the United States, united by a shared interest in supply-chain expansion – the extension of effective supply chains and the intensification of circulatory possibilities within regional transportation networks. I call these growth coalitions "infrastructure alliances." Before turning to an exploration of the place of suburban territorial politics in these alliances, in this section I briefly outline the underlying theoretical argument.

Scholarship on post-1970s North Atlantic urban governance has been profoundly influenced by Logan and Molotch's (2007) model of the city as a growth machine and by related accounts within the "new urban politics" (Cox, 1993; MacLeod & Jones, 2011). Logan and Molotch identify a common interest among a wide swath of elites – the "growth coalition" – in land-use intensification. Urban growth coalitions are typically anchored by rentiers and the real estate industry, and also include municipal governments, local media and entertainment, and utility companies among their ranks. Because growth of the built environment is the precondition of land-use intensification, infrastructure development is an important shared priority of growth coalitions. However, recent scholarship has argued that contemporary city-bound growth coalitions, despite their shared interests in promoting infrastructural development, may be splintered rather than united by the built environment: first because of neoliberal urban fiscal constraint that prevents elites from pursuing optimal strategies (Kirkpatrick & Smith, 2011), and second because of the rise of infrastructure unbundling and premium infrastructural networks that opens the possibility of secessionist development schemes (Graham & Marvin, 2001). The result is a contradiction: shared interests in land-use intensification drive local growth coalitions to attempt to direct infrastructure development consistent with those interests, but the contemporary politics of infrastructure development tend to splinter and fragment coordinated growth coalition action at the local scale.

My contention (developed at length in Wachsmuth, 2017b) is that this contradiction is now in part resolved through scale-shifting: infrastructure development in the domain of trade and logistics provides a structuring principle for new growth coalition action analogous to land-use intensification, but at a larger, multi-city scale. This structuring principle is supply-chain expansion: the extension of effective regional transportation networks and supply chains (the systems of production and circulation enabled by extant transportation and technological possibilities) and the intensification of circulatory possibilities within a regional transportation network. A new highway can increase the competitiveness of local manufacturing by providing a faster connection to regional distribution hubs, and a new logistics facility can introduce new trade-related economic possibilities into an existing transportation network. Zero-sum competition and governance coordination problems make a traditional growth machine agenda unlikely to emerge at the multi-city scale, but investment in transportation infrastructure that connects multiple city regions can become an appealing target for regional growth politics because it can (1) enhance complementarities within an existing regional division of labour; (2) improve the competitive position of an entire region within existing larger-scale spatial divisions of labour; or (3) introduce new economic development pathways by connecting a region into new divisions of labour.

Regional growth coalitions of this sort – large-scale local development partnerships oriented around supply-chain expansion – have indeed been proliferating in the United States in the past twenty years. They are a subset of a broader trend towards "competitive multi-city regionalism" in the country – an upscaling of local economic development activity in which place-bound elites coordinate place marketing, business attraction, and policy development across multiple cities and metropolitan areas spanning up to several hundred kilometres (Wachsmuth, 2017a). In the mid-1970s there were just a handful of formally institutionalized multi-city economic development partnerships, but as of the end of 2013 there were 171.

I call this subset of regional growth coalitions that are united by supply-chain expansion "infrastructure alliances." These are groups operating to strengthen functional connections within the polycentric urban fabric and to expand the extent of these connections. Their anchor, analogous to the role of real estate developers in the local growth machine, is the transportation and logistics sector. Firms in this sector, including rail companies, shippers, importers and exporters, and third party logistics operations, have the clearest direct economic interest in supply-chain expansion as a means of opening up new markets or increasing the profitability of their operations in existing markets. The key partners of logistics firms within infrastructure alliances are metropolitan chambers of commerce, which offer local business communities the opportunity to exert policy agency over the regional economic context that structures local growth possibilities.

Alongside these two major players in the infrastructure alliance are a range of supporting characters: infrastructure management and construction firms, which make money whenever there is infrastructure to be constructed, upgraded, or retrofitted; manufacturing and natural resource extraction firms, which rely for their individual competitiveness on a robust regional milieu and low-cost transportation networks; metropolitan planning organizations, which often view multi-city regional initiatives as a way of exerting influence over transportation development schemes that extend outside their jurisdictional boundaries (usually one or several counties); and local governments, for whom new investments in expanding or intensifying regional markets generally imply an infusion of new local economic activity. State governments may also be participants in regional infrastructure alliances, since they have come to play a leading role in local economic development throughout the United States (Eisinger, 1988), although regional partnerships whose memberships cross state boundaries have just as frequently been opposed as supported by state governments (Wachsmuth, 2017a).

In sum, much as land-use intensification provides a structuring principle for growth coalition action at the local scale – centred on rentiers and the real estate industry – supply-chain expansion provides a structuring principle for growth coalition action at the multi-city regional scale – centred on trade and logistics. The comparison shouldn't be overextended, however. Because the governance pathways of local growth machines map in a relatively straightforward fashion onto the institutional arenas of local politics (above all municipal governments and metropolitan chambers of commerce), while the governance pathways of multi-city infrastructure alliances are sparse and unpredictable, infrastructure alliances will not be able to achieve anything like the consistency or ubiquity of the growth machine. Still, analytically, they are amenable to the same form of structural analysis. Following, as a contribution to this task, I discuss two examples of infrastructure alliances in operation. But first I expand on the spatial implications of the concept, in particular by unpacking the distinctive territorialization of suburban space that supply-chain expansion encourages.

In-Between Territories

Up until now I have discussed infrastructure alliances in relatively aspatial terms, simply noting that supply-chain expansion encourages the formation of ground-up growth coalitions on much larger scales than are common with local growth machines. But this fact intersects with two important spatial features of contemporary US urban geography. First, the United States features many medium and large cities in relatively close proximity, but few areas of continuously dense settlement space – with the Northeast "Megalopolis" (Gottmann, 1961) and parts of the California coast standing as the main exceptions. Second, urban growth in the United States continues to be suburban growth extending out from dense cities, radiating particularly strongly along interstate highways and thus increasingly removed from city centres, but still clearly within the "gravitational" influence of cities as opposed to a formless exurban development (Nelson & Lang, 2011). The combination of these circumstances means that multi-city regionalism of the sort that supply-chain expansion encourages tends to occur over territories with multiple "centres" and multiple "peripheries." By way of illustration, figure 10.1 shows impermeable surface cover (a close approximation to the extent of the human-made built environment) in the multi-city Florida's Super Region, one of the case studies I discuss in a later section.

In what follows I want to focus on the areas of the map that are still sparsely shaded but whose colour has been filling in: the areas located along the interstate highways between major cities, which have historically been rural or unsettled but are now seeing rapid and steady population growth. I refer to these areas as "in-between territories," and my contention is that they have become key strategic points in the territorial politics of supply-chain expansion and infrastructure politics more broadly. These in-between territories have become key even though they typically feature weak and under-institutionalized local governing coalitions (as a consequence of their rapid growth from a low baseline), which are unable to effectively advocate for local developmental priorities. Even where local economic development agendas in polycentric urban regions are driven by the infrastructure priorities of the growth poles, specific suburban spaces can emerge as strategically important sites for territorial growth politics. This situation is due to (1) the imperative to unite political-economic interests across the

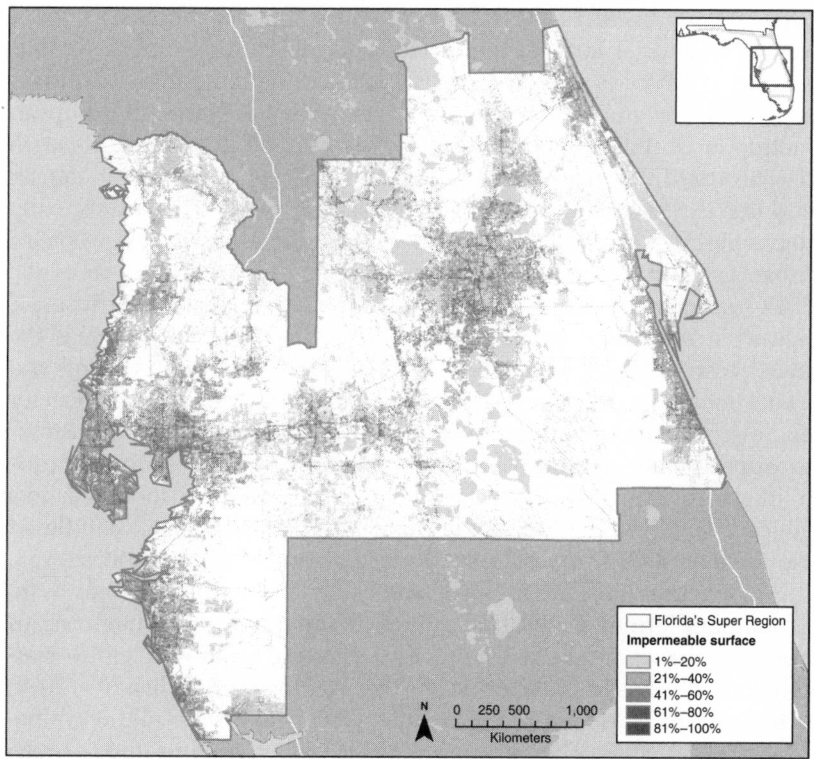

10.1 Impermeable cover in Florida's Super Region. Source: Author, with data from National Land Cover Database (2011)

entire regions; and (2) strategic opportunities for siting new infrastructure development along established corridors but outside the major growth poles. These spaces are in-between territories in the sense that their strategic significance arises from their spatial location in between growth poles along urban corridors. But describing these sites in terms of their relationship to nearby major cities is not meant to imply that they are peripheral. Indeed, while in-between territories are spatially "in-between," they are strategically central.

The concept of in-between territories builds on Sieverts's (2003) discussion of the European *Zwischenstadt*, which Young and Keil (2010) rework for a North American context as "in-between city."[1] I use the term "territory" instead to foreground the governance dimensions of polycentric urban development. Sieverts draws attention to the distinctive morphological and social form of the increasingly ubiquitous suburban landscape that serves as a kind of connective tissue within polycentric urban space. His analysis is driven by the development of Northern European metropolitan regions, which have some morphological similarity with decentralized US "edge cities" (Garreau, 1991), although within denser and less extensive overall spatial envelopes. This similarity is arguably deceiving in political terms, however, since the autarkic system of US urban governance helps produce a "localist" urban politics (Brenner, 2009), which does not have a strong European analogue. The result is that what Young and Keil (2010) call "the politics of infrastructure in the in-between city" requires further geographic specification. Young and Keil's analysis foregrounds the dual role of infrastructure as connector and disconnector of polycentric urban space; they argue that "infrastructure builds cities but it also dissolves cities as it creates centrifugal possibilities" (Young and Keil, 2010: 91). This argument is no doubt true in a general sense, but the specific centripetal and centrifugal possibilities at work are mediated through considerable institutional variation.

For my own analysis of in-between territories, the first step is to "zoom out" from the urban/suburban interface that characterizes Young and Keil's in-between city to the interface between entire metropolitan regions in polycentric or "megaregional" (Wachsmuth, 2014) urban space. These are the areas lying along interstate highways between major cities. They are rural, unsettled, or newly settled – even within the densest large-scale urban corridors, settlement space is far from continuous – and they are at the low end of the urban density gradient. They are typically the fastest growing areas of their larger regional networks, because they host suburban expansion from both poles of their urban corridor. And these areas have immature and weak political institutions, in terms of both formal government actors and non-governmental actors such as chambers of commerce and economic development agencies.

From the perspective of infrastructure alliances or other incipient multi-city growth coalitions, in-between territories are simultaneously roadblocks and opportunities. They are roadblocks because their weak governance does not allow them to pull their own weight in the partnership. They are opportunities because they have land for development, because they are located along the transportation corridors where new trade and logistics investments will likely need to be located, and because their location situates them well for compromise spatial strategies between the larger cities attempting to negotiate positive-sum developmental outcomes. In the latter case, one of the most difficult-to-overcome challenges in multi-city collaborations is the necessity to locate inward investment in one physical site. If growth coalitions in two city regions succeed in attracting a large new forward-processing plant, in whose city region will the plant – and most of the jobs – actually be located? The existence of in-between territories offers one possibility for compromise, along with the prospect of labour sheds that offer employment opportunities in both city regions.

We thus see a distinctive urban/suburban spatiality – a reworking of centre and periphery – to the infrastructure alliance, with spatial peripheries assuming a strategic centrality. Translating this spatiality into governance is by no means straightforward, but across infrastructure alliances I have observed a developmental meta-strategy that can be termed "territory to achieve network" (Wachsmuth, 2017b). Infrastructure alliances tend to pursue archetypally networked economic development goals (oriented towards extending and intensifying economic flows through, for example, new logistics facilities and trade routes) via strongly territorialized political strategies (using the large spatial reach of the necessary infrastructure development to assemble equally large territorial alliances to drive that development). Territory and network are thus not oppositional in the context of multi-city growth politics, but mutually constituting (Harrison, 2010). Reflecting on Harvey's (2006) seminal discussion of the tension between fixity and motion in the capitalist production of space, the case of in-between territories demonstrates that the need to fix some capital as infrastructure in order to let other capital circulate more smoothly does not just generate economic contradictions within existing sociospatial landscapes, as Harvey discusses, but actively generates new political geographies.

To summarize the discussion so far: supply-chain expansion offers a structuring principle for new growth coalition formation on the multi-city regional scale, based around the extension of effective regional supply chains and transportation networks, and the intensification of circulatory possibilities within a regional transportation network, but does so in a geographically uneven fashion. In particular, the in-between territories that lie along major transportation corridors between the growth nodes of the multi-city region can become important strategic lynchpins in regional economic development strategies.

Growth Politics in the Arizona Sun Corridor and Florida's Super Region

Having laid out these theoretical considerations, I now proceed to concretize them through a comparison of suburban infrastructure politics in two locations in the United States: Pinal County, Arizona, and Polk County, Florida. Both are rapidly growing suburban jurisdictions lying between two major second-tier cities (Phoenix and Tucson in the Arizona case, and Tampa and Orlando in the Florida case); both have been the sites of major new logistics-related infrastructure development or planning; and both have emerged as important lynchpins in the corridor-wide infrastructure alliances that have developed in Arizona and Central Florida. The findings here are drawn from a larger five-case comparison of competitive multi-city regionalism in the contemporary United States, which relied on document analysis, spatial analysis, and key-informant interviews conducted in 2013.

Freight Politics in the Arizona Sun Corridor

Pinal County is a rapidly urbanizing county in Arizona, located between the major growth poles of Phoenix and Tucson. In the last twenty-five years its population has increased four-fold, and much of that growth has been suburbanization from greater Phoenix along the main interstate corridor of I-10, although the edges of greater Tucson are now steadily creeping into Pinal County from the south. The county's population started markedly increasing in the 1990s, but the period of greatest sustained growth was the housing bubble of the 2000s, during which time the Phoenix–Tucson corridor was one of the fastest growing areas in the United States.

During the 2000s, private sector and public sector local growth strategies in Arizona were overwhelmingly oriented towards housing development, and regionalism was almost exclusively an oppositional perspective – articulated around the importance of resource management and smart growth in contrast to the prevailing low-density, sprawling development model (Morrison Institute, 2008). But since 2010, in the wake of the onset of the Great Recession and the complete collapse of Arizona's housing-led economy, a number of regionalism initiatives have emerged with a shared premise of uniting local elites throughout the Phoenix–Tucson corridor in pursuit of alternative growth pathways. The shared imaginary underlying all of these efforts is the "Sun Corridor" – a term for the multi-city region first proposed

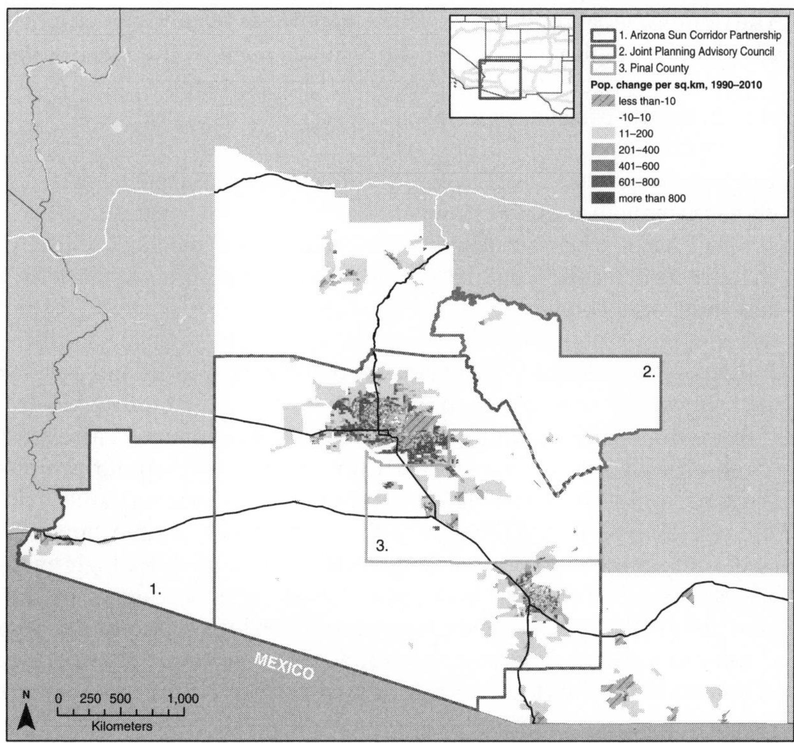

10.2 Arizona Sun Corridor Partnership and Joint Planning Advisory Council extents. Source: Author, with data from US Census (1990, 2010)

in an Arizona State University public policy course by Robert Lang and John Hall in 2006, building on the existing "Valley of the Sun" moniker for Phoenix (Lang & Nelson, 2011).

The two most significant multi-city regionalism initiatives in the Sun Corridor have been the Arizona Sun Corridor Partnership (ASCP), a local development collaboration between the five major urban economic development agencies in the region, and the Joint Planning Advisory Council (JPAC), a partnership between four of the federally designated metropolitan planning organizations in the region, which has initiated a major new freight planning process. In what follows I focus mainly on the latter initiative, although the two are to some extent intertwined. Their similar but distinct territorial extents, along with population change in the corridor from 1990 to 2010, are indicated in figure 10.2.

The defining goal of the JPAC collaboration has been to define a new logistics-led growth pathway for the Sun Corridor in the wake of the collapse of the state's housing economy. One participant laid out the rationale for the project as follows:

> You sit there and you look at freight. Freight doesn't care about borders. It doesn't care about even state boundaries or county boundaries. And so we said, hey listen, where's the population located? It's in the Sun Corridor. So let's extend that with our partners here for the JPAC, propose this idea and work together on it.

Initially, JPAC sought to define the parameters for an "inland port" to attract shipments from ports on the west coast and the Gulf of Mexico, and from Mexican maquiladoras. On its own, the plan wasn't a novelty, since JPAC was only one of several actors working on similar projects in Arizona, amidst a "scramble" of freight and trade studies in the wake of the recession (O'Dowd, 2012). The JPAC effort was notable, however, for explicitly being a corridor-scale plan. In fact, the corridor logic eventually led to the inland port being abandoned, because, despite the apparent connectivity of the Sun Corridor into various different national and international trade circuits, no single location within the corridor was adequately positioned to capture that connectivity. Moreover, the premise of broadly dispersed economic benefits throughout the corridor, which was essential to JPAC generating the political will to carry out its planning effort, was difficult to reconcile with a single large facility in a single one of the several participating jurisdictions. The result is that in 2013 JPAC, working closely with the transnational project management firm AECOM, ended up developing proposals for a series of four logistics-related infrastructure investments throughout the Sun Corridor. Two were to be located in the Phoenix area: a build-up of manufacturing and local distribution capacity around Phoenix-Mesa Gateway airport and a forward distribution centre to send exports from the region's significant manufacturing base to the west coast. And one was to be located in the Tucson region: an import distribution facility with a focus on Texan and Mexican markets.

The fourth investment was proposed to be located in Pinal County: a freight mixing centre, where large arriving shipments can be deconsolidated and sent to final destinations in smaller batches that are more amenable to precise control over delivery timing. The mixing centre has been a major priority for both the Union Pacific Railroad Company and political leadership in Pinal County since it was first proposed in 2006, but they have struggled to gain the necessary state government support for the project. So how did Pinal County, with a fraction of the population of its much more powerful neighbours and barely any governance capacity to speak of, land JPAC's support for the mixing centre? An economic development official for the county put it this way to me:

> If we were off to the side of one of these major counties, like Yavapai County is, [we wouldn't have the influence that we do] ... Pinal County is right in the middle. We're the cream filling in the Oreo cookie. They have to go through us – they don't have to go through [Yavapai]. So that puts us in the catbird seat because we've got the interstates and the rail, and a third interstate. So they're going to have to at least put up with us for the next few years.

In other words, it is Pinal County's status as an in-between territory – peripheral, suburban, and under-institutionalized, yet highly strategic to infrastructure development strategies – that has put it "in the catbird seat." This decision to empower the suburban leadership of Pinal County in corridor-wide development schemes was not free from controversy, however. Compare the following remarks by participants in the Arizona Sun Corridor Partnership. Here is the economic development representative from Pinal County again:

> We don't all quite have the same idea of what the Sun Corridor brings to us. In our county, because we're brand new at this ... we looked at this as a chance to participate with our big neighbours, which we've never had before. To be included in this – I'll be very honest – was exciting. Before we've always been treated as – well, there are two Dairy Queens along I-10 – we were always the Dairy Queen stop ... So we look at it as an opportunity to be able to make our county go forward because we don't have the population, we don't have the money, we don't have the political pull those other two counties do.

And here is a city councillor from a large, affluent inner suburb of Phoenix, speaking of Pinal County's participation in the partnership: "[Pinal county] was kind of the weak link ... It's just a rural county ... They didn't have the talent pool there."

Likewise, I was told by various JPAC participants that, while representatives from Tucson and Pinal County were supportive of the freight planning initiative, some representatives from Phoenix's suburbs were still sceptical, preferring a Phoenix-centric metropolitan economic development model in which they plausibly expect to be more influential. Still, the chief executive officer of the major Phoenix metropolitan economic development corporation has been one of the most enthusiastic supporters of corridor-wide economic collaboration, and with such key figures from the metropoles advocating for Pinal County, the latter's participation in Sun Corridor initiatives has proceeded.

The JPAC freight initiative thus demonstrates logistics-related infrastructure development driving large-scale growth coalition formation across (urban) centres and (suburban) peripheries, in part despite and in part because of the uneven development of the multi-city region. While Pinal County's strategic location between the major growth poles of Phoenix and Tucson and its wide availability of land near major transportation corridors both made the county's inclusion in the JPAC initiative a foregone conclusion, this very same peripheral location has translated into underdeveloped local governance, which has made the county's actual participation in the partnership problematic.

Logistics and Rail Politics in Central Florida

As a counterpart to the Arizona case, I now turn to another example of an infrastructure alliance mobilizing around an in-between territory: Polk County, Florida. Polk County lies along the I-4 corridor in Central Florida, equidistant between Tampa and Orlando. Like Pinal County in Arizona, Polk County has been one of the fastest growing areas of the state thanks to expansion from the two major cities it lies between, although it has grown from a larger base, with the city of Lakeland accounting for 100,000 of the now more than 600,000 residents in the county. Within the emerging multi-city regional politics in Central Florida, Polk County occupies an increasingly important location inside the functional orbits of both Tampa and Orlando. A tangible example of this imbrication is the fact that Polk County is represented by the metropolitan chambers of commerce of both cities – the only such overlap that exists in the corridor (figure 10.3).

While I will shortly discuss trade and logistics developments in the region, the story of multi-city growth coalition action in Central Florida actually begins with a different type of transportation infrastructure.

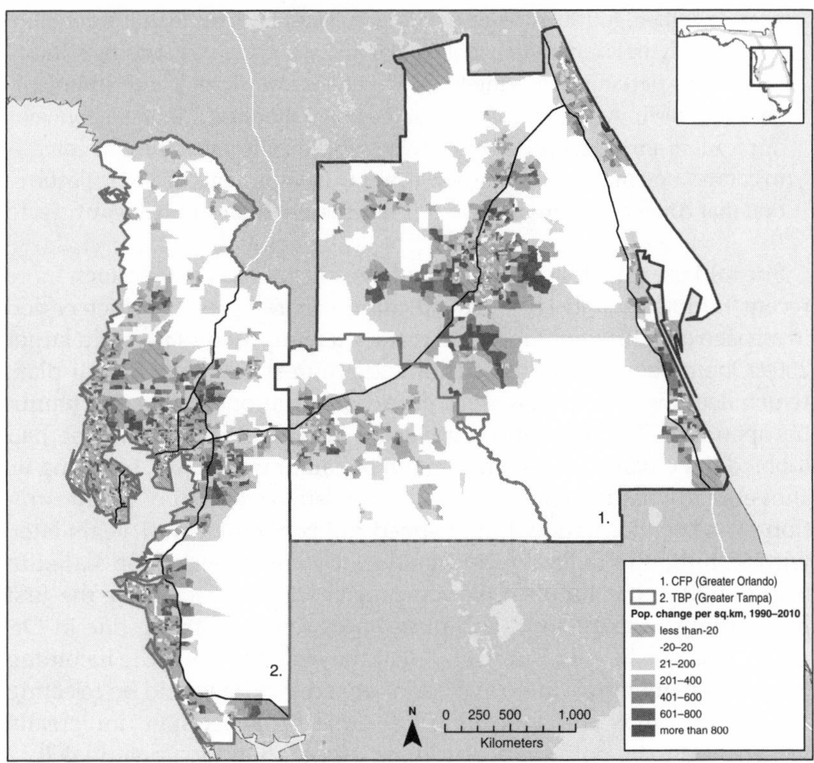

10.3 Greater Orlando and Greater Tampa metropolitan areas overlapping Polk County. Source: Author, with data from US Census (1990, 2010)

As one of the Orlando-based leaders of the Florida's Super Region economic development collaboration describes, the initial inducement to local elite collaboration in the region was the poor state of ground commuting between Tampa and Orlando:

> Back in, I'm going to say, 1997–98, there were a group of leaders in Tampa who came to us in Orlando, and said, do you know what, it's crazy that we don't bring the Olympics to Central Florida ... Let's go together, Orlando and Tampa, and put in a bid for the Olympics. So everybody was really excited about it ... and it brought our business community together, our governments were working together. The Olympic committee came down for their first review of the I-4 corridor ... that connects Tampa and Orlando ...

> And back then, fourteen years ago, it was even worse off than it is today. And the Olympic committee looked at it and went, are you kidding? This would be a parking lot. Trying to get to venues from Tampa and Orlando, back and forth. And literally, we lost the 2012 Olympics ... And so it gave our leaders a real glimpse into the twenty-first century that if we're going to compete on the global stage, we've got to have systems like transportation that connect our urban areas in a much more sophisticated way.

She told me this story as the prelude to a discussion of the much more recent Tampa–Orlando high-speed commuter rail project, which ended in a fiasco of a different kind. The route was proposed as part of a larger 2009 Obama administration nationwide high-speed rail funding plan, which itself was a component of the administration's post-crisis stimulus spending. The Republican governor at the time, Charlie Crist, had lobbied the Obama administration on behalf of the project, stressing its shovel-ready nature, and the line was chosen to be a sort of demonstration for a broader national high-speed rail program. A few years later, with $2 billion in federal funding allocated and construction slated to begin in 2011, the Tampa–Orlando project was on track to be the first line actually constructed, with three stops (one in Tampa, one in Orlando, and one in Polk County). But at the eleventh hour, the incoming Republican governor, Rick Scott, announced that he would be rejecting the federal funds and cancelling the project – a decision universally understood to be political posturing in preparation for a potential 2012 White House run, and one that drew condemnation across the political spectrum in Central Florida. (Two members of the state legislature even filed a motion in court to force the governor to accept the federal funds, although the state supreme court denied the motion and allowed the governor to reject the funds.)

Before the rail project ultimately ended in failure, it was a proximate cause for the formation of Florida's Super Region – a joint marketing and policy coordination initiative in the Tampa–Orlando corridor. The project was a major desideratum for businesses and local governments in the corridor, and they organized extensively to build in-state support for the train. The first of the Super Region leadership conferences, in 2009, was held one month after the Obama administration's rail announcement, and while the formal agenda had no mention of high-speed rail (since it had been drawn up prior to the announcement), the topic came to dominate informal conversations, according to attendees. Meanwhile, the last Super Region conference, held in 2011 before budgetary pressures led the conference to be mothballed, featured a session entitled "Overcoming Obstacles to Develop the Potential of Our Super Region" about how to regroup in the wake of the collapse of the high-speed rail project.

My Orlando-based informant described the level of mobilization that the project provoked among local elites throughout the corridor:

> I still haven't forgiven our governor for this. What was wonderful from a regional perspective is, there has never been a project that we have worked on at the super-regional level that had more dedicated leadership. We had the business community, we had every mayor from coast to coast supporting and buying into it, we had the counties on board. The State of Florida was on board – FDOT [the Department of Transportation]. And then this governor gets elected and, because of political issues with the Obama administration, killed it ... So it was very disappointing. But what it demonstrated to me was, we have leaders really organized well now to take on the next project and the next project.

The Tampa–Orlando rail project, failed though it was, demonstrates several important features of infrastructure alliances. First of all, the large-scale footprint of infrastructure investment schemes creates a similarly large-scale territory within which a territorial alliance can form, although it does not guarantee that such an alliance will be politically effective. Second, the Tampa–Orlando rail project underscores the complex and multiscalar nature of "local" infrastructure politics: local business and economic development leaders united across the entire I-4 corridor in an effectively unprecedented fashion, but pitted against a Republican governor with whom they otherwise had been very close thanks to his strongly pro-capital posture (and who appears to have been in favour of the project before he reinterpreted its politics through the national lens of a potential presidential run), and the federal transportation secretary attempting to secure side channels for the rail funding with the help of state legislators, and the oil industry apparently funding opposition to the project through a conservative think tank, and so on, and so on.

The same complexities have been at work in more recent initiatives in the trade and logistics sector, although here the institutional structure coalescing around multi-city regionalism has been quite diffuse, with a combination of state and local government agencies, public–private partnerships, and corporations all actively pursuing multi-city regionalist growth strategies. As a state economic development consultant explains:

> Particularly global trade and logistics have been a huge driver of economic development, of regions coming together in the last few years. It's created a lot of new partnerships ... The big seaports, the airports are all regional in nature. Supply chains are regional. I mean it's something that's hard to argue. The businesses that are getting involved and really helping drive that discussion really are thinking regionally, and encouraging that dialogue ... The big shippers and motor carriers, the railroads: they're looking at things like, CSX is building a significant new intermodal logistics centre in Winter Haven, which is about halfway between Tampa and Orlando, but it's really going to have impacts along that whole corridor, and so it's bringing together a lot of thoughts about, how do we work together on it.

The new logistics facility he is referring to, the Winter Haven Intermodal Logistics Center, began operations in April 2014, and is expected to process 300,000 shipping containers a year from a number of Florida's ports. It is located in the in-between territory of Polk County – seventy-five kilometres from Orlando and eighty kilometres from Tampa – and one Orlando-based economic development director I spoke with speculated that it could catalyze a process of clustering along the I-4 corridor that would open up new economic development possibilities on a regional scale.

A further complicating factor is the distributed nature of Florida's ports, which means that the centrally located Polk County has been an especially attractive destination for new logistics initiatives. The state sees one of the highest volumes of both exports and imports pass through its ports, but unlike almost all the other top shipping states it has no primate port (table 10.1). Within the Tampa–Orlando corridor alone, there are four ports and four intermodal freight terminals, including the new one at Winter Haven. The relationship between ports, urbanization patterns, and urban politics in Florida, consequently, looks little like more typical monocentric port regions (Vormann, 2015). In Florida, the politics of logistics investment are both multiscalar and polycentric, and Polk County benefits disproportionately from both of these features.

As with the Pinal County case, albeit to a lesser extent, the strategic centrality of Polk County has developed in the face of relatively weak local governance capacity. Lakeland and Winter Haven – the two

Table 10.1 Port concentration in the top eight states by freight volume

State	Total Tonnage	Number of Ports	Share of Largest Port (%)	Index of Dispersion[a]
Texas	313,691	11	47.1	59,957
Louisiana	233,311	7	41.7	29,428
California	176,423	19	39.2[b]	32,471
New York and New Jersey[c]	84,467	13	83.0	52,489
Washington	67,340	14	24.1	5,816
Virginia	67,206	3	99.2	43,760
Florida	41,779	13	26.0	4,909
Georgia	31,331	2	90.5	10,277

a: The index of dispersion is a population's variance divided by its mean. The higher the number, the more individual observations are scattered; the lower the number, the more individual occurrences are clustered. In this case, the index is calculated on total tonnage of freight per port; a higher number means port activity is concentrated unevenly in a small number of ports, and a lower number means port activity is spread out relatively evenly. Thus the very low indices for Washington and Florida show that these two states lack the port concentration that is characteristic of the other states.
b: This figure treats the ports of Los Angeles and Long Beach as separate entities, despite their close proximity. Together, they account for 64.4 per cent of California's shipping.
c: New York and New Jersey are combined into one state here, because of the difficulty separating out economic activity within the New York metropolitan region. The data does not change significantly, since New Jersey has only a very small amount of port activity outside the New York region.

Source: Calculations based on raw data from American Association of Port Authorities, 2013

largest cities in the county – neither have strong independent economic leadership nor are they strongly integrated into another city's metropolitan economy. Instead, they exist as an in-between territory in the broader regional economy of the Tampa–Orlando corridor.

Discussion and Conclusions

Since 2010, after the onset of the Great Recession, a group of metropolitan planning organizations in the Phoenix–Tucson corridor has been collaborating (as the Joint Planning Advisory Council) to develop a new freight-led economic development strategy. The centrepiece of the strategy is an attempt to create a distributed "inland port" to capture logistics activity from the Los Angeles and Long Beach ports, and from Mexican maquiladoras. Pinal County has become the lynchpin in this strategy – the "cream filling in the Oreo cookie," as one informant described it – because it has readily developable land located between Phoenix and Tucson, with good connections to two interstate highways and the Union Pacific rail line. In Central Florida, Polk County has played a similar role in corridor-wide planning and development schemes. The distributed nature of Florida's ports means that the centrally located Polk County has been an attractive destination for new logistics initiatives, while corridor-wide transportation planning has by necessity aimed to integrate this in-between territory with the growth poles in Tampa and Orlando.

Alongside the key similarities between the Arizona and Florida cases – namely, the presence of an in-between territory that is a strategic asset yet a governance detriment in corridor-wide local development schemes – there is an important difference: the extent to which the multi-city infrastructure alliances in the two cases have successfully institutionalized. In Arizona, the initial informal collaboration of the Arizona Sun Corridor Partnership has spawned a set of increasingly formal governance institutions targeting the same territory. There are now Sun Corridor economic development, freight planning, water conservation, university consortium, renewable energy, and electric vehicle initiatives, albeit operating with a wide range of efficacy. This proliferation occurred, according to my respondents, in large measure because the Arizona state government was unable or unwilling to provide regionalism leadership, thus allowing various ground-up regionalism schemes to develop in a compatible fashion. In Florida, however, despite an initially strong collaboration between Tampa and Orlando growth coalitions, the partnership has largely stalled, and has not inspired other regionalism initiatives similar to those observed in Arizona. In this case, one important culprit has been the Florida state government's competing regionalism agenda, which has attempted to establish a statewide framework for regional economic development – but a framework that does not recognize the Tampa–Orlando corridor as a single entity. The result has been governance gridlock. This contrast suggests the importance of interactions between state governments and ground-up regionalisms in their jurisdictions (Wachsmuth, 2017a).

The concept of the in-between territory, and the underlying theorization of infrastructure alliances it rests upon, is meant to draw attention precisely to this type of interplay between the (multiscalar) built environment and the (multiscalar) political economy. At the same time as urbanization and economic change create new possibilities for local development strategy, they create new contradictions and tensions within the governance/built-environment interface that can be both managed and exploited. The in-between territory offers one promising way to understand the role of suburban space within these contradictions.

From a geographical perspective, chapter 10 has introduced the most extreme form of suburbanization discussed in this volume. The "in-between" suburban phenomenon described in the chapter is indeed beyond the zone of influence of any individual metropolitan region, rising as it does between the orbits of two such regions. This in-between place is the suburb in its most disconnected form – disconnected from individual metropolitan regions as it emerges in inter-metropolitan corridors in response to the needs for centres of freight distribution at strategic points along major transportation axes. Chapter 10 stands out within the volume in another way. The predominant perspective advanced by the case studies presented in the book's chapters is that infrastructure deficiencies are in large part a consequence of insufficient institutional capacity. In this sense, chapter 6 goes as far as attributing in large part the absence of sewerage in the Indian urban area it investigates to the absence of institutional structure with the capacity to put in place such an infrastructure. Chapter 10 suggests a reverse interpretation of the relation between infrastructures and institutions: it depicts the setting up of institutions emanating from the need to create the infrastructures required for the creation and operation of the new type of suburbs it discusses.

NOTE

1 Sieverts's *Zwischenstadt* is, most narrowly, a European form of suburbanization: "the type of built-up area that is between the old historical city centres and the open countryside, between the place as a living space and the non-places of movement" (Sieverts 2003: xi). Young and Keil's (2010) "in-between city," meanwhile, is the ambiguous urban space that lies between the city proper and the suburb proper – neither truly "city" nor "suburb."

REFERENCES

American Association of Port Authorities. (2013). *US waterborne foreign trade 2013 port ranking by cargo volume*. (Data file). Retrieved from http://aapa.files

.cms-plus.com/Statistics/2013%20U%20S%20%20PORTS%20WATERBORNE%20FOREIGN%20TRADE.xls

Brenner, N. (2009). Is there a politics of "urban" development? Reflections on the US case. In R. Dilworth (Ed.), *The city in American political development* (pp. 121–40). New York and London: Routledge.

Cox, K.R. (1993). The local and the global in the New Urban politics: A critical view. *Environment and Planning D: Society and Space*, *11*(4), 433–48. https://doi.org/10.1068/d110433

Eisinger, P.K. (1988). *The rise of the entrepreneurial state: State and local economic development policy in the United States*. Madison: University of Wisconsin Press.

Garreau, J. (1991). *Edge city: Life on the new frontier*. New York: Doubleday.

Gottmann, J. (1961). *Megalopolis: The urbanized northeastern seaboard of the United States*. New York: Twentieth Century Fund.

Graham, S., & Marvin, S. (2001). *Splintering urbanism: Networked infrastructures, technological mobilities and the urban condition*. New York and London: Routledge.

Harrison, J. (2010). Networks of connectivity, territorial fragmentation, uneven development: The new politics of city-regionalism. *Political Geography*, *29*(1), 17–27. https://doi.org/10.1016/j.polgeo.2009.12.002

Harvey, D. (2006). *The limits to capital*. London, New York: Verso.

Kirkpatrick, L.O., & Smith, M.P. (2011). The infrastructural limits to growth: Rethinking the urban growth machine in times of fiscal crisis. *International Journal of Urban and Regional Research*, *35*(3), 477–503. https://doi.org/10.1111/j.1468-2427.2011.01058.x

Lang, R.E., & Nelson, A.C. (2011, November). Megapolitan America. *Places Journal*. Retrieved from https://placesjournal.org/article/megapolitan-america/

Logan, J., & Molotch, H. (2007). *Urban fortunes: The political economy of place* (2nd ed.). Los Angeles and Berkeley: University of California Press.

MacLeod, G., & Jones, M. (2011). Re*newing* urban politics. *Urban Studies*, *48*(12), 2443–72. https://doi.org/10.1177/0042098011415717

Morrison Institute. (2008). *Megapolitan: Arizona's Sun Corridor*. (Policy report). Phoenix: Arizona State University.

Nelson, A.C., & Lang, R.E. (2011). *Megapolitan America: A new vision for understanding America's metropolitan geography*. Chicago: APA Planners Press.

O'Dowd, P. (2012, 29 October). In wake of NAFTA, states eye global supply chain with envy. *Fronteras: The Changing America Desk*. Retrieved from http://www.fronterasdesk.org/content/wake-nafta-states-eye-global-supply-chain-envy

Sieverts, T. (2003). *Cities without cities: An interpretation of the zwischenstadt*. London and New York: Spon Press.

Vormann, B. (2015). *Global port cities in North America: Urbanization processes and global production networks*. New York: Routledge.

Wachsmuth, D. (2014). Megaregions and the urban question: The new strategic terrain for US urban competitiveness. In J. Harrison & M. Holyer (Eds.), *Megaregions: Globalization's new urban form?* (pp. 51–74). Cheltenham, UK, and Northampton, MA: Edward Elgar.

Wachsmuth, D. (2017a). Competitive multi-city regionalism: Growth politics beyond the growth machine. *Regional Studies*, *51*(4), 643–53. https://doi.org/10.1080/00343404.2016.1223840

Wachsmuth, D. (2017b). Infrastructure alliances: Supply-chain expansion and multi-city growth coalitions. *Economic Geography*, *93*(1), 44–65. https://doi.org/10.1080/00130095.2016.1199263

Young, D., & Keil, R. (2010). Reconnecting the disconnected: The politics of infrastructure in the in-between city. *Cities*, *27*(2), 87–95. https://doi.org/10.1016/j.cities.2009.10.002

11 Recentralization and Green Infrastructures: Seeking Compatibility between Alternatives to North American Suburban Development

SARA SABOONIAN AND PIERRE FILION

Introduction

The dispersed suburban model, characterized by near universal automobile reliance, overall low density, and rigid land-use specialization, has been the dominant form of development across North America for the past seventy years. Virtually all North American suburban areas conform to this model. But over the last decades, this development pattern has become the target of a growing opposition. Suburban dispersion is criticized for its high infrastructure requirements, its adverse quality of life sequels (especially those related to the combination of a forced reliance on the automobile for nearly all journeys and chronically congested highway networks), and, perhaps above all, its damage to the environment. Two models have emerged as possible alternatives to dispersed suburbanism and, therefore, as solutions to the ills inflicted by dispersion. There is first *recentralization*, promoting the creation of nodes within the suburban landscape. Recentralization seeks a change in suburban form and transportation dynamics by creating a network of high-density, multifunctional, and pedestrian-hospitable nodes, connected with much improved public transit services. There is presently widespread planning adherence to the recentralization alternative, which relates to the transit-oriented development model and new urbanism. The other alternative to present suburban patterns focuses on *green infrastructures*. The objective of the green infrastructure model is not so much to restructure suburban land use and journeys as it is to abate the environmental impact of suburban infrastructures.

By focusing on certain infrastructures, the model proposes interventions that are more site specific than those associated with recentralization, concentrated as it is on a reorganization of the suburb.

Present debates around suburban development are fanned by tensions between two camps – one favouring a business-as-usual scenario and the other championing alternative models. But such an opposition overlooks differences and incompatibilities among alternatives to present suburban patterns. If both the recentralization and the green infrastructure models seek environmental objectives, in practice the implementation of one model can be incompatible with that of the other. For example, while recentralization is largely about raising density, green infrastructures are often important consumers of space. The chapter will concentrate on distinctions between these two alternatives to dispersion as well as on the form they take and their respective importance in planning nodal strategies meant to move away from current suburban patterns.

The empirical substance of the chapter originates from the Toronto metropolitan region. The suburban realm of Toronto combines the North American dispersed form and automobile dependence with a relatively strong planning capacity, from the local to the metropolitan scale. Since the 1970s, planning in the Toronto region has committed to recentralization, and interest for green infrastructure strategies has grown from the mid-1990s. The chapter focuses on two planned suburban centres developed at different times and thus reflecting the evolution of planning priorities. The first, Mississauga City Centre, dates from the late 1970s and is a clear emanation of a recentralization strategy. The second, Markham Centre, was conceived later and is still under development. While also driven by recentralization objectives, it exhibits more influence from the green infrastructure perspective. The chapter is about the evolution of suburban planning, the role it gives to the creation of nodes, and the possibility of amalgamating suburban recentralization with heightened reliance on green infrastructures. The chapter thus explores the actual and potential interface between planning, recentralization, and green infrastructures. It ponders the possibility of replacing suburban dispersion by forms of development that are less damaging to the environment. With the blending of recentralization and green infrastructures, the suburban realm could benefit from raised density, walking, and public transit use, while enjoying improved quality of life and environmental conditions stemming from the protection of natural features and enhancement of urban amenities.

Dispersion, Recentralization, and Green Infrastructures

Because the circumstances leading to the emergence of the dispersed, car-oriented North American suburb have been the object of numerous accounts, we confine ourselves here to a brief historical narrative (Hayden, 2003; Knox, 2008; Marshall, 2003). The dispersed suburb model was pieced together rapidly over the two decades that followed the Second World War (Filion, Bunting, & Warriner, 1999). After fifteen years of deprivation due to the Great Depression and the war, there was a thirst for consumption. Consumption over the post-war period was directed largely at durable goods, notably the automobile. These years witnessed skyrocketing car use accompanied by a rapid decline in the reliance on other forms of transportation, especially public transit. Responding to claims for additional road space in response to rapidly rising congestion, governments engaged in the construction of arterials and, with more impact on the structure of cities and suburbs, expressways. Generalized automobile use made it possible to consume more space in low-density environments (and thereby accumulate more goods) and to scatter employment, retailing, and institutions while enforcing strict land-use specialization. The outcome was a multitude of origins and destinations, distributed in zones that are often large, rendering any alternative to the automobile impractical. Taking place in a period of accelerated economic and demographic expansion (the birth of the baby boom), fuelled by the strong demand for durable goods, the dispersed suburb grew rapidly and soon became the dominant North American urban form. It is worth noting that dispersed suburbanization was not only a consequence of the Fordist consumer-driven economic expansion, but itself contributed to the virtuous economic cycle between consumption and production characteristic of Fordism.

Planning played a major part in the advent and generalization of suburban dispersion through its effect on decisions concerning transportation networks and the key role it played in land-use functional segregation. The accessibility and land-use specialization features of dispersed suburbanism congealed around the superblock formula, which structures the morphology and journey patterns of this form of development. Superblocks are delineated by supergrids made of arterials and/or expressways, which carry the brunt of suburban traffic. The interior of superblocks, whose internal road pattern often adopts a curvilinear configuration, attracts activities avoiding high traffic flows, notably housing, and other land uses for which there is no benefit to be close to heavy circulation. Meanwhile, edges of superblocks, which front on arterials and expressways, appeal to activities depending on visibility from and ready access to these high-capacity transportation infrastructures. The superblock/supergrid configuration accounts for the multiplicity of sites offering high levels of automobile accessibility, while providing the framework wherein land-use separation takes place. Expressways and arterials indeed play the role of major boundaries separating land uses in the dispersed suburb. The superblock/supergrid configuration is therefore a factor in the dearth of multifunctionality, especially multifunctional centres, in suburban environments. Heavy reliance on the car and land-use specialization fostered by regulations, along with the superblock/supergrid formula, favours dispersion rather than the concentration of activities within centres. The superblock/supergrid configuration can be seen as the DNA of the dispersed suburb, easing its ongoing replication.

The dispersed suburb model has come under growing criticism, targeted for its development expenses, affordability, quality of life shortcomings, and environmental repercussions. Concern about development costs has to do mainly with the large financial burden of laying down infrastructures in low-density areas (Speir & Stephenson, 2002). Given the near universal reliance on the automobile, investment in the building and maintenance of roads must be substantial (Burchell, Downs, McCann, & Mukherji, 2005; Cervero, 2013). Also related to the financial aspects of dispersed suburbanism are affordability issues stemming from the unsuitability of a development formula, devised at a time when the public purse was flush and the economy and a consumerist middle class were both expanding at an accelerated pace, to the present fiscal stringency, economic stagnation, and income polarization (Abramowitz & Teixeira, 2009). Another source of criticism directed at the dispersed suburb has to do with its deficient urban amenities, street life in particular, and its overall quality of life deficiencies such as those caused by traffic congestion. Among further problems associated with this form of suburb are sedentary lifestyles resulting from wide-scale dependence on the car. But it is likely environmental considerations that figure most prominently among blames aimed at the dispersed suburb model. There are many facets to these considerations. They include the absorption of vast tracts of rural and natural land due to urban sprawl, deteriorated air and water quality, excessive energy consumption resulting from the dispersed suburb and its journey patterns, and the emission of large quantities of greenhouse gases. These reactions to dispersion have launched a search for alternative development models that could address identified flaws (Flint, 2006; Soule, 2006).

Recentralization is the dominant alternative to the dispersed suburb aired in planning documents, especially those with a metropolitan-wide focus (Filion, Kramer, & Sands, 2016). This alternative model consists of the creation of suburban nodes, which are multifunctional, dense, and walking and public transit friendly. Recentralization strategies also entail a substantial improvement of public transit services, generally in the form of new light rail transit (LRT) and bus rapid transit (BRT) services. The objective is to deliver a more compact urban form, which is less reliant on the automobile. According to the proponents of recentralization, by virtue of its higher density such a form would require less horizontal infrastructures and thus be less costly to the public sector. It would also help reduce congestion, thereby improving quality of life and economic performance, and provide a setting that promotes active transportation and affords lifestyle choices – a suburban automobile-reliant or a more urbane walking and public transit focused lifestyle. To function as intended, recentralization must generate synergy effects between the different activities constituting nodes. These effects are essential to the integration of the different functions of nodes and thus to the capacity of these districts to attract activities, a condition for their very existence. Note that in nodes, just as in the case of traditional downtowns, synergy relies on walking, hence the importance of an attractive pedestrian environment therein (Ewing, Hajrassouliha, Neckerman, Purcel-Hill, & Greene, 2015; Forsyth, Oakes, Schmitz, & Hearst, 2007; Ryan & Frank, 2009; Southworth, 2005).

The different features of recentralization are in direct opposition to dispersion. While dispersion is the outcome of an adaptation of suburban land use to a near total dependence on the car, recentralization attempts to reduce reliance on the automobile by encouraging public transit use and walking. Likewise, recentralization proposes multifunctionality in contrast with the land-use specialization inherent in dispersed suburbanism. Finally, it brings intensification to the otherwise low-density dispersed suburban realm. Essentially, recentralization is about the "urbanization" of the suburb, returning to the density and public transit orientation of pre–Second World War development patterns. Recentralization is consistent with the Green Manhattan perspective, which highlights the environmental benefits of density, mixed use, walking, and public transit by demonstrating that the per capita environmental load of Manhattan residents is substantially lower than the US average (Owen, 2009).

Across North America virtually all plans with a metropolitan focus espouse the recentralization perspective, albeit with varying degrees of commitment (Filion, Kramer, & Sands, 2016). A danger with the implementation of this model is that attention could be given exclusively to those aspects of recentralization that are easiest to achieve, the low hanging fruits. The outcome could be higher density sprawling development, with little change to prevailing modal shares, causing traffic congestion in intensified areas. To achieve its objectives, recentralization must be implemented in a systematic way. All its aspects must be coordinated. For example, intensification and public transit improvement must take place at the same time, as must multifunctionality and the creation of walking-conducive environments. Implementing one aspect of recentralization in isolation will do little to induce the sought-after departure from suburban dispersion.

There is another response to the environmental impacts of urban and suburban development: the green infrastructure model. Green infrastructures concern substituting natural processes for heavily engineered physical infrastructures, that is, for example, replacing as much as possible concrete and asphalt with green space to fulfil functions previously devoted to conventional infrastructures, with improved environmental outcomes. Green infrastructures mimic and copy the structures and functions of natural systems (Benedict & McMahon, 2006; Spatari, Yu, & Montalto, 2011). Water management in suburban areas illustrates the positive impacts of green infrastructures. A shift from conventional to green infrastructures involves daylighting streams and safeguarding their riparian zones. Such a strategy, when combined with the replacement of large impervious surfaces, reduces the occurrence and severity of flash floods while improving stream and storm water quality. The green infrastructure model also calls for the replacement of more by less environmentally damaging technologies and processes at different scales of urban and suburban settings (Gill, Handley, Ennos, & Pauleit, 2007; Young, 2009). For example, at the scale of specific buildings, the switch to green infrastructures can take the form of solar panels, green roofs, and indoor gardens, thereby lessening reliance on electricity generated by power plants, reducing water runoff and heat island effects, and improving interior air quality. Green infrastructures are responses to specific environmental issues, which often invoke local reactions, as in the case of point source water pollution. But green infrastructure strategies can also be driven by the cumulative effects of local practices, such as regional air and water quality issues and global effects of urbanization and suburbanization, most notably their contribution to the atmospheric accumulation of greenhouse gases (Demuzere et al., 2014).

There is a great deal of agreement between objectives pursued by the recentralization and green infrastructure approaches to suburban change. They both seek to improve the environmental outcomes of suburban development. It is the means they use to achieve this objective that sets them apart. Recentralization concerns the manipulation of density and transportation to reach metropolitan-scale objectives such as compact development, a shift in modal shares, and a reconfiguration of the urban structure. It thus involves a coordination of planning interventions to attain its goals. In contrast to the dependence of recentralization objectives on a careful coordination and sequencing of multiple planning interventions taking place at different scales, the green infrastructure movement can rely on specific interventions, which can take place at different scales and be independent from each other. More so than recentralization it can operate in an ad hoc fashion, outside the purview of metropolitan strategies (Firehock, 2015). It can address specific environmental problems, whether they be at the scale of a building; a particular urban feature such as a parking lot, square, or street; an urban district, as in the case of the naturalization of streams; or infrastructures strung across an entire metropolitan region. Green infrastructures can achieve positive environmental effects without the benefit of an elaborate synchronization of multiscale planning interventions required by recentralization.

All of this begs the question of the extent to which green infrastructures are compatible with recentralization. Can the abundant space required by some green infrastructures substituting natural processes to conventional infrastructures be harmonized with the pursuit of density and multifunctionality at the heart of the recentralization model? Are these two approaches to the mitigation of the environmental impacts of the dispersed suburb incompatible or can they be fused into a more comprehensive and effective alternative to dispersion? On the one hand, to operate as intended by re-urbanizing dispersed suburbs through the creation of high-density, multifunctional nodes, recentralization comes with its own assortment of adverse local environmental sequels. These relate to side effects of density such as heat islands, flash flooding, water quality deterioration, air pollution (especially concentrations of fine particles), and a limited presence of green space (Haaland & van den Bosch, 2015). On the other hand, green infrastructures can interfere with the density and synergy at the heart of the recentralization model. The protection of wetlands, riparian zones, and woodlots can impede the close-by clustering of activities required to trigger pedestrian-based synergy effects. Green infrastructures can also puncture continuous retail façades conducive to pedestrian movements essential to synergy effects. By opposing the raising of density and the achievement of urban-like environments within the dispersed suburb, on the one hand, and the preservation of natural areas and processes, on the other, the tension between recentralization and green infrastructures is not without recalling debates between new urbanism and landscape urbanism (Duany & Talen, 2013; Waldheim, 2016). Indeed, the compact development and pedestrian-conducive settings recentralization promotes are consistent with major tenets of new urbanism, while the preservation of natural features and reliance on natural processes to reduce the environmental impact of urban and suburban development advocated by the green infrastructure movement chime with landscape urbanism.

We now turn to the history of two Toronto region suburban nodes, developed at different periods, in order to examine the respective influence of recentralization and green infrastructures on their planning and development. We also draw from these examples to explore the possibility of amalgamating the two models in the search for new alternatives to suburban dispersion.

The Toronto Context and the Selection of Case Studies

Since the Second World War, Toronto region has been among the fastest growing North American metropolitan areas. It thus follows that much of the region has been developed according to dispersed suburban standards, which have predominated over the past seventy years. Yet, Toronto has deviated from the continental norm by investing in a subway in the immediate post-war years, which has contributed to foster a dual-realm metropolitan form: (1) a concentrated realm, characterized by relatively high density and elevated public transit, including the downtown, the inner city, and areas along the subway network; and (2) a dispersed realm, which embraces most of the suburban portion of the metropolitan region (Filion, 2000). Another distinctive feature of the Toronto metropolitan area stems from planning efforts, especially from the mid-1950s to the mid-1970s, at peppering the suburban landscape with sectors of high residential density. These efforts yielded a higher density suburb than the continental average, but without the continuous density patterns and multifunctionality that could have challenged suburban dispersion.

From the mid-1970s, planning adopted a suburban node strategy. Two planning agencies, with different territories, engaged in such a strategy. There was Metro Toronto, a second-tier government, whose territory coincided with the present-day City of Toronto (which was amalgamated in 1997). The Metro suburban node strategy, which became a mainstay of its 1980 Official Plan, aimed at preventing an overconcentration of employment in the downtown area that would overburden transportation infrastructures. The plan proposed the creation of nine suburban centres, seven secondary centres, and two major centres. The two major centres, North York Centre and Scarborough Town Centre, have developed largely according to plan (which was not the case for the secondary centres). Meanwhile, by the late 1960s the City of Mississauga, the largest Toronto suburban municipality, had already prepared a vision for the development on a greenfield site of the Mississauga City Centre, depicted as a downtown for this municipality (McLaughlin Group, 1969). The creation of a centre in Mississauga was in large part meant to attract land uses, in particular category A offices, that would not otherwise opt for a suburban location.

In the late 1980s, the newly established provincial government Office for the Greater Toronto Area (OGTA) commissioned a study on the future urban structure of the Toronto metropolitan region. The consulting firm proposed a nodal structure where centres would be interconnected by rapid and frequent public transit links (IBI Group, 1990). The nodal structure was incorporated in the OGTA plans, which proposed twenty-three centres (Office for the Greater Toronto Area [OGTA], 1991). However, bereft of actual planning power, the OGTA was unable to implement its multinodal strategy. In the mid-1990s, this agency was dismantled by a right-wing neoliberal provincial administration. Still, the Metro and Mississauga experience with suburban centres and the OGTA proposals motivated some suburban municipalities to create their own centres. One of these municipalities was Markham, which began planning the Markham Centre on land that was undeveloped but highly accessible.

A region-wide nodal strategy became official provincial policy when the Growth Plan for the Greater Golden Horseshoe was given force of law by the provincial legislature in 2006. As was the case with the OGTA plan, the Growth Plan identifies nodes (twenty-five urban growth centres), which are distributed across the metropolitan region and are to be connected to high-order public transit. But the province now has the power to compel municipal administrations to develop their urban growth centres in a way that respects density thresholds specified in the Growth Plan. Recentralization within the Toronto metropolitan region is now official policy. The Growth Plan is probably the most ambitious metropolitan-scale recentralization effort in North America.

This discussion is not, however, to suggest an absence of interest on the part of Toronto region in plans for green infrastructures. Two plans stand out by the exceptional attention they gave to green infrastructures. There is first the 1992 Toronto Waterfront Plan called Regeneration, which was prepared under the chairmanship of former Toronto mayor David Crombie (Royal Commission on the Future of the Toronto Waterfront, 1992). The plan intentionally overstepped its mandate by taking a water quality perspective on the planning of the entire Toronto region watershed rather than focusing exclusively on the Lake Ontario waterfront as originally intended. The focus of the plan was squarely on the reliance on natural processes and the need for coordinated environmental measures at the scale of the watershed. Reliance on green infrastructures was also stressed by York Region (the upper-tier municipality where Markham Centre, one of our two cases, is located). The York Region Official Plan, adopted in 1994, was committed to the conservation of natural areas throughout the region (York Region, 1994). A mainstay of this plan was indeed the preservation of much of the region's land in its natural or rural condition, which would result, among other things, in linear natural areas crisscrossing the region.

The Mississauga City Centre and the Markham Centre cases were selected because of their similarities and distinctions. The two centres were planned by their local and regional municipalities to become high-density multifunctional nodes. They are both designated as urban growth centres by the Growth Plan. But the planning of Mississauga City Centre began earlier than that of Markham Centre, which accounts for some of the differences between the two nodes. Mississauga City Centre was planned around an existing regional mall and its large parking lot. It is important to realize that until recently creating a planned multifunctional suburban centre was a new form of planning venture, which came with little prior knowledge. The early planning of Mississauga City Centre therefore involved an adaptation of the planning principles of the time to the novel purpose of creating a suburban centre. The situation was different in the case of Markham Centre, whose planning is more recent and whose development takes place on a greenfield site. Markham Centre was more likely than Mississauga City Centre to feel the influence of the green infrastructure movement, given the relatively recent emergence of this perspective and the fact that Markham is located in York Region, a champion of green infrastructures.

Case Study Methods

The mixed methods approach to investigating Mississauga City Centre and Markham Centre included a survey of planning documents related to the two nodes, and historical narratives documenting their evolution; direct observation; and twelve interviews with planners and other municipal and regional officials involved in the planning of the two suburban centres. Respondents were asked about the planning history of the nodes, their present condition, problems arising from this condition, their views on the Toronto region recentralization strategy as well as on green infrastructures and their present application, and their perspectives on the future of green infrastructures within the two nodes. For confidentiality purposes, quoted interviewees are identified as either "Mississauga respondent" or "Markham respondent." Each quote in the chapter is attributed to a different interviewee.

Mississauga City Centre

Morphologically, Mississauga City Centre is heavily influenced by its legacy. The centre indeed reflects the enduring effects of the planning principles that guided its early development. The core of the centre, occupied by a large regional mall (opened in 1973) and its extensive parking area, conforms to conventional suburban retailing configurations. The mall and its parking are surrounded by functionally specialized sectors occupied by offices, a civic centre, high-density residential developments, and a public transit terminal. Mississauga City Centre is traversed by high-capacity arterials and allocates a high proportion of its space to parking. In consequence, the environment of the centre is largely hostile to pedestrians. With the exception of the walking-friendly settings provided by the indoor corridors of the mall and the large civic centre plaza, pedestrians in Mississauga City Centre must negotiate the sidewalks of wide arterials bordered by parking lots and set-back building façades with few ground-level amenities. Thus, pedestrian movement in Mississauga City Centre is infrequent. Interconnection between different types of activities is impaired by these conditions, which deprive the centre of a pedestrian-based synergy, a foremost potential competitive advantage of nodes relative to other suburban locales.

The functionally specialized morphology of Mississauga City Centre, its enduring reliance on the car despite rising density, the presence of large arterials and ample surface parking space, and its resulting inhospitable pedestrian environment are major concerns for Mississauga planners. Since the mid-1990s, planning objectives have attempted to address the legacy of Mississauga City Centre's early development. A respondent described current attempts at transforming the centre in these terms:

> In the past, it started as farmers' fields. We had cows ... Our City Centre is kind of odd ... We have this big shopping centre and we are converting it into an urban centre ... The City Centre is turning into a people's place with the projects like high-rises and Celebration Square [in front of City Hall in the Civic Centre part of Mississauga City Centre]. Celebration Square attracts a lot of people and events, especially in the summer. (Mississauga respondent)

The principles presently guiding Mississauga City Centre planning are stated as follows in the Downtown Core Local Area Plan:

> The objective is to create a high quality, pedestrian friendly, human scaled environment that is a meaningful place for all citizens and also continues to attract lasting public and private investment in the downtown to support existing and planned infrastructure, particularly high-order transit. (Mississauga, 2015: 2)

The goal is to transform the centre into a pedestrian-conducive environment rich in urban amenities. Much of the attention is directed at the creation of avenues with continuous store façades and the deemphasizing of the predominant place taken by the car.

> The City has adopted a complete streets philosophy when designing and redesigning Mississauga City Centre streets, where pedestrians, cyclists and transit users will be promoted as "first" users. (Mississauga respondent)

In the same vein, plans propose the introduction of a small block grid, which lends itself better to walking than does the present Mississauga

City Centre large block layout (for example, see Mississauga, 1994). The most ambitious transformation advanced in plans involves the replacement of the shopping mall parking lot by small blocks containing multifunctional buildings with ground-level retailing aligned to sidewalks. The redeveloped parking space is to be named the "Main Street District." In the Downtown 21 Master Plan, a consultant report that encapsulates current Mississauga City Centre planning objectives, the Main Street District is presented in these terms:

> The short-term creation of the Main Street District is a crucial final step in the development of the larger downtown area. It is seen as a vital new mixed-use precinct, close to office development sites, which will incubate the transformation of the massive parking lots surrounding one of North America's major shopping centres into the nucleus of a walkable, attractive downtown community. (Glatting Jackson Kercher Anglin, 2010: 4)

Another preoccupation voiced in Mississauga City Centre plans is the difficulty since the early 1990s to attract new office space. The Downtown 21 Master Plan states:

> While some office growth has occurred in the past, it has stalled. Downtown Mississauga cannot compete with lower cost suburban locations and, currently, downtown locations do not offer tenant amenities to offset the cost differential. (Glatting Jackson Kercher Anglin, 2010: 4)

A major factor explaining the dearth of office development in Mississauga City Centre, as in other suburban centres, is that in a dispersed suburban context new office facilities prefer cheaper office park locales, with abundant room for surface parking, to more expensive and spatially constricted nodal sites (BA Group Transportation Consultants, 2009: 53; Lang, 2003; Weitz & Crawford, 2012). While Mississauga City Centre has experienced substantial retailing and residential development, evidenced by successive mall expansions and a flurry of new condo towers, the stalling of office growth jeopardizes the multifunctionality of this node. Concern about employment in Mississauga is reflected in the attention plans give to the luring of new office buildings.

Current Mississauga City Centre planning priorities mirror difficulties in achieving recentralization. In large part the centre has been successfully intensified but, as seen, is facing problems in achieving multifunctionality and in departing from the journey dynamics of the dispersed suburb. The nature of present planning proposals is consistent with these failures to provide an alternative to dispersion. These proposals, indeed, zero in on economic development measures targeted at offices, urban design interventions to improve pedestrian conduciveness, the redevelopment of parking lots, and the improvement of public transit in large part by constructing a north-south LRT and an east-west BRT, both traversing the centre. Planning interventions pertaining to infrastructures are therefore primarily aimed at achieving recentralization-related objectives.

In these circumstances, consideration given to green infrastructures is relegated to a distant backburner. Of six Mississauga City Centre planning principles listed in the Downtown 21 Master Plan, only one principle is defined specifically in environmental terms and it is ranked fourth (Glatting Jackson Kercher Anglin, 2010: 5–6). This low priority accorded to the environment is not because Mississauga City Centre has no need for a greening of its infrastructures. Notwithstanding the preservation of a creek and its riparian zone located at the eastern edge of the centre, there is little green space and an absence of reliance on natural means to reduce runoff and protect water quality within Mississauga City Centre. The surface of the centre is mostly given to buildings, roads, and parking space, precisely the type of landscape that produces the environmental sequels green infrastructures attempt to remedy.

Markham City Centre

Two major distinctions in the early planning of Markham Centre account for differences between this node and Mississauga City Centre. For one, the site selected for Markham Centre was devoid of any development, allowing the centre to begin with a clean state. For another, the 1990s, when the guidelines for this new centre were laid down, witnessed a wave of interest within the planning community for green infrastructures. As seen, such interest was especially pronounced in York Region. Therefore, from the onset of its planning process, a large proportion of the Markham Centre site was protected from development. Areas thereby safeguarded included wetlands, woodlots, and a branch of the Rouge River. Overall, 192 acres of natural land is to be preserved in Markham Centre. There was particular concern about the quality of the Rouge River water so it could support natural life as the river flows downstream through a provincial park designated in 1995 (Markham, 2014: 3.20).

A respondent from Markham explained the nature of the preoccupation about Rouge River water quality and described measures taken to protect this quality:

> We wanted to make sure that we were looking at state of the art methods to prevent downstream impacts. These included reducing water runoff, greywater, storing the water ... So there is a number of strategies devised with the Toronto Region Conservation Authority [responsible for the watersheds of rivers flowing within the Toronto metropolitan region] to try to make sure that we are not negatively impacting water quality. It's very close to the Rouge River and there is a lot of sensitivity around what happens in Markham Centre and to the Rouge River itself. The expectation is that to a large extent, the valley is kept in as a natural a condition as possible and that programming is minimized in the valley. There are some storm ponds that are being naturalized in that area and there are some natural systems we are looking forward to protect. (Markham respondent)

In contrast with the situation observed in Mississauga City Centre, among the eleven principles guiding the planning of Markham City Centre, the protection and enhancement of the Rouge River Valley is at the very top of the list. The principle is formulated as follows:

> The Rouge River and Beaver Creek waterway is a powerful influence on Markham Centre. The downtown core will be designed to protect and enhance this natural environmental system. (Markham, n.d.)

Interest in green infrastructures did not eclipse efforts at achieving recentralization. Like Mississauga City Centre, Markham Centre is planned as a high-density, mixed-use district reliant on walking and public transit use. Just as Mississauga City Centre counts on the combination of a new LRT and BRT to raise public transit modal shares, Markham Centre relies on a nearby commuter rail station and a BRT system running through the node. The influence of the recentralization concept on Markham Centre planning can also be seen as a manifestation of the long-held City of Markham interest in suburban planning innovations, notably new urbanism. According to plans, the core of Markham Centre will adhere to the tenets of recentralization by mixing different uses – office, residential, and retailing – in an area designed with pedestrians in mind. At ground level, activities will respect the rhythm of walking. Moreover, surface parking will be banned from the core parts of Markham Centre. It is thus well positioned to avoid the shortcomings stemming from the original mall focus of Mississauga City Centre and the effects on this centre of the lingering tendencies of 1970s and early 1980s planning, such as big blocks and the accommodation of large numbers of cars.

But along with a proposed recentralization-inspired high-density district, the morphology of Markham Centre includes a large natural space component embodying green infrastructure principles in vogue since the 1990s. The Markham Centre approach to green infrastructures consists primarily of large expanses of land given to the preservation of natural features, especially those relating to water. This approach thus fits within a land-use specialization planning framework. Despite strong commitment in plans to multifunctionality, portions of Markham Centre that are presently in place are functionally specialized: large preserved natural areas, a medium-density residential neighbourhood inspired by new urbanism, and a head office campus-like development. The mixed-use component of Markham Centre has proven difficult to get off the ground, doubtless in part because of the aforementioned limited interest of suburban office developers for nodal-type settings. Retailing is also a problem because the layout of Markham Centre is suited neither to large regional mall format stores nor to big-box facilities, presently two of the most popular retail configurations (Basker, Klimek & Van, 2012; Wrigley & Lowe, 2014). Therefore, in its present state, the Markham Centre model establishes a clear division between sectors that accommodate green space and those that host development.

Commitment to green infrastructures did ooze into the planning of the built-up portions of Markham Centre. In this case, however, it took the form of the environmental standards to which buildings and other aspects of Markham Centre are expected to adhere, not of the insertion of natural space within the high-density portion of the node. Here is how a respondent from Markham explained the practicality of relying on Leadership in Energy and Environmental Design (LEED) standards for Markham Centre developments:

> So, all of a sudden LEED became the sustainability initiative that we were looking for in Markham Centre. Because it was pretty consistent with our Performance Measures document, it was something that was accepted in the community and is used in a broader practice. It was much easier in terms of having the consultants, through the LEED process, certify that developments met these criteria. (Markham respondent)

Green Infrastructures and Recentralization in a Suburban Context

The case studies have exposed difficulties in creating nodes that accommodate green infrastructures while adhering to recentralization principles. The Mississauga and Markham cases confirmed the dominant status of recentralization among alternatives to the dispersed suburb, even if Markham gave more importance to green infrastructures than Mississauga did. Our case studies thus corroborate findings from research on metropolitan-scale planning documents. But these cases have also brought to light serious difficulties in achieving recentralization. For Mississauga City Centre, issues are about departing from the car-oriented and functionally specialized nature of the centre, while Markham experiences long delays in launching the multifunctional residential, administrative, and retail heart of its node. These difficulties can be attributed to the planning models that prevailed when nodes were first conceived, the enduring influence of dispersion affecting land use and journey dynamics within these centres (reflecting the insertion of centres within the superblock/supergrid suburban structure), and the predominance of the car in journeys tying centres to their suburban surroundings. Identified problems also concern the need for nodes to compete with the dispersed suburban environment to attract office employment.

Meanwhile, awareness of the benefits of green infrastructures is on the rise as evidenced by the attention the concept receives in the literature and the influence it has had on the planning of Markham Centre. But it is the obstacles confronting recentralization that have monopolized the attention of planners. As planning struggles with the implementation of recentralization, its attention is diverted away from other ways of addressing issues arising from dispersed suburbanism. Dealing with the difficult achievement of recentralization thus takes place at the expense of green infrastructures. Another recentralization-related obstacle to green infrastructures concerns tensions between the two alternatives to dispersion. These tensions explain the absence of the type of green infrastructures involving green space and surface water systems within the portions of nodes that correspond, or aspire, to recentralization-compatible forms and dynamics. Neither of the cases pointed to a successful blending at this scale of these two alternatives to dispersion. To be sure, as planners become increasingly aware of the environmental benefits of green infrastructures, we can expect attempts to integrate them to the portions of nodes that are densest and multifunctional. But in the absence of a model capable of blending recentralization objectives with the land-use thirsty forms of green infrastructures, we can expect green infrastructure policies in the core parts of nodes to concentrate on LEED building standards rather than on the preservation or restauration of natural land, rivers, streams, and other forms of wetlands.

We can speculate about the form a node combining green infrastructures and recentralization could take. At a coarse grain, it could conform to the Markham Centre model: a separation of large sectors according to whether they are devoted to natural areas or to a compact built environment. According to this model, built structures would be required to avail themselves of up-to-date environmental innovations, but land use in the built-up parts of these centres would be left unmarked by the presence of space-consuming green infrastructures. What about a fine grain coalescence of the two alternative models? It would involve marrying the presence of natural areas with high-density morphologies in a way that impedes neither the compactness of a node nor pedestrian connectivity between its different activities. Centres could be planned around preserved natural areas in a fashion that would allow the incursion of these areas within the built environment without interfering with pedestrian connectivity (Newell et al., 2013; Yngve, 2016). Such a blending would require dexterous context-specific planning. Another approach could take the form of multilevel corridors where pedestrian-hospitable settings would be designed above natural areas, or under such areas were they to be elevated as in the case of the New York High Line. Finally, it is important to note that pedestrian connectivity can better accommodate itself to natural green space and wetlands than to high traffic arterials and parking lots (Jorgensen & Gobster, 2010).

Conclusion

In large part, the chapter has addressed the transformative role of infrastructures in suburbs. It has identified two models proposing alternatives to suburban dispersion, the prevailing form of development in North America. The planning means relied upon to implement these models mix regulations and infrastructures. The chapter has emphasized the role of infrastructures, which likely reflects the uneven impact of these two categories of instruments. It is thus about infrastructures as transformative tools.

The chapter has zeroed in on the tension between the two alternative models, which makes it more difficult for them to challenge dispersed suburbanism than if we were in the presence of one comprehensive alternative model. Much of this tension has to do with differences in the nature of, and incompatibilities between, infrastructures associated with these models. Recentralization relies on a metropolitan-scale coordination of transportation networks. We have also seen that at a nodal level, recentralization attempts to replace fully car-focused transportation systems, typical of suburban dispersion, with pedestrian friendly infrastructures, such as complete streets and small blocks. The green infrastructure perspective gives less importance to metropolitan-wide infrastructures, notwithstanding naturalized water systems, which can exist at a watershed (that is, regional) scale. Most green infrastructures take the form of site-specific interventions, often independent of each other. The main incompatibility between the two categories of infrastructures relates to the tendency for recentralization-related infrastructures to promote density and pedestrian movement in a compact setting, while green infrastructures can be high consumers of space and thus hinder density and pedestrian-based synergy. There is still no model on the horizon capable of amalgamating recentralization and green infrastructures in a fashion that blends the promotion of density, multifunctionality, and transit and pedestrian conduciveness with measures capable of mitigating the adverse environmental impacts of compactness.

By exploring the relation between recentralization and green infrastructures, chapter 11 introduced the volume to the two main planning alternatives to the predominant form of North American suburbanism. In this fashion, the chapter observes the North American suburban phenomenon through the lens of present debates about how to retrofit the suburb. It opens the door on a possible future of the North American suburb – a retrofitting of the suburb to make it more sustainable. But the chapter also identified difficulties in interweaving discordant planning strategies aiming at delivering a more sustainable form of suburb.

REFERENCES

Abramowitz, A., & Teixeira, R. (2009). The decline of the white working class and the rise of a mass upper-middle class. *Political Science Quarterly*, *124*(3), 391–422. https://doi.org/10.1002/j.1538-165x.2009.tb00653.x

BA Group Transportation Consultants. (2009). *Parking strategy for Mississauga City Centre, final report*. Toronto. ON: Author.

Basker, E., Klimek, S., & Van, P.H. (2012). Supersize it: The growth of retail chains and the rise of the "big-box" store. *Journal of Economic Management Strategy, 21*(3), 541–82. https://doi.org/10.1111/j.1530-9134.2012.00339.x

Benedict, M.A., & McMahon, E.T. (2006). *Green infrastructure: Linking landscapes and communities*. Washington, DC: Island Press.

Burchell, R.D., Downs, A., McCann, B., & Mukherji, S. (2005). *Sprawl costs: Economic impacts of unchecked development*. Washington, DC: Island Press.

Cervero, R. (2013). *Suburban gridlock* (new ed.). New Brunswick, NJ: Transaction Publishers.

Demuzere, M., Orru, K., Heidrich, O., Olazabal, E., Geneletti, D., Orru, H., ... Faehnle, M. (2014). Mitigating and adapting to climate change: Multifunctional and multi-scale assessment of green urban infrastructure. *Journal of Environmental Management, 146*, 107–15. https://doi.org/10.1016/j.jenvman.2014.07.025

Duany, A., & Talen, E. (Eds.). (2013). *Landscape urbanism and its discontents: Dissimulating the sustainable city*. Gabriola Island, BC: New Society Publishers.

Ewing, R., Hajrassouliha, A., Neckerman, K.M., Purcel-Hill, M., & Greene, W. (2016). Streetscape features related to pedestrian activity. *Journal of Planning Education and Research, 36*(1), 5–15. https://doi.org/10.1177/0739456x15591585

Filion, P. (2000). Balancing concentration and dispersion? Public policy and urban structure in Toronto. *Environment and Planning C: Government and Policy, 18*(2), 163–89. https://doi.org/10.1068/c2m

Filion, P., Bunting, T., & Warriner, K. (1999). The entrenchment of urban dispersion: Residential location patterns and preferences in the dispersed city. *Urban Studies, 36*(8), 1317–47. https://doi.org/10.1080/0042098993015

Filion, P., Kramer, A., & Sands, G. (2016). Recentralization as an alternative to urban dispersion: Transformative planning in a neoliberal context. *International Journal of Urban and Regional Research, 40*(3), 658–78. https://doi.org/10.1111/1468-2427.12374

Firehock, K. (2015). *Strategic green infrastructure planning: A multi-scale approach*. Washington, DC: Island Press.

Flint, A. (2006). *This land: The battle over sprawl and the future of America*. Baltimore, MD: Johns Hopkins University Press.

Forsyth, A., Oakes, J.M., Schmitz, K.H., & Hearst, M. (2007). Does residential density increase walking and other physical activity? *Urban Studies, 44*(4), 679–97. https://doi.org/10.1080/00420980601184729

Gill, S.E., Handley, J.F., Ennos, A.R., & Pauleit, S. (2007). Adapting cities for climate change: The role of the green infrastructure. *Built Environment*, *33*(1), 115–33. https://doi.org/10.2148/benv.33.1.115

Glatting Jackson Kercher Anglin. (2010). *Downtown 21 master plan: Creating an urban place in the heart of Mississauga*. Mississauga, ON: City of Mississauga.

Haaland, C., & van den Bosch, C.K. (2015). Challenges and strategies for urban green-space planning in cities undergoing densification: A review. *Urban Forestry and Urban Greening*, *14*(4), 760–71. https://doi.org/10.1016/j.ufug.2015.07.009

Hayden, D. (2003). *Building suburbia: Green fields and urban growth, 1820–2000*. New York: Vintage Books.

IBI Group. (1990). *Greater Toronto Area urban structure concepts study: Background report no. 1, description of urban structure concepts*. (Prepared for the Greater Toronto Coordinating Committee). Toronto. ON: Author.

Jorgensen, A., & Gobster, P.H. (2010). Shades of green: Measuring the ecology of urban green space in the context of human health and well-being. *Nature and Culture*, *5*(3), 338–63. https://doi.org/10.3167/nc.2010.050307

Knox, P.L. (2008). *Metroburbia, USA*. New Brunswick, NJ: Rutgers University Press.

Lang, R.E. (2003). *Edgeless cities: Exploring the elusive metropolis*. Washington, DC: Brookings Institution Press.

Markham (City of). (2014). *Official plan*. Markham, ON: Author.

Markham (City of) (n.d.). Eleven guiding principles: The vision for Markham Centre is based on guiding principles. Markham, ON: Author. Retrieved from https://www.markham.ca/wps/portal/Markham/BusinessDevelopment/MarkhamCentre/TheMarkhamCentreStory/PlanningDevelopment/GuidingPrinciples

Marshall, A. (2003). *How cities work: Suburbs, sprawl, and the roads not taken*. Austin, TX: University of Texas Press.

McLaughlin Group. (1969). *Mississauga*. Port Credit, ON: Author.

Mississauga (City of). (1994). *City centre secondary plan, official plan amendment 90*. Mississauga, ON: Author.

Mississauga (City of). (2015). *Core local area plan*. Mississauga, ON: Author.

Newell, J.P., Seymour, M., Yee, T., Renteria, J., Longcore, T., Wolch, J.R., & Shishkovsky, A. (2013). Green alley programs: Planning for a sustainable urban infrastructures? *Cities*, *31*, 144–55. https://doi.org/10.1016/j.cities.2012.07.004

Office for the Greater Toronto Area (OGTA). (1991). *Growing together: Towards an urban consensus in the Greater Toronto Area*. Toronto, ON: Author.

Owen, D. (2009). *Green metropolis: Why living smaller, living closer, and driving less are key to sustainability*. New York: Riverhead Books.

Royal Commission on the Future of the Toronto Waterfront. (1992). *Regeneration: Toronto's waterfront and the sustainable city: Final report*. Toronto, ON: Queen's Printer of Ontario.

Ryan, S., & Frank, L.F. (2009). Pedestrian environments and transit ridership. *Journal of Public Transportation*, *12*(1), 39–57. https://doi.org/10.5038/2375-0901.12.1.3

Soule, D.C. (Ed.). (2006). *Urban sprawl: A comprehensive reference guide*. Westport, CO: Greenwood Press.

Southworth, M. (2005). Designing the walkable city. *Journal of Urban Planning and Development*, *131*(4), 246–57. https://doi.org/10.1061/(asce)0733-9488(2005)131:4(246)

Spatari, S., Yu, Z., & Montalto, F.A. (2011). Life cycle implications of urban green infrastructure. *Environmental Pollution*, *159*(8), 2174–9. https://doi.org/10.1016/j.envpol.2011.01.015

Speir, C., & Stephenson, K. (2002). Does sprawl cost us all? Isolating the effects of housing patterns on public water and sewer costs. *Journal of the American Planning Association*, *68*(1), 56–70. https://doi.org/10.1080/01944360208977191

Waldheim, C. (2016). *Landscape as urbanism: A general theory*. Princeton, NJ: Princeton University Press.

Weitz, J., & Crawford, T. (2012). Where the jobs are going: Job sprawl in U.S. metropolitan regions, 2001–2006. *Journal of the American Planning Association*, *78*(1), 53–69. https://doi.org/10.1080/01944363.2011.645276

Wrigley, N., & Lowe, M. (2014). *Reading retail: A geographical perspective on retailing and consumption spaces*. London: Routledge.

Yngve, L. (2016). Distribution of green infrastructure along walkable roads. Paper given at the Active Living Research Annual Conference, 31 January–3 February 2016, Clearwater Beach, FL.

York Region. (1994). *Official plan*. Newmarket, ON: Author.

Young, R.F. (2009). Interdisciplinary foundations of urban ecology. *Urban Ecosystems*, *12*(3), 311–31. https://doi.org/10.1007/s11252-009-0095-x

12 "Greenfrastructure": The Greater Golden Horseshoe Greenbelt as Urban Boundary?

SARA MACDONALD AND LUCY LYNCH

Introduction

"Green infrastructure" has entered into land-use planning policy discourses in Southern Ontario, Canada, in recent years. Amati and Taylor (2010) have argued that the Greater Golden Horseshoe (GGH) Greenbelt "is multifunctional and acts as green infrastructure to provide a sustainable context for future growth in the region" (147). The provincial government's 2016 Coordinated Land-Use Planning Review calls for tighter policies on infrastructure development, as well as addresses climate change through green infrastructure. If suburban infrastructures have become the most visible set of sociotechnical assemblages that stand for the ecological and financial crisis of our age, questions must be raised about how we see the value of suburban infrastructures, both grey and green, in relation to the GGH Greenbelt in light of these recent policy changes (see chapter 1, the introduction to this volume).

This chapter aims to enrich suburban infrastructure discussions by incorporating the new and contested concept of "green infrastructure" into our understanding of the role of infrastructures within processes of peripheral expansion in the GGH region. As most of the new urbanization in the world during the next generation will be peripheral, green infrastructure stands to reshape questions of infrastructural development within the suburban landscape. Moreover, as the suburbs are so often sites of uneven and contested infrastructural development, they are also a likely source of urban infrastructure innovation, which may very well come in a greener form. For these reasons, we seek to understand the growing infrastructural tensions that exist within the Greater Golden Horseshoe periphery by unpacking the implications of re-appraising the GGH Greenbelt as green infrastructure (Amati & Taylor, 2010) while also addressing existing grey infrastructural conflicts within the greenbelt. We tease out a comparative understanding of grey and green infrastructure by drawing on a range of theories and interpretations of infrastructures (Larkin, 2013; Keil & Young, 2011; Monstadt & Schramm, 2013), and we rely on Mick Lennon's (2015) comprehensive and critical assessment of the cross-continental interpretations of the "green infrastructure" concept, as well as an overview of the David Suzuki Foundation's research on ecosystem services in the greenbelt (see Tomalty & Fair, 2012; Wilson, 2008). Building on scholarship on equity and infrastructure (Tonkiss, 2014; 2015), we aim to enrich the current literature that examines negotiations of space and equity on the periphery through the lens of infrastructures in order to understand infrastructures not only as technical but also as social and political. Looking at infrastructures, both urban and green, in relation to the greenbelt allows for a richer and more nuanced understanding of the greenbelt as a contradictory and contested space.

Located within the Greater Golden Horseshoe region, the greenbelt is billed as the world's largest permanently protected countryside of 1.8 million acres that stretches 325 kilometres from the Niagara Peninsula at the American border to Northumberland County, north of Lake Ontario. As a complement to the province's Places to Grow legislation, the greenbelt provides important ecosystem services and infrastructural functions both as a conduit for major cross-cutting mobility infrastructures and as a green buffer between the city and leapfrogging suburbanizing settlements beyond. In this chapter, we will examine the ways in which the GGH Greenbelt acts as green infrastructure, while also providing a site for noxious and networked grey infrastructures necessary for the functioning of the entire GGH region.

We will provide an overview of infrastructure literature, and we will also engage emerging critiques of green infrastructure and ecosystem services in relation to the GGH Greenbelt. By providing an overview of the GGH Greenbelt, an analysis of the most recent provincial review of these policies, and an examination of emerging discourses around the valuation of the greenbelt's ecosystem services, we will demonstrate how the greenbelt functions to facilitate grey infrastructural development, furthering peripheral expansion, while also providing important green infrastructural functions that serve to mitigate climate change and counteract the adverse effects of growth and expansion. This chapter will conclude by discussing the role that green infrastructure plays in the remaking of boundaries between the suburban fringe and the surrounding non-urban areas. Inherent in these discussions about infrastructure and the greenbelt are questions about equity, distribution, and access, as well as who benefits from these greenspaces. These are the questions we will raise in the conclusion of this chapter.

Grey Infrastructure

The true story of the infrastructure explosion set to occur over the next half century is suburban, as discussed in chapter 1. By "infrastructure," we are referring to the physical systems, networks, and structures that provide the necessary means to sustain everyday urban and suburban life. These are the buildings, roads, railways, sanitation systems, hydro corridors, communication towers, and so on that "facilitate the flow of goods, people or ideas and allow for their exchange over space" (Larkin, 2013: 328). Infrastructures are technical and physical, just as much as they are social and political, and in their facilitation of connectivity and the circulation of capital, they are integral to the production of everyday urban and suburban life. Within the suburban context infrastructures have a particular role to play, holding tremendous importance for the functioning of the entire region. Through an examination of infrastructures, much is revealed about "the widespread, but spatially differentiated, phenomena of suburbanization" (Monstadt & Schramm, 2013: 86).

A Western understanding of infrastructures stems from the concept's historical roots in the Enlightenment and liberalism. Through this history, infrastructures have become synonymous with "the shaping of modern society and realizing the future" (Larkin, 2013: 332). These ties have led to the reading of infrastructures through a narrative of progress and freedom, reiterated as "copies" of infrastructures produced across the globe, creating a "common visual and conceptual paradigm of what it means to be modern" (332). As infrastructures erode over time, their crumbling realities remind us that these promises of progress and freedom are, as AbdouMaliq Simone (2015: 278) so rightly puts it, "promise[s] intended to be broken."

Infrastructures, in their operational form, function as a networked base that is intended to sustain urban life. They are "the objects that create the grounds on which other objects operate" (Larkin, 2013: 329), or, rather, they are enablers, providing the conditions that make other activities possible. As a network or system, infrastructure "comprises the architecture for circulation" (328), providing a base for the economic functioning of society. However, the "uneven economies of infrastructure put into serious question how far these can be understood as systems of collective consumption" as they are subject to elitism and exclusion (Tonkiss, 2015: 387).

Infrastructures are the "arteries" of the urban form (Tonkiss, 2014: 138) constructed by cities, while simultaneously constructing and dissolving cities (Keil & Young, 2011: 7). Often in the formal suburbs of the global north, infrastructures act as "critical preconditions" for peripheral development and expansion (Monstadt & Schramm, 2013: 86). Large-scale infrastructures have facilitated growth on the periphery, in many cases preceding residential and commercial development, functioning as a substantial fraction of fixed capital that works at the behest of larger accumulation processes and interest. Meanwhile, in the *Zwischenstadt* – a term coined in 2003 by German scholar Thomas Sieverts meaning "in-between city" – these infrastructures propelling growth on the periphery engage with this in-between landscape in an entirely different way, operating as "a thoroughput channel and staging ground for uses channeled elsewhere" (Keil & Young, 2011: 12). Alternatively, it can also be the case that infrastructure development lags behind peripheral expansion, characterized chiefly by informal settlement patterns, rapid and unequal peri-urbanization, and high degrees of social segregation. As is the case in some of the diverse peripheral regions in the global south or in neglected communities and spaces of disinvestment in the global north, "the reproduction of suburban space is shaped by the conditions of informality and self-organization defying the implementation of comprehensive master plans or reliable investments in networked infrastructure" (Monstadt & Schramm, 2013: 91).

While infrastructures appear in variegated forms across the spectrum of what we know as "suburbia," at the very core of the concept is the role of infrastructures in supporting the functioning of different aspects of society, including supplying the necessities of human life in urban environments. In this way, and through their inception, implementation, and everyday operation, infrastructures are "sociotechnical assemblages" and "social in every respect" (Amin, 2014: 138), "generat[ing] the ambient environment of everyday life" (Larkin, 2013: 328). In addition, there is a strong argument for viewing people as not only essential to the functioning of infrastructures but as infrastructures themselves (Simone, 2015; Tonkiss, 2014). As products of complex social relations and in their provision of everyday necessities and services, infrastructures raise questions of equity – infrastructures for whom? – which are front and centre in infrastructure discussions. Moreover, the often large-scale and resource-based materiality of infrastructures directly implicates the natural environment, and questions of ecology are an ever-increasing concern in this regard. And so we arrive at "green infrastructure," an emerging concept gaining increasing popularity for incorporating the environment and biodiversity into the infrastructural equation and proposed by many as a response to present-day concerns with urban sustainability.

Green Infrastructure

Much like infrastructure, the definition of "green infrastructure" is difficult to pin down. Its definitions range on a national and continental scale (Lennon, 2015). Generally, green infrastructure is accepted as naturally occurring or human-made ecological systems, networks, or sites that provide benefits to humans, which are valued as "natural capital" and measured as "ecosystem services." It encompasses a variety of material and policy-based interventions in urban and suburban environmental land use, ranging from green energy to conserved wetlands, rooftop farms, permeable sidewalks, storm water ponds, golf courses, and so on. Underlying all forms of green infrastructure, whether it is implemented for conservation, human health, recreation, aesthetics, management, or climate change mitigation, is the aim to reconcile the environment (conservation) with the economy (development), with the overarching goal of achieving urban sustainability (Lennon, 2015).

Unlike grey infrastructures, whose "path-dependent" trajectories become "annoying obstacles to change in the morphologies of cities and suburbs and the economic and political relationship between them" (Monstadt & Schramm, 2013: 86), green infrastructure is considered "multifunctional," and consequently has a much more fluid relationship with the physical environment, in many ways mimicking naturally occurring processes or preserving natural geographies, and operates as "a 'framework' to reconcile divergence between ecological conservation, economic development and social equity" (Lennon, 2015: 963). However, as Gilbert (2014) points out, there exists a "misleading balance depicted in the rhetoric of the three pillars of sustainability: economy, environment and society" where all too often "social and environmental concerns appear more easily trivialized and marginalized at the profit of economic growth" (159–60).

As a form of infrastructure, green infrastructure conceptually implies modernity or progress. In this post-modern era of "sustainability," the term "green" itself implies progress, situating green infrastructure among theories of ecological modernization because it presents "particular attractions for policy-makers pursuing economic development and place competiveness agendas as it appears to offer a way of quantifying economic benefits from green space interventions" (Thomas & Littlewood, 2010: 219). "Green" becomes a brand that overlays visuals of failing modernist infrastructures with promises of a sustainable future achieved by harnessing nature through innovation and technology. Consequently, "greening" infrastructure depoliticizes infrastructures and allows decision-makers to push their agendas with less opposition (Lennon, 2015: 964), resulting in a re-scripting of urban development narratives as *necessarily* green and normalizing green infrastructures as, like grey infrastructures, "something that we must have" (Benedict & McMahon, 2006: 2).

Like grey infrastructure, green infrastructure is presented as beneficial for society; however, as we know, these benefits are highly uneven. In the case of green infrastructure, land is valued for its natural capital, and the benefits provided by the land and its ecosystem functions are measured as ecosystem services, which are considered essential to human well-being and biodiversity. While there are various interpretations of ecosystem services, for the purpose of this chapter we will be referring to the 2003 Millennium Ecosystem Assessment definition. This definition, which has been credited with bringing the concept into mainstream policy circles (Dempsey & Robertson, 2012; Gómez-Baggethun, De Groot, Lomas, & Montes, 2010), classifies ecosystem services as provisioning, regulating, supporting, and social and cultural services (Millennium Ecosystem Assessment [MA], 2003). Green infrastructure's multifunctionality is measured through ecosystem service discourses, which, for example, might see an engineered wetland as acting as a carbon sink, as storm water management, as habitat, and as an important social and cultural destination, all of which would be considered essential to the functioning of a sustainable society, especially in the wake of climate change.

Amati and Taylor (2010: 153) see green infrastructure as "an opportunity to re-appraise" greenbelts. They argue that we can see the GGH Greenbelt as green infrastructure because it is "a multifunctional space for agriculture, nature and settlement negotiated as the local municipal level" (146), providing a "sustainable context for future growth" (147).

While they accept that "sustainability has been a double-edged sword for the green belt, in some cases even supporting arguments for development," they argue that "climate change provides a renewed urgency on the need to avoid sprawl" (152), pointing out that there is a critical "need for green space protection to be a cornerstone of urban-region planning" (153). However, as Wekerle, Sandberg, Gilbert, and Binstock (2007) caution, in the GGH "nature conservation has served as a cornerstone or lubricant for implementing a regional planning framework in the service of growth," which "has allowed the fast-tracking of a regional growth management strategy and new infrastructures projects that support growth" (34). This situation calls into question the contradictions that green infrastructure presents in the periphery as it serves multifunctional needs, and, in the case of the greenbelt, acts to both curb sprawl and climate change as well as to lubricate strategies for growth. Arguments for sustainability justify future growth while simultaneously providing the necessary grounds on which a land-use policy that prohibits growth would be established. It is through this lens of green infrastructure that the GGH Greenbelt is revealed to be a highly contradictory space, perhaps necessarily so.

The Greater Golden Horseshoe Greenbelt

Population and economic growth within Ontario is concentrated in the GGH region. This urban region covers approximately 3.2 million hectares and is made up of a range of urban developments, including large cities such as Toronto and Hamilton, and comprises 110 different municipal jurisdictions. The region was home to 9.2 million people in 2016 and is expected to grow significantly in the coming decades (calculated from Statistics Canada, 2017). The Ontario government predicts that, by 2041, the GGH region could be home to 13.48 million people, accounting for more than 80 per cent of Ontario's population growth, and will provide 6.27 million jobs (Ministry of Municipal Affairs and Housing [MMAH], 2013: Schedule 3). Throughout the 1990s and the early 2000s, there was a growing concern about the impact of rapid urban growth in Southern Ontario. In 2002, the Neptis Foundation warned that if the "business as usual land-use" conditions at the time of low-density development continued for the next several decades, quality of life conditions within the region could deteriorate, as more farmland would be consumed to meet housing demands and traffic congestion and greenhouse gas emissions would increase significantly, resulting in longer commute times and increasing the impact of climate change (Neptis Foundation, 2002). As a result, there was a growing need to address the existing sprawling patterns within the region at the time; where and how the predicted future growth would be accommodated was becoming a pressing concern. It was within this context that the GGH Greenbelt was established in 2005.

In 2005, the Greenbelt Act was passed by the Ontario provincial government; this legislation allowed for the creation of a Greenbelt Plan, which was released that year. The greenbelt was designed to safeguard farmland and protect natural heritage and water resource systems, and to support the economic and cultural activities associated with rural communities (MMAH, 2005) (see figure 12.1).

The greenbelt includes 800,000 acres of land already protected under the Niagara Escarpment Plan and the Oak Ridges Moraine Conservation Plan, and adds another 1 million acres of land known as the Protected Countryside to this landscape. The plan generally prohibits the designation of protected areas for development, prevents development close to environmentally sensitive areas, and promotes the creation of recreational spaces. The greenbelt features a diverse range of land uses and activities. Agriculture is a primary land use in the greenbelt, as 43 per cent of the total area within the greenbelt was used for agricultural purposes in 2011 (JRG Consulting Group, 2014). The greenbelt preserves some of the most significant agricultural land in Canada, and its natural heritage system covers 535,000 acres of wetlands, lakes, river valleys, and forests, and provides habitat for seventy-eight species at risk (Friends of the Greenbelt Foundation, n.d.). The greenbelt also contains more than 10,000 kilometres of trails and cycling routes, so this landscape provides many recreational and tourism opportunities to the region's residents. Municipalities are responsible for implementing the greenbelt policies and are required to bring their planning documents into conformity with the Greenbelt Plan. The Ministry of Municipal Affairs and Housing is responsible for these policies, and they will conduct a review of the Greenbelt Plan every ten years to assess its effectiveness. The first ten-year review started in 2015; it is only during this review process that the minister can make amendments to protected areas within the greenbelt, and those changes are not allowed to decrease the total area of the greenbelt (MMAH, 2005: 45).

The greenbelt works in conjunction with the Places to Grow legislation, as the greenbelt policies dictate where growth cannot occur, while the Places to Grow specifies where development can happen and under what conditions. In 2006, under the Places to Grow Act, a growth plan for the

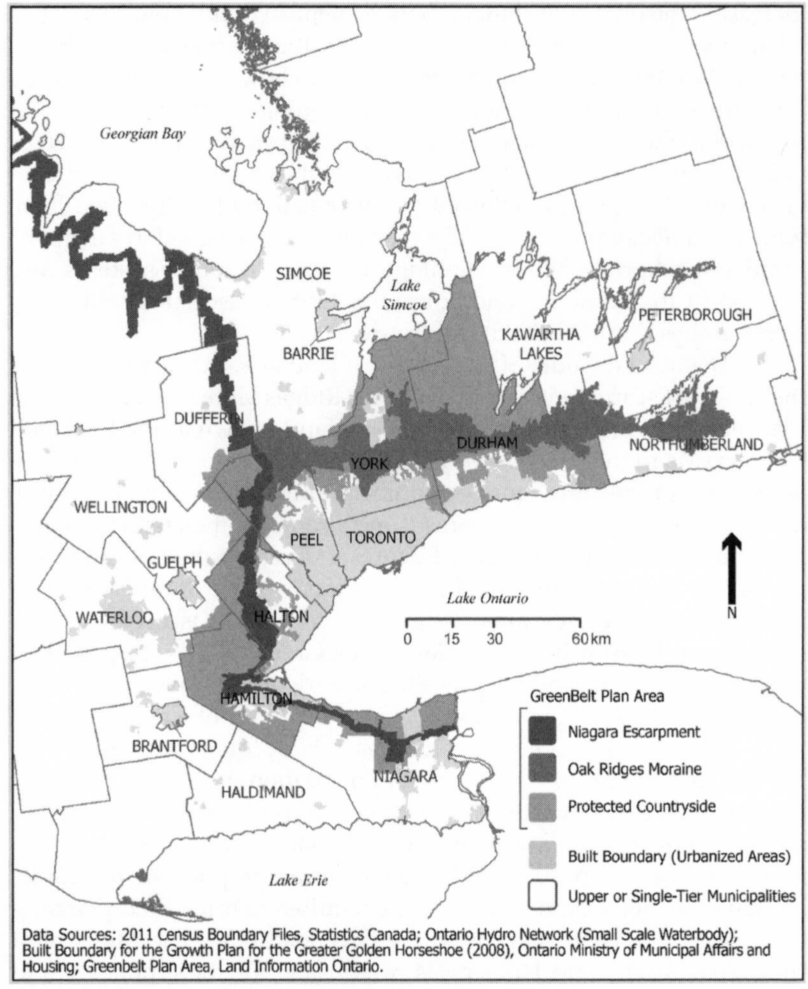

12.1 Greater Golden Horseshoe Greenbelt. Source: Keil & Macdonald, 2016

GGH region was released by the province (called interchangeably Places to Grow and Growth Plan), which was designed to manage growth in the region for twenty-five years, promote intensification, and direct new development away from greenfield sites and towards existing urban areas (see Ministry of Public Infrastructure Renewal, 2006). Similarly to the Greenbelt Plan, the Growth Plan is to be reviewed by the provincial government every ten years. The review of the Niagara Escarpment Plan and Oak Ridges Moraine Plan was, however, delayed to sync up with the review of the Greenbelt Plan and Growth Plan in 2015.

For the Coordinated Land-Use Planning Review of all four plans, the province appointed an advisory panel and asked them to make recommendations for improvements to these policies. Their findings were released in December 2015. In May 2016, the province introduced proposed changes to all four plans, and, after a series of public consultations, will release amendments to these policies. At the time of the writing of this chapter, the proposed plans had only been recently released. However, an initial analysis of the proposed Greenbelt Plan and GGH Growth Plan shows a number of significant revisions that were made to the original policies. The greenbelt will be expanded by adding twenty-one major river valleys and seven coastal wetlands to the plan's "urban river valley" designation, while also allowing for the future expansion of the greenbelt into areas of hydrological significance that are facing development pressures (MMAH, 2016b: 13). There is a clear focus on increasing urban intensification and directing growth into transit-accessible areas, as the current density and intensification targets are raised in the proposed Growth Plan (MMAH, 2016b: 7). Also, the policies now include a strong focus on addressing the impacts of climate change, and the language of ecosystem services, green infrastructure, and resiliency is now featured prominently throughout the proposed Greenbelt Plan (MMAH, 2016b: 15). However, the Neptis Foundation cautions that, while many of the proposed amendments in the Growth Plan align with a sustainable view of regional planning, it is only if these policies are fully implemented that they have the potential to direct growth into built-up urban areas (Neptis Foundation, 2016: 1–2). Since the success of the Greenbelt and GGH Growth Plans are dependent upon each other, then "getting the Growth Plan right is critical, as the greenbelt alone will not stop sprawl" (Neptis Foundation, 2016: 1).

Grey among the Green Infrastructure within the Greenbelt

Chapter 1 discusses the importance of suburban infrastructure for the functioning of the entire region and, in particular, the urban core. It addresses the consequences of the location of necessary services such as airports, golf courses, landfills, and water treatment facilities in the urban periphery. In the case of the GGH region, these noxious uses or networked spaces can be located within the greenbelt, so it is important to examine the implications of these suburban infrastructures. Within the GGH context, the Greenbelt Plan allows for the creation and expansion of major infrastructure projects by way of the following policies:

> All existing, expanded or new *infrastructure* ... is permitted in the Protected Countryside ... provided it meets one of the following two objectives:
>
> a. It supports agriculture, recreation and tourism, rural settlement areas, resource use or the rural economic activity that exists and is permitted in the Greenbelt; or
> b. It serves the significant growth and economic development expected in southern Ontario beyond the Greenbelt by providing for the appropriate *infrastructure* connections among urban growth centres and between these centres and Ontario's borders. (MMAH, 2005: 30)

It is the second objective that will allow for these infrastructures within the greenbelt, provided they support the significant regional growth expected in the coming years. This objective has been called a loophole in the plan, as the language is vague enough that it will allow for projects such as new highways, airports, quarries, landfills, and golf courses to be built in the greenbelt. A number of changes have been introduced to the infrastructure policies in the proposed Greenbelt Plan, including increasing the resiliency of infrastructure, ensuring its financial viability, and encouraging the use of green infrastructure (MMAH, 2016a: 40–1). The original policies' loopholes allowing major infrastructure projects within the greenbelt still remain in the proposed amendments. However, those gaps have been tightened up by new policies stating that new or expanding infrastructure shall avoid specialty crop and prime agricultural land unless no reasonable alternative is available, and if so, an impact assessment will be necessary for the development to proceed (MMAH, 2016a: 43). Wekerle and colleagues argue that allowing this type of large-scale infrastructure in the greenbelt suggests that growth is prioritized over conservation, and assert that conversation can only happen if growth is also supported (Wekerle, Sandberg, Gilbert, & Binstock, 2007: 29).

One significant threat to the greenbelt is that of the creation or expansion of infrastructure, in particular highways (figure 12.2).

Some of the most used highways in Southern Ontario pass through the greenbelt; given the substantial growth predicted for the region, it is expected that either new highways will need to be built or existing ones expanded. There is currently a new regional highway proposal being

12.2 Highway construction within the GGH Greenbelt. Photo: Courtesy of Sara Macdonald

planned by the Ministry of Transportation. The Greater Toronto Area (GTA) West Corridor (or Highway 413) is a four- to six-lane highway that would connect north Vaughan to Milton. Some of the potential routes for this highway would cut through the greenbelt in Vaughan and prime farmland in Caledon, and pave over 2,000 acres of prime farmland in the greenbelt (Javed, 2015, 2 January).

A common argument from environmental organizations against highway development in greenbelts is that it enables leapfrog development, facilitating new development beyond the existing urban boundaries. Also, highways can undermine the environmental integrity of the greenbelt because the roadways cut through environmentally sensitive areas, disrupt wildlife corridors, destroy the habitat for endangered species, and threaten the carbon storage capacity of the greenbelt (Tomalty & Fair, 2012: 28–9). Environmental organizations have been protesting this project, saying that highways represent an outdated model of growth and that the province would be better to invest the billions of dollars needed for this project into public transit infrastructure instead (see figure 12.3) (Javed, 2015, 30 January).

12.3 Campaign to stop Highway 413. Source: Environmental Defence (www.environmentaldefence.ca)

In contrast, some regional municipalities and developers are putting pressure on the province to go forward with Highway 413, as they counter that the expected population growth cannot be accommodated without new highway development and that the movement of goods and services must be considered (Javed, 2016, 28 April). The result is conflicting visions of what type of growth and infrastructure is needed for the region in the coming decades. This controversial highway proposal was scuttled by the Liberal provincial government in February 2018, but Doug Ford, the newly elected Progressive Conservative premier, has committed to take a second look at the project.

While these suburban infrastructures are vital for the functioning of the region, questions need to be raised about who benefits from, and who would be disadvantaged by, the creation or expansion of grey infrastructure through these suburban areas, because when highways are proposed to run through the greenbelt, there are different implications for a variety of stakeholders including farmers, urban residents, and tourists. Looking at infrastructure, both grey and green, in relation to the greenbelt provides new avenues for understanding the greenbelt as a contested and contradictory space. How can we understand the greenbelt as green infrastructure, while also addressing suburban infrastructure conflicts within the greenbelt? If we are to accept the greenbelt as green infrastructure, what does this endorsement mean for grey infrastructure within the greenbelt? How can we quantify the ecosystem services that green infrastructure such as the greenbelt provides, while also allowing for infrastructural development such as highways, aggregate mining, soil dumping, or even industrial-scale agriculture occurring in this landscape? These contradictions are highlighted by examining infrastructure discourses associated with the greenbelt.

The Greenbelt: A Working Countryside

The GGH Greenbelt was designed by the province to be more than a park; rather, it is a working countryside with medium-sized cities within it and a policy regime that allows for a variety of land uses (Provincial planner, personal communication, 30 March 2015). However, there are often different visions of this protected landscape, which can create conflicts between its users. Keil and Shields (2013) have argued that the use of infrastructure within the greenbelt is not driven by agricultural demands anymore, but rather by the interests of urbanities in both real and imagined green landscapes. Instead of agricultural uses, we are now seeing "para-agriculture" (for example, trails, parks, golf courses, wineries, cycling, horse farms, and apple picking) replacing the more traditional agricultural activities that had dominated the greenbelts of the past (76).

Recreational and tourism opportunities in the greenbelt such as the cycling routes, wine tours, and nature trails have become popular in the past several years, in particular with the region's urban residents. Conflicts can erupt around issues related to tourism within the greenbelt, as, for example, in the case of wineries in the Niagara Region of Southern Ontario, which offer wine tours and host weddings that cause tension because of increased traffic congestion from tour buses and noisy parties (Municipal politician, personal communication, 8 October, 2014). This discussion raises key questions related to infrastructure and equity: Who are the users of these para-agricultural services? What types of demands are being placed on both green and grey infrastructures within the greenbelt as a result of these activities? It is important to balance the region's growing demand for recreational opportunities with conservation goals. There are no limits to the number of recreational facilities allowed within the greenbelt; converting forests or wetlands to new facilities can result in the loss of ecosystem storage capacity, while certain types of activities such as golf courses can place a strain on the greenbelt's hydrological systems (Tomalty & Fair, 2012: 30). Allowing for unlimited recreational infrastructures within the greenbelt can also be seen as a contradiction to the plan's policies; this loophole within the policies could have adverse impacts on the capacity of the greenbelt to act as a green infrastructure, as the hydrological systems of the greenbelt can be seen as part of that system of green infrastructure.

Conflicts can also arise as suburbanization approaches the greenbelt and creates issues about how to separate suburban development from adjacent agricultural operations. As the countryside becomes gentrified with more urban residents moving into rural communities, there may be complaints from these new residents about the noise, dust, traffic, and smells associated with farming operations, while farmers might have their own concerns about trespassing and vandalism on their properties (Farming non-governmental organization [NGO] representative B, personal communication, 27 August 2014). Part of the challenge is that the number of urban residents who have a direct connection to farming is decreasing over time, and, as a result, they may complain about normal aspects of farming operations and question why open fields are not available for public use, when in fact that land is often privately owned (Farming NGO representative A, personal communication, 26 August 2014). In the proposed Greenbelt Plan, there is now a stronger focus on policy changes that would reduce conflicts between agricultural and non-agricultural uses, supporting local food production and minimizing the impact of new infrastructure projects on agricultural operations, which reflects an attempt by the province to address some of the concerns from farmers about the original greenbelt policies (MMAH, 2016a: 14–15). The infrastructures that have been laid out in rural communities have been set up to facilitate agricultural production (Keil & Shields, 2013: 75). However, today's greenbelts have to be seen as "conjugations of postsuburban relationships," and the previously mentioned examples highlight that "the use of greenbelts today are taking shape not *against* but *in conversation with* the push of the suburban expansion that surrounds those greenspaces" (Keil & Shields, 2013: 76).

Ecosystem Services and the Greenbelt

In recent years, the language of ecosystem services has been used to re-script relations between the GGH Greenbelt and its suburban boundaries by framing the greenbelt as green infrastructure, necessary for preserving and linking ecosystems and ensuring sustainable growth and development in the region. In Ontario, ecosystem service discourses have been most widely disseminated by the David Suzuki Foundation, reporting that the wetlands, forests, and agricultural land found within the GGH Greenbelt provide significant ecosystem services, such as carbon and water storage, flood control, erosion control, waste treatment, water filtration, and pollination (Wilson, 2008: 1). According to research conducted by the David Suzuki Foundation, the greenbelt's forests, wetlands, and agricultural soils store 86 million tonnes of carbon, the economic value of which is estimated at $4.5 billion or $366 million per year over a twenty-year period (Tomalty & Fair, 2012: 22). Moreover, additional research argues that the GGH Greenbelt has multiple roles to play in a climate change adaptation strategy for the region: providing protection against the loss of biodiversity, protecting farmlands to ensure food security in the wake of climate change and collapse of global food systems, combating urban heat island effects, and providing a refuge for urban population, as periods of extreme heat and poor air quality are expected to increase in the future (Tomalty & Komorowski, 2011: 24).

There is widespread concern about the impact of climate change becoming more pronounced within the coming decades, and, as the GGH region continues to grow, one of the most problematic trends could be increasing greenhouse gas emissions associated with worsening traffic congestion around the region. Viewing the GGH Greenbelt through the lens of ecosystem services opens the possibility of regarding these one million acres of protected countryside as having "untapped potential" that could become increasingly important for urban and suburban areas in the face of these uncertain global changes. However, framing the GGH Greenbelt as green infrastructure and valuing the land for its ecosystem services is not a politically neutral project and is problematic for the ways in which it presents opportunities for measuring nature and ecologies according to their exchange value, as well as for how it employs the technocratic language of experts, effectively hijacking land-use planning questions that should involve the entire public.

Infrastructure discourses occur in three forms – technical, political, and critical – all serving to shape the decision-making processes and public perceptions regarding infrastructures. All too often debates around infrastructures end up being dominated by a technical overview, taking advantage of the fact that, for most people most of the time, infrastructures are taken for granted and seen as the exclusive purview of experts, effectively depoliticizing decision-making. While the ecosystem services concept is, as of now, mainly confined to experts, it is important to realize that this concept should not inherently be about the marketization of nature and should not necessarily imply framing the environment in neoclassical economic terms.

Originally a pedagogical concept linked to biodiversity conservation, ecosystem services have departed significantly from this perspective due to anthropocentric reinterpretations of these services (Gómez-Baggethun et al., 2010: 1209).While Dempsey and Robertson (2012) admit that "market strategy clearly hybridizes well with state strategy in ES [ecosystem services] policy-making," the concept "also hybridizes well with scientific agendas, the strategies of civil society actors, and even with the distinctly non-market impulses of traditional state Keynesianism" (760). The concept has proven to be effective in "deciding how to allocate the resources provided by nature among alternative desirable ends, whether or not the end receiving highest priority is monetary value, quality of life, or the preservation of nature for its intrinsic values" (Farley, 2012: 48).

As one Suzuki Foundation report points out, there is an "exclusion of natural capital in our current measures of progress and decision-making" (Wilson, 2008: 12). While we recognize the problematic nature of discussing the greenbelt through the lens of ecosystem services and natural capital, this discourse allows us to have a broader discussion and gain wider recognition of the contributions that maintaining a healthy ecosystem in the greenbelt has to offer, which ultimately provides important grey and green infrastructural value for the GGH region. However, Gilbert (2014) reminds us that "the discourses of sustainability and greening do not necessarily take us to the future but rather to the status quo if redressing inequality is not at the core of such agendas" (168).

Conclusion

If suburbs are to be sites of infrastructural innovation as suggested in chapter 1, what kind of prospects does the greenbelt offer as a site for new and old infrastructures, as well as a form of green infrastructure itself? Throughout this chapter, we have examined the various ways in which the greenbelt acts as green infrastructure, while also negotiating provincial policies that allow for networked and noxious infrastructures to exist within this suburban growth boundary. We have examined the implications of the contradictory reality that the greenbelt is green infrastructure, a discourse likely to be reinforced by the provincial government's Coordinated Land-Use Planning Review, as well as a working landscape in which highways and other mega-infrastructure projects exist. We argue that we see the value of suburban infrastructures, both grey and green, in relation to the GGH Greenbelt in constant negotiation with one another, acting in a "multifunctional" manner not only as a suburban boundary that curbs sprawl, connects hydrological systems, and provides important ecosystem services in this era of changing global conditions but also as a site for grey infrastructures necessary for the functioning of the GGH region.

All of this discourse raises important questions around equity. Whether it is grey or green, infrastructure "reproduces and generates relations of economic and spatial inequity" (Tonkiss, 2015: 387). Access to the many benefits provided by ecosystem services are not always distributed evenly, raising questions of both whom this infrastructure is intended to serve and who stands to gain as a result of its implementation. As Wolch, Byrne, and Newell (2014) point out, access is often highly stratified based on income, age, gender, ethnicity, and other factors; however, "despite a growing literature, there is no consensus among scholars about how to measure green space access" (235–6). As Angelo and Hentschel (2015) argue, "the fight about infrastructure is really a fight about who has the right to place" (311). In the greenbelt context, there are many conflicts over infrastructure that directly raise issues of how greenbelt lands are to be utilized, and for whom.

While research demonstrates that green space is often linked to gentrification, both within the core (Wolch et al., 2014) and on the periphery (Sandberg & Wekerle, 2010), environmental gentrification is not an inherent outcome of green space planning (Curran & Hamilton, 2012). When discussing a green space such as the GGH Greenbelt, which operates as a site for recreation and leisure as well as a working landscape that holds grey infrastructures while simultaneously providing important ecosystem services that benefit an entire region, quantifying equity is complex. However, one only needs to go as far as the closest stables or apple orchard to see that these new sites of consumption are places of privilege, accessible mainly by private vehicle by those who can afford to visit them, both in terms of time and income. Thus, infrastructures that serve existing communities are redirected to address the needs of visitors to the greenbelt, which, as we discussed earlier in this chapter, could be a source of conflicts and create challenges for these rural communities within the greenbelt.

One significant factor regarding infrastructure and equity within the greenbelt is the issue of access. As a result of its sheer size and much of its land being privately held, the GGH Greenbelt can be seen as hampered by a lack of proximity to potential users, especially those without access to a private vehicle. In contrast, European greenbelts, such as those in Frankfurt, Germany, have well-developed transit connections linking these greenspaces to the urban core. How could a better infrastructure of recreation be built within the GGH Greenbelt to make it more accessible and inclusive? How could the provincial and local governments enhance the recreational infrastructures that already exist, and what lessons could the jurisdictions with more established greenbelts offer the Ontario context?

Since the GGH Greenbelt is only just over a decade old, and the first review of the greenbelt policies are happening as this chapter is being written, there is a unique opportunity right now to rethink how infrastructures within the greenbelt can be made more equitable, inclusive, and innovative. Moreover, with this integration of the language of green infrastructure into the land-use planning policy context in Southern Ontario, new opportunities for understanding the greenbelt present themselves. In 2010, Amati and Taylor said that green infrastructure presents "an opportunity to re-appraise" greenbelts, "in some cases link[ing] them within a larger network of green infrastructure" (153). However, they also pointed out that the successful implementation and management of a twenty-first century greenbelt is clearly reliant on public support (150). Just over half a decade later, polls show that public support for the greenbelt is high, with more than 90 per cent of Ontarians favourable to its existence (Friends of the Greenbelt Foundation, n.d.). While these polls are promising, as support for the greenbelt is crucial for its long-term success, they also mean that there is a lot at stake. There is immediate pressure on the provincial government to ensure that they get these new proposed changes right; otherwise, we risk continued low-density sprawling development patterns that have plagued the region for decades at the expense of the greenbelt and the many benefits it holds.

Just like chapter 11, the present chapter deals with green infrastructure. It does so at a different scale, however, by concentrating on the largest green infrastructure present within certain metropolitan regions: greenbelts. Chapter 12 introduces us to the Toronto regional planning strategy of the Ontario government, one of the most ambitious such policies in North America. As with all planning inspired by sustainable development, the Ontario Growth Plan policy is riddled with contradictions between environmental and development objectives. In this sense, the chapter goes to the root of tensions affecting sustainable development. It shows how one of the main aspects of the Growth Plan, the GGH Greenbelt, provides a vivid illustration of such tensions. The chapter indicates how infrastructures required to support ongoing development outside the greenbelt and a desire to tap the recreational potential of this vast undeveloped area, which borders a large concentration of population and economic activity, imperil the environmental integrity of the greenbelt. Chapter 12 thus brings to light how planning contradictions inherent in sustainable development affect region-wide planning efforts in the Toronto periphery.

REFERENCES

Amati, M., & Taylor, L. (2010). From green belts to green infrastructure. *Planning, Practice & Research*, *25*(2), 143–55. https://doi.org/10.1080/02697451003740122

Amin, A. (2014). Lively infrastructure. *Theory, Culture & Society*, *31*(7–8), 137–61. https://doi.org/10.1177/0263276414548490

Angelo, H., & Hentschel, C. (2015). Interactions with infrastructure as windows into social worlds: A method for critical urban studies: Introduction. *City*, *19*(2–3), 306–12. https://doi.org/10.1080/13604813.2015.1015275

Benedict, M.A, & McMahon, H. (2006). *Green infrastructure: Linking landscapes and communities.* Washington, DC: Island Press.

Curran, W., & Hamilton, T. (2012). Just green enough: Contesting environmental gentrification in Greenpoint, Brooklyn. *Local Environment*, *17*(9), 1027–42, https://doi.org/10.1080/13549839.2012.729569

Dempsey, J., & Robertson, M. (2012). Ecosystem services: Tensions, impurities, and points of engagement within neoliberalism. *Progress in Human Geography*, *36*(6), 758–79. https://doi.org/10.1177/0309132512437076

Farley, J. (2012). Ecosystem services: The economics debate. *Ecosystem Services*, *1*(1), 40–9. https://doi.org/10.1016/j.ecoser.2012.07.002

Friends of the Greenbelt Foundation. (n.d.). About the Greenbelt. Retrieved from http://www.greenbelt.ca/about_the_greenbelt

Gilbert, L. (2014). Social justice and the "green" city. *urbe. Revista Brasileira de Gestão Urbana, 6*(2), 158–69. https://doi.org/10.7213/urbe.06.002.se01

Gómez-Baggethun, E., De Groot, R., Lomas, P.L., & Montes, C. (2010). The history of ecosystem services in economic theory and practice: From early notions to markets and payment schemes. *Ecological Economics, 69*(6), 1209–18. https://doi.org/10.1016/j.ecolecon.2009.11.007

Javed, N. (2015, 2 January). Proposed highway to pave parts of Greenbelt proving controversial. *The Toronto Star*. Retrieved from http://www.thestar.com/news/gta/2015/01/02/proposed_highway_to_pave_parts_of_greenbelt_proving_controversial.html

Javed, N. (2015, 30 January). *GTA West highway: Forward thinking or retro mistake? The Toronto Star*. Retrieved from http://www.thestar.com/news/gta/transportation/2015/01/30/gta-west-highway-forward-thinking-or-retro-mistake.html

Javed, N. (2016, 28 April). GTA West highway decision could determine future of the region. *The Toronto Star*. Retrieved from http://www.thestar.com/news/gta/transportation/2016/04/28/gta-west-highway-decision-could-determine-future-of-the-region.html

JRG Consulting Group. (2014). *Agriculture by the numbers: Understanding the Greenbelt's unique advantages*. Friends of the Greenbelt Foundation. Retrieved from http://www.greenbelt.ca/agriculture_by_the_numbers_2014

Keil, R., & Macdonald, S. (2016). Rethinking urban political ecology from the outside in: Greenbelts and boundaries in the post-suburban city. *Local Environment, 21*(12), 1516–33. https://doi.org/10.1080/13549839.2016.1145642

Keil, R., & Shields, R. (2013). Suburban boundaries: Beyond greenbelts and edges. In R. Keil (Ed.), *Suburban constellation: Governance, land and infrastructure in the 21st century* (pp. 71–8). Berlin: Jovis.

Keil, R., & Young, D. (2011). Introduction: In-between Canada – The emergence of the new urban middle. In D. Young, P. Burke Wood, & R. Keil (Eds.), *In-Between Infrastructure: Urban Connectivity in an Age of Vulnerability* (pp. 1–18). Kelowna, BC: Praxis (e)Press. Retrieved from http://www.praxis-epress.org/IBT/inbetween.pdf

Larkin, B. (2013). The politics and poetics of infrastructure. *Annual Review of Anthropology, 42*, 327–43. https://doi.org/10.1146/annurev-anthro-092412-155522

Lennon, M. (2015). Green infrastructure and planning policy: A critical assessment. *Local Environment, 20*(8), 957–80. https://doi.org/10.1080/13549839.2014.880411

Millennium Ecosystem Assessment. (2003). *Ecosystems and human well-being: A framework for assessment*. Washington, DC: Island Press.

Ministry of Municipal Affairs and Housing (MMAH). (2005). *Greenbelt plan, 2005.* Toronto: Queen's Printer for Ontario.

Ministry of Municipal Affairs and Housing (MMAH). (2013). *Amendment 2 (2013) to the Growth Plan for the Greater Golden Horseshoe, 2006* Retrieved from https://www.placestogrow.ca/index.php?option=com_content&task=view&id=398&Itemid=14

Ministry of Municipal Affairs and Housing (MMAH). (2016a). *Proposed Greenbelt plan 2016.* http://www.mah.gov.on.ca/AssetFactory.aspx?did=14951

Ministry of Municipal Affairs and Housing (MMAH). (2016b). *Shaping land use in the Greater Golden Horseshoe.* Retrieved from http://www.mah.gov.on.ca/AssetFactory.aspx?did=14910

Ministry of Public Infrastructure Renewal. (2006). *Places to grow: Better choices, brighter future – Growth Plan for the Greater Golden Horseshoe 2006* Retrieved from https://www.placestogrow.ca/images/pdfs/ggh/en/2006_growth-plan-for-the-ggh-en.pdf

Monstadt, J., & Schramm, S. (2013). Beyond the networked city: Suburban constellations in water and sanitation systems. In R. Keil (Ed.), *Suburban constellation: Governance, land and infrastructure in the 21st century* (pp. 85–94). Berlin: Jovis.

Neptis Foundation. (2002). Toronto-related region futures study: Interim report: Implications of business-as-usual development. Retrieved from http://www.neptis.org/sites/default/files/ibi_reports/neptis_bau_final_report.pdf

Neptis Foundation. (2016, 17 May). Province must embrace its role as regional planner for Growth Plan to succeed. Retrieved from http://www.neptis.org/latest/news/province-must-embrace-its-role-regional-planner-growth-plan-succeed

Sandberg, L.A., & Wekerle, G.R. (2010). Reaping nature's dividends: The neoliberalization and gentrification of nature on the Oak Ridges Moraine. *Journal of Environmental Policy & Planning, 12*(1), 41–57. https://doi.org/10.1080/15239080903371915

Simone, A.M. (2015). Afterword: Come on out, you're surrounded: The betweens of infrastructure. *City, 19*(2–3), 375–83. https://doi.org/10.1080/13604813.2015.1018070

Statistics Canada. (2017). Census Metropolitan Areas: Table 1.1: Population and demographic factors of growth by CMA, Canada. (91-214-X). Retrieved from https://www150.statcan.gc.ca/n1/pub/91-214-x/2017000/tbl/tbl1-1-eng.htm

Thomas, K., & Littlewood, S. (2010). From green belts to green infrastructure? The evolution of a new concept in the emerging soft governance of spatial strategies. *Planning, Practice & Research, 25*(2), 203–22. https://doi.org/10.1080/02697451003740213

Tomalty, R., & Fair, J. (2012). *Carbon in the bank: Ontario's Greenbelt and its role in mitigating climate change*. Vancouver, BC: David Suzuki Foundation. Retrieved from http://www.greenbelt.ca/carbon_in_the_bank_ontario_s_greenbelt_and_its_role_mitigating_climate_change

Tomalty, R., & Komorowski, B. (2011). *Climate change adaptation: Ontario's resilient Greenbelt*. Toronto, ON: Friends of the Greenbelt Foundation. Retrieved from http://www.greenbelt.ca/climate_change_adaption_ontario_s_resilient_greenbelt

Tonkiss, F. (2014). *Cities by design: The social life of urban form*. Hoboken, NJ: John Wiley & Sons.

Tonkiss, F. (2015). Afterword: Economies of infrastructure. *City*, *19*(2–3), 384–91. https://doi.org/10.1080/13604813.2015.1019232

Wekerle, G.R., Sandberg, L.A., Gilbert, L., & Binstock, M. (2007). Nature as a cornerstone of growth: Regional and ecosystems planning in the greater golden horseshoe. *Canadian Journal of Urban Research*, *16*(1), 20–38.

Wilson, S. (2008). *Ontario's wealth, Canada's future: Appreciating the value of the Greenbelt's eco-services*. Vancouver, BC: David Suzuki Foundation.

Wolch, J.R., Byrne, J., & Newell, J.P. (2014). Urban green space, public health, and environmental justice: The challenge of making cities "just green enough." *Landscape and Urban Planning*, *125*, 234–44. https://doi.org/10.1016/j.landurbplan.2014.01.017

13 Building on Quicksand: Infrastructural Megaprojects in China

XUEFEI REN

Introduction

Since the early 1990s, China has constructed more megaprojects than any other country in the world, and many of these are infrastructural projects built on or extended to the urban periphery. Local governments are the biggest investors in infrastructural megaprojects, and municipal governments across the country have invested massively in roads, bridges, tunnels, parks, ports, airports, and subways. Private capital has also quickly flowed into the construction sector: private investment, often in partnership with local government, has spurred new megaprojects, especially in residential and commercial property development. Although the 2008 global recession may have led one to expect a significant decline in megaproject investment, in China the economic downturn has given the central government a legitimizing tool in the form of a stimulus plan to direct investment into infrastructural megaprojects. The spending on fixed-asset investment – a measure that captures building activities such as real estate and infrastructure construction, among other things – reached an alarming 70 per cent of the national gross domestic product (GDP) in 2013, and infrastructure spending has surpassed foreign trade as the biggest contributor to national economic growth (Barboza, 2011). China used 6.4 gigatons of cement in the three years of 2011, 2012, and 2013, more than the United States did in the entire twentieth century (Swanson, 2015).

How can we explain this unprecedented level of investment in infrastructural megaprojects in Chinese cities? To date the scholarship on urban megaprojects has mostly focused on post-industrial cities in North America and Europe. This literature has identified two major themes in the current wave of megaproject development since the early 2000s. First, many have pointed out the temporal variations of megaprojects: compared to the "great mega-project" era of the 1950s and 1960s, urban megaprojects today tend to be more inclusive, less disruptive, and more flexible for addressing different needs of local communities (Altshuler & Luberoff, 2003; Diaz Orueta & Fainstein, 2008). Second, several authors identify megaproject development as part of a broader strategy of entrepreneurial urban governance (Del Cerro Santamaria, 2013; Flyvbjerg, Bruzelius, & Rothengatter 2003; Olds, 2002). Noting that economic restructuring over the past several decades has intensified inter-city competitions for investment, they argue that urban governance has become increasingly dominated by high-profile projects that rely heavily upon private financial contributions.

These observations on urban megaprojects in the West, however, have limited explanatory power for China's megaproject boom. First, this shift from the "great mega-project era" to "do no harm" (Altshuler & Luberoff, 2003) is nowhere to be seen in Chinese cities, which since the early 1990s have been experiencing their own phase of large-scale and highly disruptive urban renewal. Urban megaprojects significantly changed the textures of Chinese cities. Traditional neighbourhoods were bulldozed to make way for new expressways, subway lines, and office and shopping complexes, and millions of residents were displaced from their homes in inner cities to the urban fringes. Second, the prevalence of megaprojects in Chinese city building calls into question whether they are an entrepreneurial strategy consciously pursued by local officials – an enquiry that can only be resolved through empirical fieldwork. Different from Western cities, urban megaprojects have become the "rule" instead of the "exception" for city builders in China.

Building on the literature on regional variations of urban megaprojects (Bezmez, 2008; Haila, 2008; Lehrer & Laidley, 2008; Ren & Weinstein, 2012), this chapter examines the economic and sociopolitical conditions that gave rise to China's megaproject era of the 1990s and 2000s. The chapter discusses both the deregulatory measures and financial innovations behind the massive boom of megaproject construction and the contestations and mixed legacies of China's megaproject boom. Megaprojects are conceived here broadly to include infrastructural projects (such as expressways, subways, sports stadia, airports, utilities) as well as mixed-use commercial and residential projects. Instead of focusing on particular kinds of megaprojects, this chapter examines the common conditions that underlie the planning and implementation of megaprojects, such as the regulatory reforms in the land, housing, and finance sectors.

The distinction between urban and suburban infrastructure is not particularly meaningful in the Chinese context. Unlike in the West, the governance structure of metropolitan regions in China is centralized around municipal government. What are sometimes referred as "suburbs" in the urban China literature are not independent municipalities but peripheral districts subordinate to municipal governments. These outlying districts exhibit a heterogeneous mix of built environment features different from the North American suburb, encompassing high-rise residential new towns, industrial parks, and agricultural villages. The Chinese word for "suburb"– *jiaoqu* – often entails negative meanings, such as backwardness and lack of development (Ren, 2017). Therefore, settlement occurring around the metropolitan edge that may be described as suburb in a Western context is better referred to as "urban periphery" in a Chinese context.

A series of deregulatory reforms in land, housing, and finance in China have significantly empowered municipal governments, which are the largest undertakers of infrastructural megaprojects. The land reform in the 1980s and 1990s separated the public land ownership from privatized land-use rights, thus making land a commodity to be transacted in the market. The public land ownership gives municipal governments enormous power to acquire land for megaprojects with far below market rate compensations to its current users. Since the early 1990s, the central government has also promulgated a series of regulations over housing demolitions and evictions, and these regulations have paved the way for large-scale residential displacement. Even with land acquired and resident populations evicted, megaproject construction would not be possible without financing. The rise of local investment corporations (LICs), a system of quasi–state enterprises set up by local governments, solved the problem by borrowing directly from the market, especially from state banks, on behalf of local governments. Thus, the various market-oriented regulatory reforms have created optimal conditions for the coming of China's great megaproject era.

Securing Land for Infrastructural Megaprojects

Infrastructural megaprojects often require acquisition and consolidation of land parcels, and securing land for megaprojects is no easy task in many countries. In China, land acquisition for urban megaprojects has been facilitated by a number of major legal and regulatory changes in the land sector since the beginning of the market reform. These regulatory changes have significantly empowered municipal governments in land acquisitions, which in turn have triggered strong resistance and contestations across the country.

Land was not a commodity in socialist China between the 1950s and 1970s, and it was allocated by the state to *danwei* (state enterprises) for specific industrial projects. Article 10 of the 1982 Constitution of the People's Republic of China specifies that no organization or individual may seize and sell land or make any other unlawful transfer of land (Yeh & Wu, 1996). In the early days of the market reform in the 1980s, local governments lacked capital to build infrastructure and attract foreign investors, and some coastal cities such as Shenzhen began to experiment with privatization of land-use rights. In the early phase of the land reform, before the mid-1980s, the main agenda for local governments was providing cheap land to foreign investors to induce them to help build infrastructure. The experiment in the south went well, and investors from Hong Kong and Taiwan flocked to the special economic zones in the Pearl River Delta to set up factories and take advantage of the cheap land.

To expand the land reform to other cities, the constitution was amended in 1988 by the central government with the clause, "the right to use land may be transferred in accordance with the provisions of law" (Yeh & Wu, 1996). Thus, the constitution for the first time recognized market transactions of land-use rights. Privatized land-use rights were treated differently than land ownership, which remained public. Individual and institutional investors could obtain land-use rights for up to seventy years for residential property development, and fifty years for commercial property development.

The agenda of local governments for privatizing land-use rights has changed since the late 1990s, from giving incentives to investors to provide infrastructure, to generating local government revenues. Under current fiscal policies, revenues from land leasing are counted as "extra-budgetary" and do not have to be shared with the central government. As land is leased by local governments to private parties, and transacted among investors, land-leasing and transaction fees have become a major source of municipal government revenue. Governments use the revenue obtained from land leasing to improve urban infrastructure, which in turn can open up more land for development.

China's hybrid land market, comprising both public ownership and private land-use rights, has been operating with a multitrack system in which land is transacted through both public auctions and under-the-table negotiations between the government and private investors (Lin, 2009; Lin & Ho, 2005). Because local governments commonly transfer land-use rights through negotiations, the central government has sought to improve market transparency by urging that all commercial land be transferred publicly through auctions to the highest bidder. But as a practical matter, due to the intertwined interests of local governments and investors, under-the-table negotiations remain the dominant method of land leasing. Many officials and their relatives are shareholders in real estate companies and construction projects, and local governments commonly lease out land at prices far below market value to these well-connected enterprises. Local governments prefer negotiations to auctions because of the flexibility to lower land prices as a means of attracting investors (Xu, Yeh, & Wu, 2009). Good connections with local governments are crucial for individuals and companies trying to obtain prime land at a fraction of its market price.

State expropriation of land takes many forms, from violent means such as mobilization of the armed forces and forced demolitions to administrative means such as manipulation of the national land ownership classification scheme, or *hukou* system.

In China, urban land is owned by the state and rural land by village collectives. The state has the right, according to the current constitution, to acquire rural land from farmers and urban land from its current users in the name of the public interest. In most cases, farmers and urban residents are compensated by local governments at rates far below the market prices that investors would pay to acquire the same land, and the local governments can easily pocket huge profits through land transactions – by acquiring land cheaply and then leasing it to investors.

The protest over land acquisition in Wukan village in the Pearl River Delta marked a turning point in Chinese peasants' resistance against land grabs. In 2011 villagers stormed into government offices, chased away local officials, and blocked the roads after they learned that the village government had leased out about a third of the village land to an outside investor. One of the representatives chosen by the villagers to negotiate with the village government had died in police custody. The standoff between the villagers and security forces sent by local authorities continued for months. Finally, in 2012, the Guangdong provincial government intervened. The village officials were dismissed, the land deal was put on hold, and the villagers were allowed to hold elections to choose new leaders (Jacobs, 2012).

Land expropriation can also be accomplished by changing the *hukou* status of farmers. *Hukou* divides the population into categories of urban and rural. Rural *hukou* holders, by law, are entitled to the land in the villages where they reside. In 2004, Shenzhen converted 235 square kilometres of rural land to urban land in Bao'an and Longgang districts by reclassifying the *hukou* status of villagers from "agricultural" to "urban," or "non-agricultural." After becoming "non-agricultural" *hukou* holders, villagers are no longer entitled to rural land, and village collectives are abolished. Farmers are often compensated at rates far below market prices, and local governments apply most of the compensation to buying pensions and medical insurance for the farmers and the rest to building community facilities. The strategy of acquiring farmers' land by making them urban *hukou* is widely pursued by many large cities (Xu et al., 2009).

To summarize, the current land market in China – a legacy of both the Communist revolution and the market reform – is unique, and it makes land acquisition for megaprojects much faster, more disruptive, and undemocratic as compared to other countries. The public ownership over urban land gives municipal governments a monopoly to acquire land without having to offer adequate compensation to its current users, and the privatization of land-use rights – in contrast to land ownership – has further strengthened municipal governments by channelling land-leasing revenues to municipal coffers. The ready availability of land has prepared the ground for the coming of China's megaproject era.

Demolitions and Evictions

Public land ownership alone, however, cannot free up sufficient land for megaprojects because most urban land parcels are used by communities and enterprises in the densely populated Chinese cities. Building urban megaprojects in China almost inevitably requires demolition of existing housing stock and eviction of residents. Large-scale urban renewal programs, designed to make inner-city land available for real estate speculation and megaproject constructions, began in the early 1990s and peaked in the mid-2000s. As with regulatory reforms in the land sector, the central government has implemented a series of policies to legalize housing demolitions and speed up the pace of land acquisition for profit-driven urban megaprojects.

Large-scale housing demolitions began in the early 1990s in Beijing and Shanghai first, as these two cities adopted ambitious plans to remake themselves into global cities. For instance, in 1990 Beijing's municipal government launched an Old and Dilapidated Housing Renewal Program to redevelop three million square metres of old urban housing stock within fifteen years. The target of this first wave of redevelopment was inner-city neighbourhoods. In 1993, the then Xicheng district government began to develop a financial district within the second ring road – the heart of historic Beijing – razing many *hutongs* (traditional neighbourhoods with courtyard houses and narrow alleys). Driven by inter-district competition, other district governments soon followed suit to build their own flagship megaprojects. Oriental Plaza, a massive shopping and office complex investment by the Hong Kong business tycoon Li Ka-Shing, underwent construction in 1993, and more *hutong* neighbourhoods vanished in spite of opposition from residents and preservationists (Johnson, 2005; Wang, 2003; Zhang & Fang, 2004). The 2008 Beijing Olympics further fuelled the machinery of urban renewal, and millions of residents were displaced from inner-city neighbourhoods to the outskirts without fair compensation (Ren, 2011). By 2010 the work of urban renewal was mostly completed. Ordinary citizens, unable to afford the skyrocketing housing costs and rents, were relegated to the urban periphery of the continually sprawling Beijing. The "growth machine" of local governments and private developers was driving urban renewal programs, exacerbating the power imbalance that divided residents and the private–public growth coalition (Chen, 2009; He & Wu, 2005; Zhang, 2002).

To understand the massive scale of forced housing demolitions in the 1990s and 2000s, one has to examine, again, the regulatory changes made by the state to enable such practices. Until 2011, forced demolitions were legalized with the Urban Housing Demolition Regulation enacted successively in 1991 and revised in 2001 (State Council, 1991; 2001). Both versions of the regulation formally legalized the practice of redeveloping neighbourhoods without residents' consent and carrying out forced demolitions by administrative order. Residents could appeal to the People's Court, but the court could not issue an order to stop demolitions, and the regulation explicitly prohibited any delays in the demolitions during appeal if residents were provided some form of compensation or temporary housing. Mass demolitions in the 1990s led to widespread protests, and the central government eventually revised the demolition regulation in 2001. The 2001 regulation tightened the control over developers, but the fundamental power relationships did not change: municipal governments were vested with great power to acquire housing from residents by administrative order, and residents were excluded from the decision-making process of urban renewal programs.

The State Council replaced the 2001 regulation with new legislation in 2011. The most significant change was forbidding the eviction of residents through violence or otherwise forcing them out through practices such as cutting off water, heat, and electricity (State Council, 2011). Departing from previous laws, the 2011 Urban Housing Acquisition and Compensation Regulation requires consent from the majority of residents before any urban renewal project can proceed (Article 11). It specifies the standard of compensation that includes the market value of the acquired property, relocation costs, and other losses incurred by the affected residents and businesses (Article 17). The 2011 regulation requires an independent third-party real estate agency to assess the market value of the property, and these assessors are subjected to fines of between 50,000 to 200,000 yuan for flawed practices. It orders both monetary compensation and in-situ resettlement, or a combination of the two, for affected residents to choose.

The 2011 regulation reflects the changing power relations among the main stakeholders in redevelopment processes. Although municipal governments are the key decision-makers over urban renewal programs, their power is no longer unchecked. Rights to housing are better protected today than in the 1990s and 2000s. Large-scale urban renewal programs can no longer be initiated without consent from residents, as the state is increasingly concerned with social unrest. The Shanghai bylaw of the 2011 regulation, for example, requires 90 per cent of residents' consent for any urban renewal project to proceed, after which 80 per cent of the residents must agree to the compensation terms in order for demolition to take place (Shanghai Municipal Government, 2011).

Housing demolitions in China, as in other countries, are highly contested, and forced demolitions have sparked widespread protests and demonstrations across the country dating back to the early 1990s. The changes ushered in by the 2011 regulation over housing demolitions, including majority consent requirements and market rate compensation for residents, can be interpreted as a response from the state to the two-decades-long housing-rights movements waged across the country. The pace of urban redevelopment and demolitions in China has clearly slowed down after 2010, partly because residents need to be compensated at the market rate for their housing, and developers and local governments can no longer afford megaprojects that involve relocating a large number of residents in central locations.

Financing Infrastructural Megaprojects

How local governments can invest so much in infrastructure often puzzles both specialists and casual observers of Chinese cities. The funding for local governments to invest in infrastructure megaprojects comes from a variety of sources, the most important of which are the local investment corporations (LICs), quasi–state enterprises set up by municipal governments to borrow directly from the market to fund infrastructural megaprojects.

Municipal governments have long replaced *danwei* (state enterprises) as the main provider of infrastructure. Most municipalities do not have sufficient tax revenue – for example, they do not collect property taxes – to finance their ambitious infrastructure projects. Most properties are still owned by the public sector, and local governments are reluctant to tax their own assets and share the tax revenue with the central government. Moreover, the propertied middle class resists any policy experiments to impose property taxes. Given the limited local tax revenues and diminishing subsidies from the central government, extra-budgetary revenues and borrowing from the capital market have become major sources of infrastructure financing. Wu (2010) finds that the Shanghai city government charges more than twenty-eight kinds of fees to real estate investors and that this extra-budgetary income accounts for up to 25 per cent of infrastructure investment funds. Income from land leasing also counts as extra-budgetary revenue, and has become the most significant source of revenue for local governments. In 2013, the Shanghai city government pocketed more than 200 billion yuan from leasing land to developers – income that did not have to be shared with the central government – and land-leasing revenues from the twenty largest cities in the country reached 1,500 billion yuan, a 63 per cent increase over the previous year (Xinhua Net, 2014).

Moreover, borrowing from domestic and foreign capital markets through bonds, equities, loans from state-run banks, stock market listings, and joint ventures has become a new way to raise funds, accounting for more than 30 per cent of infrastructure investment capital (Wu, 2010). Among these types of borrowing, loans from state-run banks have become the main method for municipal governments to finance their ambitious infrastructure projects. As the central government has tightened requirements for state-run banks' lending to municipal governments, special-purpose investment platforms including LICs have been set up to evade these regulations. According to some estimates, by

2011 more than 10,000 LICs in capital markets had been set up by local governments to raise funds for investing in infrastructure (Wong, 2011). Bank loans do not go directly to municipalities, but instead go into the accounts of these investment companies that finance the construction of roads, bridges, subways, and government offices. City-owned land valued at high prices serves as collateral for these loan deals. In this way, the debt incurred does not show up on the municipalities' balance sheets. As a result of using off-balance-sheet vehicles to raise funds, the size and structure of local government debt are opaque.

Build-operate-transfer (BOT) is a widely adopted model for building infrastructural megaprojects. The costs are shared between private investors and local governments; after the initial investments are recovered through toll charges, the ownership rights are transferred to local governments. Although BOT has proved to be effective at tapping into the private market to finance infrastructure – by charging exorbitant fees to users – it has turned various infrastructure projects into profit-making machines for local governments and their private partners. For instance, in 2012 China possessed 100,000 kilometres of the 140,000 kilometre toll roads throughout the world, and by 2015 China had increased its toll roads to 162,600 kilometres (Bloomberg News, 2015; Wu, 2012). Owning and driving a car is a symbol of middle-class living in China, but it has also become costly with the endless toll road charges. The surge in toll charges for roads and bridges has also contributed to the inflation of food prices, as meat and vegetables have to be trucked into cities via toll roads on a daily basis.

Chinese cities have made remarkable progress in building urban infrastructure by tapping into private capital from both domestic and foreign sources. The massive investment in infrastructure has been facilitated by public–private partnership and non-transparent LICs for fund-raising. Many infrastructure megaprojects are profit-making machines for both local governments and private investors. But the growing inability of local governments to repay their infrastructure loans threatens to derail the growth model based on infrastructure investment.

Consequences of Building Big: Municipal Debt

Soon after the 2008 Olympics, the economic recession set in, but in China the recession was used as a legitimizing tool to direct more investment into infrastructural megaprojects. The recession alarmed China's top leaders, who worried that the country would be dragged into economic stagnation, with adverse consequences in terms of political stability. The central government responded in 2009 by rolling out a stimulus plan of 4 trillion yuan – the world's largest, and three times larger than Obama's package – among which 1.18 trillion would be provided by the central government with the rest to be matched by local governments through bank loans (Wong, 2011). The stimulus package amounted to 12.5 per cent of China's GDP in 2008 and was to be spread out over a period of twenty-seven months. The top three priority areas earmarked for the spending were transport and power infrastructure, post-earthquake construction, and affordable housing, with the rest extending to rural infrastructure, environmental protection, technology, health, and education. This large fiscal injection triggered deregulation in the financial sector, especially regarding bank lending, and also a relaxation of fiscal rules over local debt.

Local governments are the key players entrusted with implementing the 2009 stimulus program, and specifically with identifying particular infrastructure projects to be funded. To help local governments raise capital and co-finance these projects, the central government made a series of deregulatory moves (Wong, 2011). For example, the State Council approved a 200-billion-yuan treasury bond on behalf of local governments, and the Ministry of Finance relaxed the standards on what counterpart funds could qualify for stimulus projects – basically all sources at the discretion of local governments could be used for co-financing.

One particular deregulatory measure has been the widespread use of LICs. Since local governments themselves are not allowed to borrow from the market, they set up LICs to borrow from banks and raise funds through corporate bonds in order to invest in infrastructure and enhance city revenue in general. The LICs bundle together bank loans and other funds raised, and use municipal assets as equity and collateral. Due to the public ownership of urban land, land has become the principal asset backing LICs, and city governments have used income from land leasing to service LIC debt. No national agencies, not even the Ministry of Finance, can oversee the tens of thousands of LICs in the country set up by local governments. During the stimulus-spending period from 2008 to 2010, local governments borrowed massively through LICs and accrued a large amount of local debt. Servicing the debt has become a problem for many cities struggling with less robust economic growth and lower tax revenues.

Because of the overinvestment in infrastructure brought about by the stimulus, Chinese city governments have accumulated a large amount of debt since 2008. The local government debt in the country doubled from 7 trillion yuan in 2009 to 14 trillion yuan at the end of 2010 (Wong, 2011). The soaring municipal debts have to be explained by both long-term structural tendencies in municipal finance, such as the mismatch of revenues and responsibilities, and short-term policy programs, such as the 2009 stimulus program of the central government for which local governments were encouraged to borrow massively in order to match central subsidies and invest in infrastructures.

An urban fiscal crisis has been in the making in China since 2008. The rise of LICs has significantly enhanced the fiscal capacity and autonomy of local governments, but it has also created an unsustainable level of municipal debt because of binge borrowing by LICs from state banks. The deregulatory measures taken by the central government have significantly expanded the borrowing power of local governments and spawned an escalation of municipal debt. To rein in local debt, the central government is increasingly turning to the public–private partnership (PPP) model for financing. In May 2015, the central planning agency, the National Development and Reform Commission, announced a list of 1,043 projects totalling 1.97 trillion yuan (about USD 317.75 billion) for which it is inviting private investors to help build and operate. But as long as the state banks continue to lend to local governments, it is questionable whether the shift to the PPP model can help to reduce local government debt (Goh, 2015).

Conclusion: Lessons from China

The construction of infrastructural megaprojects has significantly altered the structure of Chinese cities and the way people live. Chinese cities in the socialist era were fairly compact, and the boundary between cities and the countryside was clear. The compact urban form of socialist cities was gradually dismantled with the construction of infrastructural megaprojects such as subways, expressways, airports, and massive-scale new towns targeting the middle class, who cannot afford to live in inner cities anymore. Old neighbourhoods were bulldozed to make way for shopping malls and luxury housing, and agricultural land on the periphery was acquired to build development zones and more high-rise residential buildings for the middle class. In the process, older urban fabrics were destroyed and replaced by non-human–scale megablocks, and millions of urban households and peasants were either relocated or simply evicted. The displaced would find accommodation on the periphery where real estate prices are cheaper. Today, the average urbanite in a top-tier Chinese city lives in a high-rise apartment building on the urban periphery and commutes to work by subway or private car. The massive investment in public transit has improved the connectivity between the centre and periphery, but the mismatch between housing, mostly on the periphery, and jobs, mostly at the centre, makes a long commute inevitable. The sprawled urban form resulting from property development and infrastructural projects has changed how Chinese live, work, and socialize. Future research is needed to study the new Chinese urbanism as a way of life in the aftermath of the megaproject boom.

This chapter has examined the economic and sociopolitical conditions in China that have made the massive-scale construction of megaprojects possible. Since the beginning of the market reform in 1978, governments at all levels have initiated a series of market-oriented regulatory reforms in sectors of land, housing, and infrastructure financing. These regulatory changes explain why China has been able to build more infrastructural megaprojects in the past two decades than any other country in the world.

China's experience in infrastructural investment offers two major lessons for other cities. First, the provision of urban infrastructure is strongly conditioned by local state capacity. China offers an extreme case of strong municipal institutions operating under a non-democratic but highly decentralized system. Power, authority, and resources have been devolved to local governments since the 1990s, and decision-making over infrastructural investment is fast and centralized within municipal government in the absence of electoral politics. The empowered local state is behind most infrastructural megaprojects. China's story offers a counter-example to the narrative of cities in the global south that emphasizes informality and inadequacy. As shown in this chapter, most infrastructural projects in China are "formally" provided by the local state or public–private partnerships, and the infrastructural sector shows clear signs of overinvestment rather than lack of funding. Second, China's example also illustrates the financial consequences of overinvesting in infrastructural megaprojects. Today, many local governments find themselves in deep debt from overinvestment in infrastructural megaprojects. Since the local government debt had reached about 40 per cent of the GDP in 2015, it is unsustainable, and many local governments and investment vehicles will default on their debt without debt restructuring. In 2015, the Ministry of Finance introduced a refinancing program for local governments, which will be allowed to swap about 3 trillion yuan (USD 160 billion) of high-interest debt for lost-interest bonds (S.R. [Shanghai], 2015). The economic slowdown and the mounting municipal debt signal the end of China's megaproject era.

The chapter has focused on the regulatory, institutional, and, especially, financial conditions for the massive urban and suburban infrastructure boom that has taken place in China. It thus concentrates on "infrastructures for infrastructures," instruments that enable infrastructures, just as infrastructures themselves enable different forms of urban and suburban development and activities taking place in these environments. The preoccupation of chapter 13 with the financial conditions of suburbanization echoes the object of chapter 4, which examined financial mechanisms making the existence of the Canadian, and by extension the North American, suburb possible. A comparison of the chapters reveals similarities and differences between how the two suburban contexts have been financed. In both cases, the development of suburbs is perceived as a macroeconomic instrument intended to stimulate the national economy in an attempt to compensate for other economic sectors that are underperforming. And in the two contexts, spending in and on suburbs relies on debt. But these two factors are where similarities between the two cases end. The North American suburban context, seen through the lens of chapter 4, is depicted as dependent on personal debt needed to support the consumption of housing and other durable goods. Meanwhile, chapter 13 has demonstrated how in China the focus of debt financing is on large-scale urban and suburban infrastructures, and the burden of this debt falls mostly on the shoulders of municipal administrations.

NOTE

This chapter is adapted from Xuefei Ren, Biggest infrastructure bubble ever? City and nation building with debt-financed megaprojects in China, in *The Oxford Handbook of Megaproject Management*, edited by Bent Flyvbjerg, 137–53, Oxford: Oxford University Press, 2017.

REFERENCES

Altshuler, A., & Luberoff, D. (2003). *Mega-projects: The changing politics of urban public investment*. Washington, DC: Brookings Institution Press.

Barboza, D. (2011, 6 July). Building boom in China stirs fears of debt overload. *New York Times.* Retrieved from http://www.nytimes.com/2011/07/07

/business/global/building-binge-by-chinas-cities-threatens-countrys-economic-boom.html

Bezmez, D. (2008). The politics of urban waterfront regeneration: The case of Haliç (the Golden Horn), Istanbul. *International Journal of Urban and Regional Research, 32*(4), 815–40. https://doi.org/10.1111/j.1468-2427.2008.00825.x

Bloomberg News. (2015, 5 August). China considers plan to bolster toll roads in search for growth. *Bloomberg News*. Retrieved from http://www.bloomberg.com/news/articles/2015-08-05/china-considers-plan-to-bolster-toll-roads-in-search-for-growth

Chen, X. (Ed.). (2009). *Shanghai rising: State power and local transformations in a global megacity*. Minneapolis, MN: University of Minnesota Press.

Del Cerro Santamaria, G. (Ed.). (2013). *Urban megaprojects: A world view*. Bingley, UK: Emerald Group Publishing.

Diaz Orueta, R., & Fainstein, S. (2008). The new mega-projects: Genesis and impacts. *International Journal of Urban and Regional Research, 32*(4), 759–67. https://doi.org/10.1111/j.1468-2427.2008.00829.x

Flyvbjerg, B., Bruzelius, N., & Rothengatter, W. (2003). *Megaprojects and risk: An anatomy of ambition*. New York: Cambridge University Press.

Goh, B. (2015, 25 May). China invites private investors to help build $318 billion of projects. *Reuters*. Retrieved from http://www.reuters.com/article/2015/05/25/us-china-economy-infrastructure-idUSKBN0OA07R20150525

Haila, A. (2008). From Annankatu to Antinkatu: Contracts, development rights and partnerships in Kamppi, Helsinki. *International Journal of Urban and Regional Research, 32*(4), 804–14. https://doi.org/10.1111/j.1468-2427.2008.00824.x

He, S., & Wu, F. (2005). Property-led redevelopment in post-reform China: A case study of Xintiandi redevelopment project in Shanghai. *Journal of Urban Affairs, 27*(1), 1–23. https://doi.org/10.1111/j.0735-2166.2005.00222.x

Jacobs, A. (2012, 1 February). Residents vote in Chinese village at center of protest. *New York Times*. Retrieved from http://www.nytimes.com/2012/02/02/world/asia/residents-vote-in-chinese-village-at-center-of-protest.html

Johnson, I. (2005). *Wild grass: Three stories of change in modern China*. New York: Pantheon Books.

Lehrer, U., & Laidley, J. (2008). Old mega-projects newly packaged? Waterfront redevelopment in Toronto. *International Journal of Urban and Regional Research, 32*(4), 786–803. https://doi.org/10.1111/j.1468-2427.2008.00830.x

Lin, G. (2009). *Developing China: Land, politics and social conditions*. London: Routledge.

Lin, G., & Ho, S. (2005). The state, land system, and land development processes in contemporary China. *Annals of the Association of American Geographer, 95*(2), 411–36. https://doi.org/10.1111/j.1467-8306.2005.00467.x

Olds, K. (2002) *Globalization and urban change: Capital, culture, and Pacific Rim mega-projects*. New York: Oxford University Press.

Ren, X. (2011). *Building globalization: Transnational architecture production in urban China*. Chicago, IL: University of Chicago Press.

Ren, X. (2017). Lost in translation: Names, meanings, and development strategies of Beijing's periphery. In R. Harris & C. Vorms (Eds.), *What's in a name? Talking about urban peripheries* (pp. 316–33). Toronto, ON: University of Toronto Press.

Ren, X., & Weinstein, L. (2012). Urban governance, megaprojects, and scalar transformations in China and India. In T. Samara, S. He, and G. Chen (Eds.), *Locating right to the city in the global South: Transnational urban governance and socio-spatial transformations* (pp. 107–26). New York: Routledge.

Shanghai Municipal Government. (2011). *Shanghai Urban Housing Acquisition and Compensation Regulation.*

S.R. (Shanghai). (2015, 11 March). China's local government debt: Defusing a bomb. *The Economist*. Retrieved from http://www.economist.com/blogs/freeexchange/2015/03/china-s-local-government-debt

State Council. (1991). *Urban Housing Demolition Regulation.*

State Council. (2001). *Urban Housing Demolition Regulation.*

State Council. (2011). *Urban Housing Acquisition and Compensation Regulation.*

Swanson, A. (2015, 24 March). How China used more cement in 3 years than the U.S. did in the entire 20th century. *The Washington Post* (Wonkblog). Retrieved from http://www.washingtonpost.com/news/wonkblog/wp/2015/03/24/how-china-used-more-cement-in-3-years-than-the-u-s-did-in-the-entire-20th-century/

Wang, J. (2003). *Beijing record*. Beijing: Sanlian Press.

Wong, C. (2011). The fiscal stimulus program and public governance issues in China. *OECD Journal on Budgeting, 11*(3), 1–21. https://doi.org/10.1787/budget-11-5kg3nhljqrjl

Wu, W. (2010). Urban infrastructure financing and economic performance in China. *Urban Geography, 31*(5), 648–67. https://doi.org/10.2747/0272-3638.31.5.648

Wu, Y. (2012, 25 April). Roads take an unbearable toll. *China Daily*. Retrieved from http://usa.chinadaily.com.cn/business/2012-04/25/content_15137576.htm

Xinhua Net. (2014, 3 January). Land leasing revenues in 20 cities increased by more than 60 percent. *Xinhua Net*. Retrieved from http://news.xinhuanet.com/fortune/2014-01/03/c_125954247.htm

Xu, J., Yeh, A., & Wu, F. (2009). Decoding urban land governance: State reconstruction in contemporary Chinese cities. *Urban Studies, 46*(3), 559–81. https://doi.org/10.1177/0042098008100995

Yeh, A., & Wu, F. (1996). The new land development process and urban development in Chinese cities. *International Journal of Urban and Regional Research, 20*(2), 330–53. https://doi.org/10.1111/j.1468-2427.1996.tb00319.x

Zhang, T. (2002). Urban development and a socialist pro-growth coalition in Shanghai. *Urban Affairs Review, 37*(4), 475–99. https://doi.org/10.1177/10780870222185432

Zhang, Y., & Fang, K. (2004). Is history repeating itself? From urban renewal in the United States to inner-city redevelopment in China. *Journal of Planning Education and Research, 23*(3), 286–98. https://doi.org/10.1177/0739456X03261287

14 Retrofitting Obsolete Suburbs: Networks, Fixes, and Divisions

REBECCA INCE AND SIMON MARVIN

Introduction

British suburbs were largely created under a modernist growth logic that separated home from work and encouraged high-consumption lifestyles. However, these prevailing conditions have changed, with global pressures such as climate change, fears over resource security, and economic crisis creating new priorities around domestic energy use of carbon regulation, economic resilience, and energy security (Hodson & Marvin, 2009). These combined pressures have made the reduction of energy use a political and social priority, thus highlighting the need to change the way that energy efficiency in suburban housing is addressed (Sims, 2012) and to develop new capacity for domestic retrofit as an urban infrastructure.

This chapter views retrofit responses as attempts to create new urban infrastructures that tackle multiple entwined global and local pressures. Looking at retrofit in this way shows how an apparently mundane act of installing insulation, boilers, and other energy-efficient technologies into people's homes connects global challenges with suburban lives. These isolated actions problematize the individual suburban household; yet, the infrastructural solution for global challenges relies on the buy-in and decisions of millions of individual homeowners. Furthermore, an analysis of retrofit illustrates the place-specific, messy, and complex interrelations between "soft" (that is, social) elements of an infrastructure and "hard" (that is, material and technical) elements, providing a lens for investigating the multiple relational facets of a complex sociotechnical issue. It particularly illuminates how suburban spaces are positioned as sites of infrastructural experimentation through policy, and focuses attention on the lived reality of these experiments for those involved in them.

This chapter examines the issues involved in developing retrofit as an integrated urban infrastructure in the particular context of Haringey, North London. It explores the challenges of creating a retrofit infrastructure under conditions of austerity governance in which the state prioritizes selective market-based approaches, and critically examines the effects on the local suburban context. We ask the following questions: What enabled or hindered the creation of this infrastructure? To what extent did the national policy context condition the local response? What was produced by this infrastructural experimentation, and what effects did it have on the local context? And ultimately, did this form of deliberate place-based experimentation successfully produce a working infrastructure system?

Constructing Retrofit as a New Urban Infrastructure

There are approximately twenty-two million households in England, and 83 per cent of those properties are in urban or suburban areas (Department for Communities and Local Government [DCLG], 2014). The vast majority of properties that stand now – around 90 per cent – were built more than twenty years ago (DCLG, 2014), at times when English suburbs were almost exclusively *not* built with energy efficiency as a priority. With the energy use of existing housing stock making up over a quarter of the total energy use of the United Kingdom (around the same as road transport, more than heavy industry, and far more than air transport) (HM Government, 2011), poor domestic energy efficiency has become a problem. Furthermore, with a demolition rate of domestic properties at approximately 0.3 per cent per year, most of those existing properties – at least 75 per cent – will likely still stand in 2050 (Ravetz, 2008). Thus, any attempt to significantly reduce the energy demand of domestic properties must address the existing housing stock.

The housing itself has not changed, but the energy efficiency standards by which it is assessed have: there is "slippage" between the original logic upon which previous housing was built and the new priorities – particularly targets of reducing carbon emissions by 80 per cent by 2050. This slippage renders existing housing obsolete, and improved energy efficiency at the household level, through various policy measures intended to increase retrofitting, is now seen as the "fix" to deal with this obsolescence (Theobald & Shaw, 2014). Simply put, domestic retrofit means people making structural or technological alterations to an existing domestic property for the specific purpose of improving that property's energy efficiency performance by reducing its energy use/demand. This reduction involves a range of activities and alterations either concerning the building fabric or the building's systems (Swan, 2013). Retrofitting the building fabric includes insulating walls, lofts, and floors; replacing or improving areas of glazing; and draught proofing or improving air tightness to reduce heat losses from the property and thus reduce the energy required to heat it. Retrofitting the building's systems includes modifying mechanical and electrical systems such as heating/air conditioning, lighting, small power (that is, power to kitchen equipment and sockets), and water (particularly hot water) to reduce their energy demand. These modifications can include replacing boilers and central or electric heating with more efficient systems; altering the fuel source for these systems (for example, from gas/electricity to biomass); installing microgeneration technologies such as photovoltaic (solar) panels or heat pumps to provide electricity, air heating/cooling, or hot water; and replacing light fittings with more efficient ones.

Domestic retrofit has the potential to reduce carbon emissions, decrease dependency and vulnerability to energy supply and pricing issues at both individual household and national scales, protect people against ill health and financial hardship as a result of fuel poverty (that is, households spending a disproportionate amount of their income on energy), and create a new industry that will boost employment and economic growth (Guertler, 2012; Kelly, 2009; Williams et al., 2013).

However, there is no existing conceptual framework for a suburban retrofit infrastructure that adequately captures the multiple issues and factors at play. Domestic retrofit is not solely a "hard" infrastructure like public transport, water, or waste services, and it is not solely a "soft" service infrastructure like care services. It relationally comprises elements of both: The hard elements are the material or technical elements of the infrastructure such as the properties themselves, and the heating and lighting systems, products, and technologies such as insulation, boilers, windows, solar panels, or underfloor heating that might be installed in the retrofitting process. The soft elements of the infrastructure comprise the multiple networks of organizations, people, and processes, such as finances, surveys, information, and advice.

The primary function of this potential urban infrastructure is to effectively channel the products and technologies into private homes through the relational network. Key elements of stable, working relational infrastructure services (Graham, 2010; Graham & Thrift, 2007;

Kimbell, 2015; Star, 1999) include their embeddedness and familiarity (invisible unless broken), their embodiment of social norms and conventions of practice, their knitting with habits of the service users' normal lives – that is, their user-friendliness – and their equitability and inclusiveness. The decision to retrofit involves a range of motivations and considerations (Tweed, 2013) that take into account lifestyles, cultural values, temporal events, and attitudes towards home improvements. The complexity of domestic retrofit, with multiple technologies and material elements and numerous actors and contributing factors, have to be assembled as an effective and stabilized sociotechnical system (Geels 2002; 2005; 2010). This perspective allows us to understand a domestic retrofit infrastructure as the sum of its component parts, which range from technologies or products to the people involved in their development and implementation. A complete retrofit infrastructure would require at least four working system areas: the technical (products, systems, and building components and fabrics); production (installers, manufacturers, skilled workers, and so on); user practices (behaviour and preferences in multiple housing sectors); and government (policy and regulations) and non-government (for example, insurances, contracts, and warranties) regimes (Swan, 2013; Vergragt & Brown, 2012).

A sociotechnical perspective also helps us to understand the complexity of transitioning from an existing regime based on domestic property repair and maintenance according to existing energy performance to the systemic reshaping of energy performance around new low carbon priorities. This reshaping requires both change in the existing norms and practices that guide domestic retrofit and the establishment of a new sociotechnical system for delivering retrofit services. Change in sociotechnical systems occurs when broad "landscape" pressures, along with niche-level experimentation with both products and approaches, create significant instability in the existing regime, thus opening up opportunities for reconfiguring or transforming existing norms and practices around domestic retrofit. However, existing regimes are notoriously resistant to change (Seyfang & Smith, 2007; Smith, 2007), and viewing existing housing and domestic energy use as an obdurate regime reminds us why widespread changes in domestic energy efficiency may be difficult to achieve.

Despite these difficulties, there have been attempts to envision more wide-ranging notions of retrofit through integrated regeneration strategies for the reuse and adaptation of suburban spaces as a whole

(Dunham-Jones & Williamson, 2011). This approach substantially broadens the concept of retrofit from focusing on energy-saving technologies to encompassing multiple suburban infrastructures and systems, such as the retrofit of a deserted mall becoming a focal part of a connected and walkable urban development. This vision of an incrementally "retrofitted suburbia" is dependent upon more systemically planned suburban transformation at a metropolitan scale, incorporating civil society, developers, and state institutions in a unified approach. Creating an effective urban infrastructure system for domestic retrofit therefore requires a holistic approach that considers all relevant technologies, actors, and factors, and is embedded and integrated with suburban life and other suburban redesign processes and designed thoughtfully and inclusively with householders in mind. The issue then becomes how and whether a retrofit system is constituted as a stable and effective place-based infrastructure through sociotechnical experimentation by state and urban authorities.

State Retrofit Responses: "Able to Pay" and "Fuel Poor"

Urban authorities have been undertaking local forms of experimentation to develop retrofit as an infrastructure service but have been doing so in heavily conditioned and prescribed national policy contexts. Austerity conditions require local governments to compete for national funding in order to experiment with the development of local responses. State funding and support is heavily regulated by national priorities that have attempted to frame retrofit as a market response for those households able to pay through the Green Deal and a subsidized response for the fuel poor through the Energy Company Obligations (ECOs).

The Green Deal was a financial mechanism that enabled householders to access loan finance to pay for retrofit measures and to pay back the cost through energy savings by means of a loan attached to their energy bill rather than to themselves as individuals. It was intended to overcome the significant perceived barrier to retrofit of the high upfront cost – particularly for solid wall insulation – and to open up finance options to those who may otherwise not be able to access them (for example, those with poor credit ratings). Additionally, it provided a framework for supporting and upskilling a supply chain for retrofit; establishing a standardized retrofit service provided by professionals such as assessors, finance providers, and installers; and coordinating Green Deal Providers. Service professionals were all accredited to an industry standard (PAS 2030 for installers) in order to create trust and

Figure 14.1 The Green Deal process

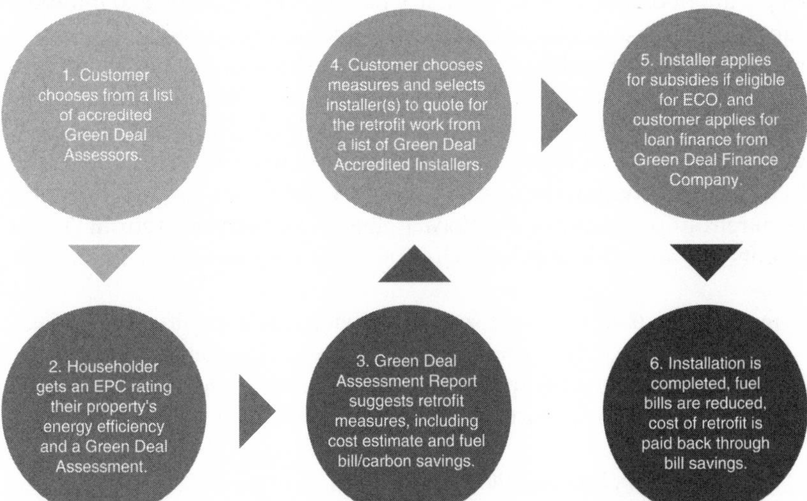

ECO = Energy Company Obligation; EPC = Energy Performance Certificate

confidence in them and avoid the inclusion of "cowboys" in the industry. The process for accessing Green Deal is outlined in figure 14.1.

The Energy Company Obligation (ECO) is a three-part legal obligation on energy companies to provide funding for energy efficiency measures. They must provide funding for specific households that are vulnerable to fuel poverty (Home Heating Cost Reduction Obligation), funding for specific areas considered deprived and in need of energy efficiency upgrades across an area (Carbon Saving Communities Obligation), and funding for specific household types that are considered "hard to treat," that is, homes requiring difficult cavity wall insulation, homes on off-grid locations, or, most commonly, solid walled homes that need costly and difficult internal or external wall insulation (Carbon Emissions Reduction Obligation). The Carbon Emissions Reduction Obligation component of the ECO has the most impact for private sector homes.

Developing a Holistic and Integrated Response

Using the two quite different policy mechanisms, the intention was to develop a systemic retrofit response at the urban level. Green Deal was "soft" launched in October 2012 (the first assessments took place), with the energy companies spending their ECO alongside it. In July 2013, the Department of Energy and Climate Change (DECC) made £20 million of capital funding available for local authorities to use to deliver key retrofit measures of solid wall insulation and boilers in a program called Green Deal Communities, of which Haringey was awarded £6.7 million in May 2014. This program was part of a broader localism agenda, in which urban authorities were being encouraged to define their own approaches to social and economic issues, including retrofit. DECC created this short-term retrofit program for the "local delivery" of domestic retrofit "experiments" in various localities (including Haringey), and its vision was to

> regenerate whole areas and see them having a real facelift and looking nicer, for people to be warmer and paying less for their bills but for that to be integrated with other systems like district heating: it needs to be holistic and integrated so not just a retrofit market on its own, but insulation is crucial to this because it has to be fabric first. (DECC official, personal communication, February 2015)

The Green Deal Communities program had explicit experimental priorities designed to (1) "kick start the market" for retrofit by offering subsidies, particularly for external wall insulation, which would increase the number of installations and socially "normalize" retrofit; (2) "create partnerships" or different local models for delivering retrofit (for example, collaborations between local authorities and large contractors, energy companies, or multiple delivery partners); and (3) "test the 'street-by-street' approach" in which households of varying tenures and circumstances could be offered a retrofit service of some kind. While this particular policy context and the conditions of the Green Deal Communities program created space and possibility for localized experimentation and an expectation of variation between places, it also fixed certain aspects of the retrofit process. It required use and blending of Green Deal and ECO, prioritized solid wall insulation and boilers as technological fixes with funding for additional measures conditional upon installing these, and provided only a short timescale of just over a year for experimentation.

A Partial Infrastructure?

The Green Deal/ECO framework embodied in the Green Deal Communities program represented an effort to create a new infrastructure for retrofit upon which local authorities could build an urban service.

However, while there were clear areas of focus in the national policy approach such as developing the supply chain and incentivizing market-pull through subsidies and loan finance, there were also significant deficits. First, the system is characterized by distinctly separate provisions and approaches for those households that are able to pay and those that are not. UK policy assumed that ECO subsidies would address households in fuel poverty, and thus stimulating a retrofit "market" would address privately owned and occupied households, but often these groups are not mutually exclusive. Second, the emphasis on particular technological "fixes" such as external wall insulation ignored fundamental technical, social, and cost issues associated with those technologies. Third, experimental and short-term streams of funding offered to local consortia (often led by local authorities) positioned suburbs as spaces of experimentation, responsible for coordinating fragmented policy. The consequence of this setup was a set of conditions and principles imposed on places, which were charged with "finding a solution," but these principles were inherently divisive of households in different socioeconomic circumstances. Policy encouraged the reconfiguring of household energy efficiency choices around market logic, and the focus on local experiments and partnerships in finding solutions created temporary, fluid networks rather than consistent, stable governance, as well as geographically uneven activities and retrofit "hotspots." In July 2015, funding for the Green Deal Finance Company was cut by the new Conservative government, with no replacement scheme.

The next section uses the example of Haringey, London, to illustrate, from the perspective of the network of actors and practitioners involved, the lived reality of trying to create an infrastructure for domestic retrofit in a specific urban context under these difficult policy conditions.

Haringey, London: Constructing Retrofit as a Developmental Opportunity

Haringey's particular response to the issue of domestic retrofit spans the period between autumn 2012, when retrofit was established as a political priority in anticipation of the Green Deal, and September 2015. During this time, Haringey Council provided funding to set up a retrofit cooperative in July 2013, which would provide retrofit services for the local area. In May 2014, Haringey Council, in partnership with five neighbouring boroughs, was also awarded £6.7 million to deliver a Green Deal Communities scheme running until September 2015, delivering over 1,000 retrofits to private sector housing, with installations continuing into the spring of 2016. This section explores how national and local priorities interacted through this time period to create a place-specific infrastructure with particular local effects.

Haringey has a growing population of around 225,000 people housed in 97,000 residential dwellings. As a suburb, it exhibits elements of both high-consumption suburban living in the west and inner-city deprivation in its eastern part. Haringey's particular socioeconomic mix makes it the most unequal borough in London: five of its eastern wards are among the poorest 10 per cent of all London wards (Northumberland Park, Tottenham Green, Tottenham Hale, West Green, and White Hart Lane), while four of its nineteen wards in the west are among the richest 10 per cent (Alexandra Palace, Highgate, Muswell Hill, and Crouch End). There are also significant differences in life expectancy between the boroughs (Carbon Commission, 2012). This inequality is even starker when viewed in terms of fuel poverty: Haringey's Joint Strategic Needs Assessment shows significantly higher earnings in the west of the borough, while higher incidences of fuel poverty occur in the lower earning east of the borough (Haringey Council, 2015).

Consequently, Haringey's priorities mirror, to some extent, landscape pressures around climate change, economic crisis, and resource scarcity, but in this specific context, economic and employment issues represent a more pressing and urgent issue. The borough's efforts to reduce inequality do not fit with national priorities around market-making and divisions between able-to-pay and fuel-poor households. On the one hand, there are clearly issues with fuel poverty and poor living conditions, particularly in the east of the borough, but the west houses wealthier communities that are highly informed and engaged with domestic retrofit and energy efficiency as a climate change issue. There are urgent local priorities around job creation, with boosting a local retrofit industry in order to tackle unemployment in the east of the borough seen as a clear task for any emerging retrofit infrastructure.

Carbon Commission – RetrofitWorks Co-operative

As a result of these issues specific to Haringey, the borough developed a strong local political will around climate change and the local economy, bringing together councillors, community leaders, and green and ethical business leaders, as well as the New Economics Foundation, to produce its 2013 Carbon Commission report. This report identified a range of key priorities for the borough, centred on a local target of reducing carbon emissions by 40 per cent by 2020 (labelled "40:20 target") and utilizing this target as an economic and social opportunity – particularly to reduce inequality – for the borough. The development of a domestic retrofit infrastructure was positioned in multiple ways: as a tool for reducing carbon emissions from the borough's homes; as an opportunity to build a local industry around retrofit and create jobs to tackle unemployment; and as an opportunity to develop alternative business models (such as cooperatives) to deliver retrofit that would reinvest revenue and wealth in the borough instead of losing it to external corporations.

Haringey produced a number of creative responses as a result, particularly its investment in the development of the RetrofitWorks Co-operative, a network of local retrofit installers, with community group members representing households and generating leads for the installers. RetrofitWorks was established in July 2013 after consultation with Haringey Council, local tradespeople, and community organizations, with around twenty North London–based installer members. Its intention was to create a network connecting and coordinating the retrofit process, offering holistic retrofit services integrated with other home improvements. RetrofitWorks also developed its own quotation and project management software, and generated revenue through project management fees, membership fees, and fees for using their software and quoting system. Any profit would be reinvested into the local network, enabling it to eventually offer discounted retrofits to those less able to pay. It intended to generate its own revenue and be an alternative to Green Deal rather than relying on government funding.

Smart Homes–Green Deal Communities

However, governance of retrofit in Haringey shifted from the Carbon Commission rationale to a top-down approach dictated by policy at the beginning of the Green Deal Communities scheme in May 2014, which made use of the £6.7 million funding from DECC. Haringey partnered with five neighbouring North London boroughs: Hackney, Enfield, Islington, Waltham Forest, and Camden, which provided geographical scale to meet high target numbers of retrofitted properties as well as existing resources and skills such as an energy advice line in

Islington and retrofit supply chain development projects in Enfield. The thirteen-month timescale of this project and the targets associated with it – retrofitting 1,210 buildings across the six boroughs, of which 718 were domestic owner occupied, 420 privately rented properties, and 72 non-domestic buildings – created the impetus for an immediate, larger scale domestic retrofit response with installations heavily subsidized by grant funding and supported by ECO funding. Under conditions of austerity governance, the local authority was under pressure to take on this very challenging project and its targets to try to sustain local capacity; without it, an entire team from within the council would have been without funding for their posts. Significantly, the only member of the team responsible for maintaining the Carbon Commission agenda and priorities was made redundant.

The scheme – entitled "Smart Homes" – offered the householder a grant of 75 per cent, up to a maximum of £6,000 per household, for solid wall insulation or hard-to-treat cavity walls, and, for a short period, up to £3,000 for heating upgrades (boiler replacements) and double glazed windows. In order to be awarded the maximum grant, eligible householders had to be living in a generally solid walled property and be able to make a contribution to the cost of the works. A Green Deal assessment was a condition for qualifying for grant funding before householders were put in touch with the main delivery partner, independent advisor, and specific installers. At the close of the Smart Homes scheme, 1,060 grants were approved for upcoming installations across the six boroughs. Retrofitted properties included a multiple occupancy block of flats, 506 privately rented properties, and a range of housing types. Approximately a quarter of these also received ECO funding.

The arrival of Green Deal Communities funding signalled a move towards creating a domestic retrofit infrastructure for Haringey based less on local priorities and more on the technical, ideological, and economic parameters and principles dictated by national priorities. The financial conditions for the scheme directly targeted the able-to-pay householders in the area rather than economically vulnerable households or the fuel poor. The measures supported by Smart Homes – external wall insulation and boilers – promoted the national technological "fixes."

The time constraints on Smart Homes also produced a complex, rapidly assembled network. Contract regulations in the United Kingdom require all work funded by local authorities over certain thresholds to be put out to competitive tender and advertised in the *Official Journal of the European Union* (OJEU) to procure partners – a notoriously long and arduous process. In order to create a partnership in time to deliver the works within thirteen months, Haringey Council opted for an alternative contract form – the "concessionaire" – which allowed them to procure a partner without a full OJEU process. Although Haringey Council's own investment, RetrofitWorks, was written into the original funding bid as the main delivery partner for Smart Homes, by the time the scheme started it did not meet certain conditions attached to DECC's funding, namely, being a registered Green Deal Provider with access to ECO or being registered with the Financial Services Agency (FSA) and able to access Green Deal Finance. InstaGroup, a large corporate insulation manufacturer and installer that did meet these conditions, was instead procured as the main delivery partner, and RetrofitWorks had to become one of many subcontractors in order to remain involved. The role of RetrofitWorks therefore changed from being the single domestic retrofit response in Haringey to one of many at a crucial point in their business development. In addition, Haringey employed a London-wide outreach organization (London Sustainability Exchange, or LSx) to promote the scheme and an independent "Smart Advisor" (ECD Architects) to advise both householders and installers. With six local authorities, a main delivery partner, various subcontractors including RetrofitWorks and its own network, an architect, an outreach organization, and various others, the infrastructural network was quickly populated with multiple actors and intermediaries.

Smart Homes presented an opportunity to experiment and learn about different forms of retrofit infrastructure. It provided local businesses with a considerable volume of work for a period and illuminated North London's particular strengths in the supply chain: having a large number of accredited retrofit installers and good business support services. The financial incentive from Smart Homes also gave people, who would not otherwise have been able to, access to solid wall insulation. More than 1,000 installations over a suburban area of varying property types contributed to the social normalizing of retrofit and began to tackle domestic carbon emissions as per the Carbon Commission's original priorities. In particular, the Smart Homes program produced detailed learning about the role of retrofit intermediaries in providing professional advice. The inclusion of the Smart Advisor in the scheme was intended to protect and educate householders and installers, prevent hygrothermal issues such as mould and damp occurring as a result of poorly installed measures, and ensure good thermal comfort levels and energy savings. A number of householders felt the Smart

Table 14.1 Comparison between the RetrofitWorks and the Smart Homes approaches

	RetrofitWorks – Local Response	Smart Homes – National Response
Technical Approach	Bespoke retrofits, multiple measures (insulation, windows, heating, solar PV, solar hot water) integrated with home improvements (e.g., kitchens, redecorating)	Primarily solid wall insulation, secondarily boilers, other measures conditional on solid wall insulation
Network	Local SMEs and community groups, slowly developed around existing capacity	National and local organizations, quickly assembled, new relationships
Governance	Cooperative model – local network governs itself	Governed top down – testing national policy and meeting targets
Economic Approach	Long-term cooperative business model returning profits to local area	Short-term subsidies for particular measures

PV = photovoltaic; SME = small and medium-sized enterprise

Advisor's presence offered reassurance that there was someone protecting their interests, and their presence generated significant technical conversations among installers in Haringey and its neighbouring suburbs. However, the system that was set up also produced a number of significant issues and tensions, illuminating the disparity between national and local priorities and raising questions about how to manage this divide at the local level.

Two Competing Suburban Infrastructure Governance Rationales

The responses emerging from Haringey – RetrofitWorks and Smart Homes – became combined in one network. However, with dramatically different visions and rationales for retrofit, the two approaches were in tension with one another. RetrofitWorks was slowly developing, with modest targets and long-term aims that focused on local small and medium-sized enterprises (SMEs) and the local economy, as well as a holistic technical approach to retrofit incorporating multiple measures and broader home improvements. Smart Homes, on the other hand, embodied national policy priorities and was time limited, with high target numbers of retrofits in a short-term period and a narrow technical approach. Table 14.1 shows the key differences in rationale between the two approaches. A consequence of the two responses occurring simultaneously was that national priorities dominated local ones due to the high project targets, their demand on local capacity and resources, and the conditions attached to the funding given to the local authority.

Complex Networked Relations

The network that was created had multiple actors with often overlapping and competing roles, producing a very complicated customer journey with unclear touchpoints for the householder and fractious relationships between organizations in the network.

Figure 14.2 shows the complex network created by Smart Homes. Contracts were unclear and did not appropriately define the roles and responsibilities of the various actors. The three retrofit intermediaries, InstaGroup, RetrofitWorks, and ECD Architects – all responsible for coordinating, project managing, and advising on similar parts of the process – each felt they were not able to perform their "normal" roles properly because of the presence of the other intermediaries. The network also included an outreach intermediary (LSx), which managed other local community groups, as well as Haringey Council in its strategic leadership and network managing role. Multiple intermediaries created complex relationships and communication channels. Tools such as software systems were not compatible between partners in the network, and reporting processes and timescales were not developed "on the job," which meant that information about customers was not transferred effectively. Rules for assigning leads to local contractors were not written into the concessionaire contract because of the risk of contravening competitive procurement rules; there were no sanctions for partners who did not provide timely customer services; and providing a consistent schedule of rates for installers to guide costing and quotations was very difficult when dealing with a large number of installers. The Smart Advisor's standard approach was meant to enable consistently high-quality installations, but adherence to it by installers in the network was variable.

As a result, the customer journey in Smart Homes was protracted and complicated. Householders experienced delays between registering for the scheme and receiving any contact from InstaGroup to arrange a Green Deal assessment, delays between the assessment and the Smart Advisor visit, and delays in visits or quotations from installers. Furthermore, there was a lack of clarity about who to contact if there were problems, and who to contact at different stages of the process.

Figure 14.2 Haringey's multi-stakeholder retrofit network

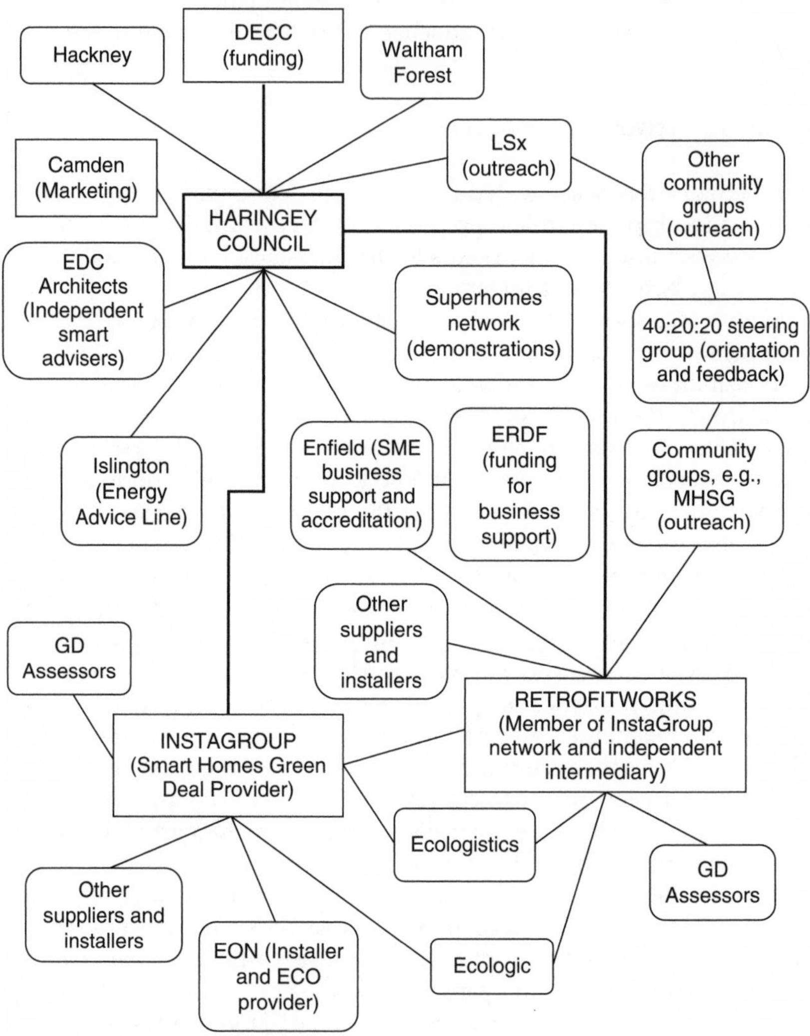

DECC = Department of Energy and Climate Change; ECO = Energy Company Obligation; ERDF = European Regional Development Fund; GD = Green Deal; LSx = London Sustainablility Exchange; MHSG = Muswell Hill Sustainability Group; SME = small and medium-sized enterprise

The experience of one householder illustrates how a family opted for the independent installer route after a series of setbacks:

> We applied to the scheme in September but nobody contacted us for months. Then somebody came out but we weren't happy with their price and we weren't really happy having strangers or people that weren't recommended to us ... We ended up going with MakeMyHomeGreen, because they were recommended to us by someone at Muswell Hill Sustainability Group, and an electrician who was recommended to us by our roofer, because we trusted him ... Then the approval process was so long and all the time we never knew who to contact so we kept phoning the advice line, the installer, the council, the smart advisor, it was so frustrating. We had to pay for a second set of scaffold because of all the delays ... But it's really improved how we feel about the house – we couldn't live with the cold. Our electricity bills are down to £15 a month. (Householder, personal communication, June 2015)

Inequitable Access for Households

The structure of Smart Homes excluded a large number of households on socioeconomic, spatial, and technical grounds. Residents living in social housing – including local authority housing – were ineligible for the scheme. It was assumed that there were other programs providing energy efficiency upgrades to those properties through the ECO or through existing regeneration projects in the borough. However, despite efforts to coordinate with these programs and point social housing tenants towards a more appropriate service, there was nothing to coordinate – no social housing retrofit program existed to connect with – and there was no relationship between the retrofit agenda and regeneration activities.

Those households in privately owned and occupied homes to whom the scheme was targeted, but who were unable to make a financial contribution to "top-up" the grant on offer, were unable to access the new infrastructure at all and were filtered out upon their first contact with the Smart Homes advice line. Many homes in Camden and Hackney, which have large proportions of conservation housing, were unable to make use of the offer even if they could contribute financially, because planning permission would not be given for external wall insulation on conservation properties. Those households whose homes were not solid walled were largely ineligible due to the Smart Homes grant being conditional upon installing solid wall insulation. Furthermore, despite a clear link between the worst private rented sector (PRS) properties and an increased likelihood of experiencing fuel poverty and energy vulnerability, these properties proved extremely difficult to access, with tenants lacking confidence in their landlords to do the work and promotions only reaching landlords who were already engaged with energy efficiency as a priority. Consequently, the PRS portion of the Smart Homes target was largely met by an independent installer and maintenance manager for multiple landlords who were likely to already be aware of the environmental and economic advantages of energy efficiency.

The infrastructure service established was therefore inaccessible to many households, including those that were most vulnerable. Only those who could pay, lived in solid walled homes not in conservation areas, *and* in boroughs whose planning departments would permit external wall insulation could access it.

Inequitable Access for Retrofit Professionals

DECC, along with procurement rules and the short timescale of the scheme, created significant difficulties for Haringey in balancing the needs of DECC and its local partners. The concessionaire contract with InstaGroup brought existing partnerships with E.ON (a multinational energy giant who would provide ECO funding) and a "snug network" of five other installers, most of whom were not local SMEs. Smart Homes overwhelmed the retrofit agenda in North London, and because RetrofitWorks was registered neither as a Green Deal Provider nor with the FSA, it had to become a member of InstaGroup's "snug network" in order to be given leads from Smart Homes and continue to develop its business. Other local retrofit installers were being approached directly by householders about the Smart Homes grant offer, but were unable to access it if they were not part of InstaGroup's network. After the Haringey Council was lobbied, eventually an "independent" route to retrofit was allowed, which meant that householders did not necessarily have to use InstaGroup to access the grant. Similarly, with competition rules and DECC's conditions inherently favouring large organizations and their economies of scale over smaller local organizations, contracts for "outreach" and householder engagement were tendered, but not won, by local community groups. Rather, they were awarded to a Greater London "umbrella" organization, which then recruited those same local community groups to do the outreach on its behalf. Trust between RetrofitWorks and Haringey Council was badly eroded by the unmet expectations that

RetrofitWorks would be the main delivery partner. Existing relationships were significantly damaged by this process as well: there was suspicion between local installers and both the council and InstaGroup due to difficulties in meeting objectives, and trust between local community groups and Haringey was eroded due to the employment of LSx for outreach.

Compromised Local Priorities

Despite being clear local priorities, unemployment and inequality became less important than short-term targets and market-building, eroding the political will that originally drove the retrofit agenda. Priorities around economic development and job creation through developing the local retrofit industry were not addressed due to a fragmented and inconsistent policy context. Installers shared a view that subsidies ruined what was an emerging market because people began to expect retrofit measures for free, and were subsequently less willing to pay for services. Furthermore, the short-term nature of schemes such as Green Deal Communities produce inconsistent volumes of work, which create difficulties for SMEs in financial planning, such that they are less able to employ extra staff. Despite the intentions of DECC to create an industry around retrofit, and Haringey's commitment to create jobs for its poorer, unemployed residents, the structure of Green Deal training and accreditation undermined these priorities. PAS2030 accreditation – the industry standard for retrofit installers – is a quality management reporting system for businesses, not a technical standard, thus providing no skills training to the unemployed, nor creating any retrofit-specific jobs. It is also clear that the new infrastructure system was not accessible to vulnerable households, which needed it the most. Due to the disconnect between the fuel poverty agenda and the able-to-pay agenda created through the structure of the ECO, there is an available supply chain and products for the able-to-pay market, but these parts of the system are not available to the fuel poor or to those in social housing. Thus, the issue of economic inequality in Haringey is exacerbated rather than ameliorated, and east-west divisions are reinforced.

Conclusion: Retrofit – An Incomplete and Imposed Infrastructure

Exploring and analysing the products and effects of this new suburban infrastructure, including the process of its creation, show us how the deliberate positioning of suburban spaces as sites of experimentation did successfully develop some components of a domestic retrofit infrastructure system. However, overall, the process created a partial, incomplete system rather than a fully functioning one. Negotiating inconsistent, market-focused and strongly technoeconomic policy with short-term funding support meant that many features of the new infrastructure were divisive and exclusive – particularly towards North London's most vulnerable residents. Conditions placed upon participating in or using the infrastructure created inequitable access for both householders and retrofit practitioners, and its existence and thus many of its potential positive effects were short lived.

Pressure from state priorities produced a partial and incomplete infrastructure that was characterized by three features. First, dominant and large-scale socioeconomic interests played a key role: policy stipulations and procurement rules encouraged partnerships with large businesses (InstaGroup/E.ON) over small local organizations. Second, it produced highly selective sociospatial responses: temporary, fluid suburban alliances and relationships that were quickly assembled around short-term priorities excluded both areas and people from the infrastructure. Third, it was characterized by coercive control: the precariousness of austerity governance meant the local authority was heavily dependent upon the policy funding connected with the retrofit infrastructure, accepting conditions on funding that significantly compromised local priorities of tackling inequality and supporting the development of the local retrofit industry.

In terms of the sociotechnical systems framework described earlier (Geels, 2002; 2005; 2010; Swan, 2013; Vergragt & Brown, 2012), what this situation has produced is a partial system. Some elements are present and well developed, such as professional know-how and technical competence and a well-developed supply chain. But many touchpoints, processes, and tools in the system are ineffective or absent, such as software and contracts, consistent policy and governance, a coherent, inclusive local vision, and a regulatory framework from government. In infrastructural terms (Graham, 2010; Graham & Thrift, 2007; Kimbell, 2015; Star, 1999), this system does not represent a stable, working infrastructure. It is not user-friendly, as it has been designed around policy targets and priorities rather than around clients. It is also a temporary infrastructure that will struggle to last beyond one event, has been forcibly assembled rather than incrementally developed, and, through the conditions it embodies, is inequitable and not inclusive. In terms of retrofitting suburbia, the Haringey system does not represent an integrated, coordinated response – it stands alone, lacking vision, and does not incorporate other principles around regeneration or connect with other infrastructures such as social housing or services for the vulnerable. *What has been produced is an incomplete and highly selective infrastructure – a partial attempt at creating a service or a system.*

The partial creation of an infrastructure illuminates missing pieces at the level of its principles: The positioning of the individual household and technological fixes as the focal point of a suburban domestic retrofit infrastructure is not enough to build a functioning and stable system. The struggle at the local level between national and place-specific priorities creates an incoherent vision and rationale for retrofit, and the inequity of the emerging infrastructure demonstrates that domestic retrofit in the United Kingdom has not been designed as an inclusive metropolitan infrastructure.

The chapter has covered a wide extent of the infrastructure field by considering hard and soft infrastructures – material infrastructures and the institutional and financial arrangements enabling the implementation of material infrastructures. Chapter 14 thus intersects with the object of many chapters in this volume, which investigate both suburban infrastructures and the institutional and financial conditions essential to their existence. It thus demonstrates, like those chapters do, that the form infrastructures take, including their effectiveness and social biases, emanates from the conditions that undergird their existence. The present chapter has also addressed two of the main preoccupations of the book: the environmental dimension of infrastructures and the social inequity of infrastructure delivery. Chapter 14 blends these two themes by showing how a program that was originally intended to aid households most in need of infrastructure upgrading to reduce their home energy consumption was subverted and transformed into a program targeted at the middle class.

REFERENCES

Carbon Commission. (2012). *The Haringey Carbon Commission report: A sustainable new economy.* London: Haringey. Retrieved from https://neweconomics.org/uploads/files/4cbbc5b78668f89032_utm6vr961.pdf

Department for Communities and Local Government (DCLG), UK. (2014). *English housing survey 2012: Energy efficiency of English housing report*. London: Author. Retrieved from https://www.gov.uk/government/statistics/english-housing-survey-2012-energy-efficiency-of-english-housing-report

Dunham-Jones, E., & Williamson, J. (2011). *Retrofitting suburbia, updated edition: Urban design solutions for redesigning suburbs*. Hoboken, NJ: John Wiley & Sons.

Geels, F.W. (2002). Technological transitions as evolutionary reconfiguration processes: A multi-level perspective and a case study. *Research Policy*, *31*(8–9), 1257–74. https://doi.org/10.1016/s0048-7333(02)00062-8

Geels, F.W. (2005). *Technological transitions and system innovations: A co-evolutionary and socio-technical analysis*. Cheltenham, UK: Edward Elgar Publishing.

Geels, F.W. (2010). Ontologies, socio-technical transitions (to sustainability), and the multi-level perspective. *Research Policy*, *39*(4), 495–510. https://doi.org/10.1016/j.respol.2010.01.022

Graham, S. (Ed.). (2010). *Disrupted cities: When infrastructure fails*. London: Routledge.

Graham, S., & Thrift, N. (2007). Out of order understanding repair and maintenance. *Theory, Culture & Society*, *24*(3), 1–25. https://doi.org/10.1177/0263276407075954

Guertler, P. (2012). Can the Green Deal be fair too? Exploring new possibilities for alleviating fuel poverty. *Energy Policy*, *49*(C), 91–7. https://doi.org/10.1016/j.enpol.2011.11.059

Haringey Council. (2015). *Joint strategic needs assessment (JNSA) Environmental Section (transport, fuel poverty, green and open space, CO2 emissions and fluvial flood)*. London: Haringey Council. Retrieved from http://www.haringey.gov.uk/social-care-and-health/health/joint-strategic-needs-assessment/other-factors-affecting-health/environmental-factors

HM Government. (2011). *The Carbon Plan: Delivering our low carbon future*. London: Department of Energy and Climate Change. Retrieved from https://www.ukgbc.org/sites/default/files/3702-the-carbon-plan-delivering-our-low-carbon-future.pdf

Hodson, M., & Marvin, S. (2009). "Urban ecological security": A new urban paradigm? *International Journal of Urban and Regional Research*, *33*(1), 193–215. https://doi.org/10.1111/j.1468-2427.2009.00832.x

Kelly, M.J. (2009). Retrofitting the existing UK building stock. *Building Research & Information*, *37*(2), 196–200. https://doi.org/10.1080/09613210802645924

Kimbell, L. (2015). *The service innovation handbook: Action-oriented creative thinking toolkit for service organizations*. London: Department for Business, Innovation and Skills.

Ravetz, J. (2008). Resource flow analysis for sustainable construction: Metrics for an integrated supply chain approach. *Proceedings of the Institution of Civil Engineers: Waste and Resource Management*, *161*(2), 51–66. https://doi.org/10.1680/warm.2008.161.2.51

Seyfang, G., & Smith, A. (2007). Grassroots innovations for sustainable development: Towards a new research and policy agenda. *Environmental Politics, 16*(4), 584–603. https://doi.org/10.1080/09644010701419121

Sims, B. (2012). Breakdown, obsolescence and repair: Drivers of change in urban infrastructure. Presented at the *International Retrofit Conference*, Manchester, UK, 2012.

Smith, A. (2007). Translating sustainabilities between green niches and socio-technical regimes. *Technology Analysis & Strategic Management, 19*(4), 427–50. https://doi.org/10.1080/09537320701403334

Star, S.L. (1999). The ethnography of infrastructure. *American Behavioral Scientist, 43*(3), 377–91. https://doi.org/10.1177/00027649921955326

Swan, W. (2013). Retrofit innovation in the UK social housing sector. In W. Swan & P. Brown (Eds.), *Retrofitting the built environment* (pp. 36–52). Oxford: Wiley.

Theobald, K., & Shaw, K. (2014). Urban governance, planning and retrofit. In T. Dixon, M. Eames, M. Hunt, & S. Lannon (Eds.), *Urban retrofitting for sustainability: Mapping the transition to* 2050 (pp. 87–98). Abingdon, UK: Routledge.

Tweed, C. (2013). Socio-technical issues in dwelling retrofit. *Building Research & Information, 41*(5), 551–62. https://doi.org/10.1080/09613218.2013.815047

Vergragt, P.J., & Brown, H.S. (2012). The challenge of energy retrofitting the residential housing stock: Grassroots innovations and socio-technical system change in Worcester, MA. *Technology Analysis & Strategic Management, 24*(4), 407–20. https://doi.org/10.1080/09537325.2012.663964

Williams, K., Gupta, R., Hopkins, D., Gregg, M., Payne, C., Joynt, J.L., & Bates-Brkljac, N. (2013). Retrofitting England's suburbs to adapt to climate change. *Building Research & Information, 41*(5), 517–31. https://doi.org/10.1080/09613218.2013.808893

15 The Uneven Outcomes of Sustainable Transport Infrastructure Planning: The Case of Montreal and Vancouver Commuters

MARKUS MOOS AND JONATHAN WOODSIDE

Introduction

There is growing interest locally to provide more sustainable transportation options for urban residents. This focus is in part due to growing concerns over the environmental, health, and public finance issues associated with auto-dependence (Ewing, Bartholomew, & Winkelman, 2008). Even though sustainability is a loosely defined concept, it has been interpreted in very specific ways in the North American planning context (Gunder, 2006). For the most part, urban planners have taken sustainability to mean increasing development densities, constraining urban sprawl, and investing in public transit, cycling lanes, and other improvements to the public realm to promote more active modes of transportation such as walking, cycling, and transit usage (Filion & Bunting, 2010; Filion & Kramer, 2011). The intent is to produce a more sustainable transportation infrastructure following what has broadly been called the "sustainability-as-density" approach (Quastel, Moos, & Lynch, 2012). Electric vehicles and car sharing are sometimes part of sustainable transport planning, but have raised suspicion among some planners who worry that these will not alleviate broader concerns associated with car dependence such as congestion and liveability (Toderian, 2014). While there are other interpretations of sustainability in planning (for instance, ecosystem planning, slow growth movements, or local food systems), sustainability-as-density has arguably been one of the more prevalent approaches.

Cities such as Vancouver, British Columbia, and Portland, Oregon, are often celebrated in the planning community for their success in reducing automobile use by relying largely on policy tools that would fall within the broad concept of sustainability-as-density (Kenworthy, 2006). However, there are also reasons to be critical of this type of infrastructure planning. For instance, planning for alternative modes of transport and the compact urban form required to support them have been associated with gentrification (and increasing land/housing values in general), romanticizing European urban forms, and neglecting the existing social geographies of cities that could reinforce and produce new forms of exclusion (Burton, 2000; Dale & Newman, 2009; Danyluk & Ley, 2007; Gilderbloom, Riggs, & Meares, 2015; Grube-Cavers & Patterson, 2015; Jones & Ley, 2016; Rayle, 2015; Revington, 2015). The impacts resulting from lower income earners being displaced and excluded from walkable neighbourhoods near transit has been referred to as "eco-gentrification" (Quastel, 2009).

It is often argued that this model of implementing sustainability in cities has been successful in part because it relies on the private market to develop high-density housing where price premiums command it. The process squeezes out households that cannot afford this sustainability premium and circumvents normative values of the "just city" (Fainstein, 2010; Lefebvre, 1968; 1991). The important question at this juncture is therefore not only whether the development of sustainable transport infrastructures lessens environmental burdens, although that is an important line of enquiry as well, but also who benefits from what amounts to a new sustainable transportation infrastructure (see also Marcuse, 1998).

We use a novel approach to examine outcomes of sustainability-as-density planning by comparing commuting decisions for otherwise similar households using multivariate statistical analysis. The intent is to demonstrate empirically that the ability to access what are now often deemed more sustainable commute modes is not equally available to all types of households. While prior literature has demonstrated the implications for low-income earners, we add a more general demographic dimension to the question of who benefits by considering implications for household size and the presence of children, and by considering the impacts in two different kinds of metropolitan areas with different planning and housing market contexts.

Unevenness in Sustainable Transport Infrastructure

Sustainability-as-density has also become a form of cultural capital (Quastel, 2009). The presence, density of coverage, and quality of networks are often lowest among dispersed populations of low socioeconomic standing (Dupuy, 2008). Mobility, therefore, can be determined by financial capital but also by proximity to network infrastructures that some households may attain by virtue of particular household attributes – non-family households, for instance, that can access high-density areas by fitting into smaller apartments. As Quastel, Moos, and Lynch (2012) note, barriers to access sustainable transport infrastructures take on greater meaning when considered alongside the increasing value of mobility networks in general and the uneven abilities to access these networks.

Here, it is useful to draw on Kaufmann's (2002) and Urry's (2007; 2012) concepts of "motility capital" and "network capital," respectively, to conceptualize these effects. Since mobility is essential to daily life but may also be a burden on one's time and commitments, they argue that the capacity to access various kinds of mobility systems and "networks" is a form of capital that is "relatively autonomous from economic, social and cultural capital" (Urry, 2007: 39). In addition, Quastel, Moos, and Lynch (2012) state:

> Propinquity is a class-based badge sewn not on the sleeves of residents but rather on the very landscape where they live and consume. The result is that walkability meshes easily with the construction of the "consumer city" – a focus on the provision of attractive and diverse consumer opportunities as a means to lure and retain a capital-intensive new middle or creative class endowed with a cosmopolitan identity." (Quastel, Moos, & Lynch, 2012: 1,069)

The framing of the ability to pursue more sustainable commuting patterns as a function of network and cultural capital allows us to draw out plausible consequences for exclusion of environmentally motivated policy. However, whereas traditional environmental justice concerns have dealt with unequal exposure to pollutants, access to natural areas or parks, and the burden of the costs of environmental protection (Buzzelli, 2008), unevenness in sustainable transport infrastructure is produced by inhibiting some from living in proximate locations (figure 15.1). Sustainable mobility requires access to public spaces, such as transit, sidewalks, or cycling lanes and trails, in order to travel to work, shops, school, or recreational activities. The price of private property abutting public spaces or transit systems thus shapes who can make use of this infrastructure, and the need to be near becomes the element of exclusion. This issue plays out along several

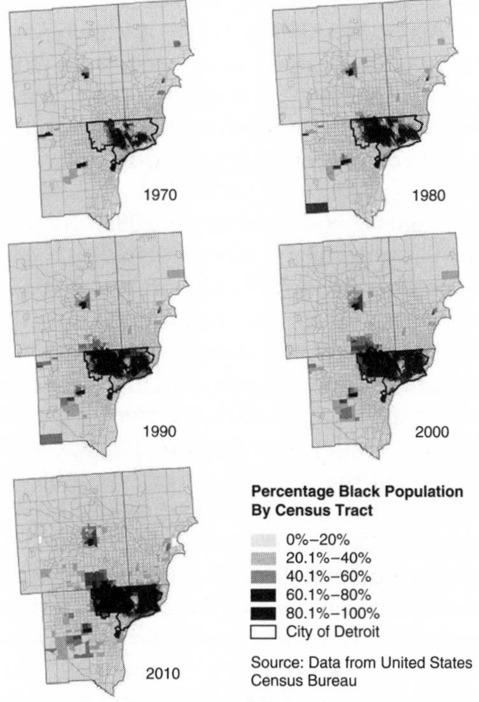

15.1 Different forms of environmental injustices

dimensions of sociospatial differentiation such as class, gender, race, but increasingly also age and household type.

The mass production of the automobile, government investment in automobility infrastructure, and the growth of the middle class under Fordism meant that an increasing share of the population saw access to mobility networks increase (Walks, 2015). Today, the development of most of our built forms to accommodate automobile-based life-styles has at the same time had an isolating effect, and those with low levels of economic capital often find themselves dependent on particular forms of mobility such as transit or walking. In North America, this circumstance can also be seen, somewhat counter-intuitively, in the day-to-day dependence on automobility in suburbs, where local alternatives to work are more limited, particularly in the context of some declining suburbs (Hanlon, 2010; Hanlon & Vicino,

2007; Hulchanski, 2010; Kaufmann, 2002; Kaufmann, Bergman, & Joye, 2004; Lucy & Phillips, 2006).

The "roll-out" of neoliberalism in the search for efficiencies in public sector operations (Leitner, Peck, & Sheppard, 2007) and what we might call "token" environmentalism have become intertwined in complex ways. This combination has produced a vision of the future that is advertised as greener and leaner but actually has quite unclear environmental benefits (Krueger & Gibbs, 2007; Quastel, 2009). It plays out in the Smart Growth debate, as municipalities advocate for more compact development and investment in alternative transportation modes to reduce public sector expenditures that would otherwise be required to extend road and utility network infrastructures into evermore sprawling suburbs and exurbs (Sewell, 2003).

The result is an expanding sustainable transport infrastructure that consists of networks of sidewalks, walking paths, cycling lanes, public transit, and bike and car sharing services that facilitate mobility without car ownership. Like for all infrastructures, there is spatial variability in terms of access, but, in the case of sustainable transport infrastructures, these variabilities are potentially going to be much more pronounced since provision is so closely tied to the density of the built form and proximity to amenities. The added variability is because these policies and infrastructures are produced by way of market strategies, and they are affected by market forces and feedback into the market, translating spatial variables into economic inequalities – and the reverse. In other words, there is the potential for new kinds of inequalities determined by varying abilities to access this emerging sustainability infrastructure that favours high-density living. We cannot walk to work, comfortably, if we cannot afford to live near work. As Kern (2010) notes, "the ability to take advantage of urban amenities ... condominium ownership," and shorter, less auto-oriented commutes are "connected to sites of privilege, including class status, education [and] family status" (375).

Under neoliberalism, where we find the withdrawal of state spending on housing and infrastructure, and a concentration on market-based regulation, localities almost exclusively rely on the private sector real estate market to attain higher building densities through condominium ownership (Hulchanski, 2004; Kern, 2010). Reliance on private sector investment to increase densities necessitates higher land values, as height is usually only prioritized where land prices exceed building costs (Skaburskis & Moos, 2008). While most new high-density housing is now based on the condominium ownership model (versus purpose-built rental, for instance) (Harris, 2011), some apartments in new buildings are purchased by investors and rented at market rates. Either way, the whole system of delivering sustainable transport infrastructures, in a market framework, is in some ways premised on new (generally higher priced) housing in urban locations (figures 15.2a and 15.2b).

Collaborative car and bike sharing platforms could actually provide greater access to sustainable transport systems as they reduce capital investment needs. Yet, redistribution is often not integrated into the operations. In interviews with bike share operators, for instance, Shaheen, Martin, Chan, Cohen, and Pogodzinski (2014) report that equity was only mentioned in nine of twenty-one bike share programs when choosing the location of docking stations, and only three of nineteen programs had a goal of serving low-income populations. Indeed, while broad surveys proclaim access in all major demographic subgroups (Neilsen, 2014; Owyang, Samuel, & Grenville, 2015), emerging reports focusing on low-income groups in marginal neighbourhoods suggest that some sharing economy networks may not expand equally across metropolitan areas, aligning more or less with those areas of high-density urbanism and landscapes of gentrification (Dillahunt & Malone, 2015; McLaren & Agyeman, 2015; Thebault-Spieker, Terveen, & Hecht, 2015).

As sustainable transport networks are being "rolled out," and some current neighbourhoods, due to their development in the pre-automobile era, are being "discovered" by the upwardly mobile, inequalities are reinforced and created. The kinds of infrastructures that facilitate walkability contribute to unequal geographies of class (Quastel, 2009), and, increasingly, household types as demography and densification (and urbanization) are closely interconnected (see, for instance, Champion, 2001). As Harvey (2008 [1973]) notes, "certain groups, particularly those with financial resources and education, are able to adapt far more rapidly to a change in the urban system, and these differential abilities to respond to change are a major source in generating inequalities" (56).

An Empirical Case Study of Two Metropolitan Areas

In the remainder of this chapter, we put the question of who benefits from sustainable transport infrastructures to an empirical test. We ask, "sustainable for whom?" in terms of being able to access more sustainable commuting modes. We analyse the changing social geographies of two Canadian metropolitan areas, Montreal and Vancouver, to assess

15.2a and 15.2b Private sector real estate markets have played an important role in implementing sustainability-as-density policies. Photos: Courtesy of the authors

inequities that may arise from transport infrastructure planning via the sustainability-as-density model (which implies planning for density primarily through private housing market development). The particularities of how the changes play out at the household level are further refined by focusing on young adults, aged twenty-five to thirty-four, who, because of their early stage in the housing market, are a good indicator of specific housing decisions in a local context (Moos, 2016; Myles, Picot, & Wannell, 1993). Their housing decisions are more likely still a reflection of current constraints and housing market conditions than those of more established households.

As part of this test we compare commuting modes and distances of young adults in the Montreal and Vancouver census metropolitan areas, two urban regions with quite different housing markets and approaches to sustainability planning. Vancouver is among the most expensive metropolitan regions in North America and a prime example of the sustainability-as-density approach to planning. Montreal is more affordable, has a much larger stock of rental housing, and density has traditionally not been a primary objective in policymaking in the same way it has been in Vancouver. Although Montreal certainly has also planned for bike lanes, public transit, and walkability, its involvement with sustainable transportation planning has been less pronounced than in Vancouver.

We report on a logistic regression analysis of Statistics Canada census data that has been conducted as part of prior work; the details of the methods are explained elsewhere (Mendez, Moos, & Osolen, 2014; Moos, 2012). The analysis considers commuting mode and commuting distance as functions of demography for the two metropolitan areas. The metropolitan area is used as an interaction dummy variable, allowing us to empirically examine whether or not certain demographic groups or household types have longer or more auto-oriented commutes in Vancouver as compared to Montreal.

There is a large literature on how demographic and household variables relate to commuting characteristics (see Molin, Mokhtarian, & Kroesen, 2016; Shearmur, 2006; Walks, 2015), but here we are primarily interested in the relative effect of these in two metropolitan areas with different housing and sustainability planning characteristics. It is the differences in commuting patterns between Montreal and Vancouver among otherwise similar households that reveal the ways the local urban context, including planning, shapes travel behaviour.

As explored in more detail elsewhere (Moos, 2012; 2013; 2014), in Vancouver where there has been heightened emphasis on neoliberalization and a more pronounced shift towards a post-Fordist occupational structure, households spend more on housing and face higher affordability burdens than similar households in Montreal. Montreal was perhaps at its peak during the Fordist period, its heavy manufacturing base supporting an expanding middle class locating in suburban areas. Although the metropolitan area is seeing recent growth in the new economy sectors, Montreal continues to struggle in terms of overall economic growth (Germain & Rose, 2000).

In contrast, Vancouver developed as a regional service centre for the resource economy. Despite being a gateway city for Pacific trade and immigrant flows, Vancouver's global connections remain limited in the sense of having head offices or a financial sector, yet its labour force and urban economy are archetypical of the new economy as it has emerged since the 1980s (Barnes, Hutton, Ley, & Moos, 2011; Hutton, 2004; 2008; McGee, 2001; Walks, 2010; 2015). The urban context in Montreal is perhaps aptly described as post-Fordist in a "Fordist spatial canvass" (Kesteloot, 2000), displaying gentrification in a declining industrial inner city, surrounded by expanding suburbs (Germain & Rose, 2000). This pattern is an urban form that reflects elements of the traditional Keynesian welfare state in terms of housing and labour market policy, an emerging quaternary sector, and an urban entrepreneurialism marketing the central city as a space of consumption and private real estate. It should be noted, however, that the Fordist period was also unmistakably different in Montreal in that rental housing played, and continues to play, a much larger role than anywhere else in North America (Choko & Harris, 1990).

As inner cities again became more desirable places, gentrification increasingly took the form of large-scale condominium apartment developments built on former industrial lands, something much more common in Vancouver than in Montreal. In fact, because condominiums have "become the principal form of residential property ownership in the inner city," any "account of Vancouver's transformation would be incomplete without considering the role of condominiums" (Harris, 2011: 24). In Vancouver, the housing context is thus unmistakably characterized by high-rise, central city condominiums that have become symbols of neoliberal urbanism and are an extension of the home-ownership ideals to the central city (Harris, 2011; Kern, 2010). Higher density housing forms also extend into the suburban areas along transit lines, which are outcomes, in part, of proactive regional planning policies and higher housing costs (Harcourt & Cameron, 2007; Newman & Kenworthy, 1999).

Vancouver differs from Montreal in that the former had already begun to lay the policy groundwork for more compact urban development during the 1970s in what came to be seen as examples of more sustainable urban development patterns later on. The establishment of an Agricultural Land Reserve that contained urban growth and the Liveable Region Strategic Plan that identified a series of growth nodes connected by transit corridors helped in the mid-1970s to place the ideals of sustainability into planning policy documents (Harcourt & Cameron, 2007).

Conversely, in Montreal local politicians argued in the early 1980s for the strengthening of the downtown and the continuation of suburban development, and opposed the kind of nodes and corridor system pursued in Vancouver. As Léonard & Léveillé (1986: 94) note, local politicians "objected to the creation of what they called 'artificial growth centres' in the suburbs," and so density remains more centralized than in Vancouver. Montreal also did not make a direct effort at regional and sustainability motivated urban planning until the mid-2000s (Brown, 2006; Filion & Bunting, 2010; Fischler & Wolfe, 2012).

A regional growth boundary actually existed for some time in Montreal, but it ultimately has had little effect because of the large amount of undeveloped land still available inside the boundary. Germain and Rose (2000) argue, citing Frisken (1994), that "the Montreal metropolitan region has never been able to successfully integrate the development of transportation and land use planning" (Germain & Rose, 2000: 108). The policies during the 1980s that aimed to attract people to the inner city in the face of decline, such as Opération 20,000 logements, actually increased single-family and townhouse developments in the inner city (Germain & Rose, 2000), contrasting with Vancouver's active attempts to raise development densities in the downtown and the growth nodes (Hutton, 2004).

While both cities have a history of planners trying to bring more families into the core by planning for amenities and housing diversity and affordability, in both cases market dynamics have put price pressures on central areas in the context of declining government involvement in the housing market (Germain and Rose, 2000; Kern, 2010). The interplay of planning for sustainability and reduced government assistance in housing have made provision of larger, family-sized apartments more difficult, particularly in Vancouver where price pressures are more pronounced.

A relative success in the context of North American cities dominated by sprawl and the automobile, Vancouver has been highly successful in attracting residential development to central areas and made walking, biking, and public transit realistic alternatives to driving a car (Filion & Bunting, 2010; Newman & Kenworthy, 1999). However, Montreal, because of its longer history of urban density, rapid transit provision, and lower incomes, has long seen higher transit usage and walking and cycling in the central city. For instance, over 60 per cent of commuters living in the City of Vancouver travelled by automobile in 2001 as compared to just under 49 per cent in the City of Montreal (Danyluk & Ley, 2007: 2203).

In the Vancouver case, one should also not overlook the particulars of the neoliberalization of policy that shifted attention to the housing market as a means to attract foreign capital (Mitchell, 2004) or the post-Fordist restructuring, facilitated by local planning policies, that transformed the inner city from industrial to residential space (Hutton, 2004; 2008). While both metropolitan areas have witnessed increases in housing costs, it is in Vancouver where central city "interconnectivity" is not only "fast becoming a very expensive place to live" (Kern, 2010: 375) but has also been deemed unaffordable for some time. In contrast, Montreal, due in part to government policies, a culture of renting, and a sluggish economy, has retained a relatively larger stock of affordable rental housing near high-density areas (Germain and Rose, 2000).

We can therefore expect to see impacts on sustainable mobility outcomes in a more pronounced way in Vancouver than in Montreal, as the former has a more expensive housing market and has pursued sustainable transport infrastructures more aggressively in tandem with residential development than in Montreal. We suggest that differences in housing prices and market dynamics, regional containment strategies, intervention of the state in housing markets, and the size of the rental stock in central areas will result in relative differences in terms of who can live near transit, or in walkable neighbourhoods, or near bike lanes in Montreal versus Vancouver (see Moos, 2012). The two-city comparison is a natural experiment of sorts exploring the impacts of a sustainability-as-density planning approach and tight housing market conditions on commuting patterns.

Who Commutes Using Sustainable Transport Infrastructures?

The empirical analysis shows that access to more sustainable transport infrastructures is not equal – "sustainable transport" defined coarsely here as those commuting by public transit, walking, or cycling, and those commuting shorter distances (tables 15.1 and 15.2). The changing socioeconomic context facilitated a demographic and economic transition that made higher density locations appealing to young, non-family households in quaternary sector occupations, and this effect is more pronounced in Vancouver than Montreal. Thus any celebration of Vancouver's sustainability planning ought to consider that decreases in commuting distances and automobile commuting have been in part due to the changing demographic composition of the population.

As documented in more detail elsewhere (Mendez et al., 2014; Moos, 2013; 2014), Vancouver, the metropolitan area where urban sustainability has been a predominant objective in urban planning policy, has made some gains in housing densities and proximity to transit among young adults. However, it is evident from prior research that these gains have come at a cost in terms of housing affordability – as young adults face a more expensive housing market context in Vancouver than in Montreal (Moos, 2012). It is also apparent that in Vancouver, immigrants, larger households, and those with children have longer and more auto-oriented commutes than their counterparts in Montreal, which speaks to the structuring effect of the expensive, high-density housing market near transit in Vancouver.

Another important finding is that walkability commands higher housing prices (see also Gilderbloom et al., 2015), because walking requires access to a public infrastructure facilitating proximity and accessibility. Walking infrastructure also has a particular aesthetic – not walking along an arterial road but along "nicely" designed places. These public amenities likely get capitalized into the housing market. The ability to easily walk places is in essence an externality of agglomeration economies in cities that in the past was reaped primarily by low-income residents – walking was even seen as a lower class activity (Amato, 2004; Solnit, 2001). The equalizing feature of walking whereby motility capital acts independent of finance capital is being eroded by the marketization of housing that creates uneven geographies. Walking to work (at least within a reasonable distance) increasingly requires the financial capital to buy into the places where public walking infrastructure is accessible (see also Grube-Cavers & Patterson, 2015; Rayle, 2015).

Perhaps the findings should not come as a surprise. Twenty to thirty years of neoliberal restructuring have resulted in a more competitive urban housing market context (Hulchanski, 2004; 2010), where housing

Table 15.1 Young adults' commute distance as a function of demography

Variables	Coefficients
Income ($1,000)	.015***
Income^2 ($1,000)	–3.E-08***
Health	.093
Social sciences, arts, culture	–.171*
Sales and services	–.195*
Clerical	.039
Manual	.201*
Primary sector	–.859
Vancouver	–.224
Health* Vancouver	–.161
Social sciences, arts, culture* Vancouver	.124
Sales and services* Vancouver	.192
Clerical* Vancouver	.186
Manual* Vancouver	.178
Primary sector* Vancouver	1.832**
Immigrant	–.418***
Immigrant* Vancouver	.265**
Child present	.243***
High school	.135
College or trades	.256**
University degree	.031
Male	.158***
Non-family	–.492***
Housing cost per room	–7.E-05**
Housing cost per room* Vancouver	–3.E-05
LR Chi^2	803.500***
Log likelihood	–14395.054
Cut 1	–.385
Cut 2	.737
Cut 3	1.513
Cut 4	2.192

Variables	Coefficients
Cut 5	2.819
Cut 6	3.421
N-cases	8823

Note: Ordered logistic regression is used because commute distance is only available as a categorical variable. Resulting coefficients reveal the propensity of workers being in categories with shorter versus longer commuting distances. The model includes the young adult labour force, with individual workers as the unit of observation. The independent variables describe the characteristics of the worker and households. Included is a dummy variable that distinguishes Montreal from the Vancouver metropolitan area. The metropolitan dummy variable is included as an interaction with the variables describing the characteristics of the worker (and retained only in cases where coefficients were significant). It reveals the relative difference between the two metropolitan areas for a given variable. Base for the regression are the young adults in Montreal, managerial occupations, non-immigrants, less than high school education, no children present, female, and in multiple-person households. ***p<0.0001, **p<0.01, *p<0.05. Categories for commuting distance are <5km, 5–9.9 km, 10–14.9 km, 15–19.9 km, 20–24.9 km, 25–29.9 km, >30 km.

Source: Calculated using Statistics Canada 2006 census data

in highly accessible locations is built at high densities according to economic theories of "highest and best use." Whereas low housing demand in Montreal may have helped the city avoid the wholesale implementation of this approach, the outcome as seen in Vancouver's urban landscape are high-rise condominium towers consisting largely of studio and one-bedroom apartments that at 400 to 600 square feet would rarely, according to North American standards, make them suitable for larger households with children. The households with children, and larger households in general, that are moving near sustainable transport infrastructures are therefore generally those with higher incomes who can afford more space in expensive locations (see Moos, 2012).

The cities that have been successful in pursuing sustainable transport infrastructures are also young cities. Smaller, non-family households without children are more likely to take transit, cycle, or walk as opposed to drive to work. Young adults are also more centralized, and downtowns are increasingly marketed as "youthful spaces," resulting in "youthification." However, in Vancouver there is also evidence of some age

Table 15.2 Young adults' commute mode as a function of demography

Public Transit	Coefficients
Income ($1,000)	–.013***
Health	–.106
Social sciences, arts, culture	–.047
Sales and services	–.114
Clerical	.307**
Manual	–.921***
Primary sector	–1.494***
Immigrant	.969***
Vancouver	–.075
Immigrant* Vancouver	–.558***
Children present	–.592***
Children present* Vancouver	–.049
High school	.009
College or trades	.135
University degree	.459***
Male	–.193***
Non-family household	.568***
Housing cost per room	1.E-05
Constant	–.773***
Bicycle	**Coefficients**
Income ($1,000)	–.013***
Health	.405
Social sciences, arts, culture	.800**
Sales and services	.260
Clerical	.216
Manual	–.348
Primary sector	.106
Immigrant	.278
Vancouver	.300
Immigrant* Vancouver	–.469
Children present	–.553*
Children present* Vancouver	–.619
High school	–.227
College or trades	–.076
University degree	.497
Male	.764***
Non-family household	.485***
Housing cost per room	2.E-05
Constant	–3.857***

Walking	Coefficients
Income ($1,000)	−.020***
Health	−.378
Social sciences, arts, culture	−.048
Sales and services	.126
Clerical	−.139
Manual	−1.283***
Primary sector	−2.537*
Immigrant	.515***
Vancouver	.470***
Immigrant* Vancouver	−.458*
Children present	−.315*
Children present* Vancouver	−.883***
High school	−.104
College or trades	−.259
University degree	.226
Male	−.128
Non-family household	.814***
Housing cost per room	1.E-04***
Constant	−2.066***
LR Chi^2	1549.860***
Log likelihood	−7822.145
N-cases	9816

Notes: A multinomial regression is used to analyse the factors associated with different commuting modes of young adults in the two metropolitan areas. Commuting modes are divided into four categories: automobile (including driving alone or as a passenger), public transit, cycling, and walking. Commuting by automobile is the base category, meaning coefficients show the likelihood of using a particular mode for each independent variable relative to the base (i.e., driving). Base is young adults commuting by automobile, Montreal, managerial occupations, non-immigrants, less than high school education, no children present, female, and multiple-person households. ***$p<0.0001$, **$p<0.01$, *$p<0.05$.

Source: Calculated using Statistics Canada 2006 census data

bifurcation as young and older populations reside in the smaller downtown apartments. But overall, there is evidence of increasing generational segregation within Canadian metropolitan areas (Moos, 2016).

Generational segregation is nowhere near at the same levels as other kinds of segregation such as ethnicity or income, but it has increased over time. Whether generational (or age) segregation contributes to heightened intergenerational conflicts remains to be seen, but clearly there is already evidence that the nightlife dimension of inner city living is causing intergenerational tensions due to noise and safety concerns among the elderly (Bromley, Tallon, & Thomas, 2005; Chatterton & Hollands, 2002). At present, the sustainable city is also a heavily generationed one.

It is important to note that the presence of children in young adult households continues to be associated with longer and more automobile-oriented commutes, despite the highly centralized residential geography of young adults as a whole. This finding points to the importance of life cycle stages in shaping location and commuting decisions. As van Diepen & Musterd (2009) note, "household structure and urbanity are indisputably related to each other" (344). The non-family households are more likely to take transit, cycle, or walk as opposed to drive to work (see also Jarvis, 2001).

But again, in terms of metropolitan differences, the negative effect of the presence of children on the likelihood of walking to work is stronger in Vancouver than in Montreal. This observation points to the strong association between densification and price appreciation in Vancouver that would make it more difficult to find housing for families with children in central locations where densification has been highest. The sustainable city is an inherently expensive one, as it favours and promotes proximity to sustainable transport infrastructures and amenities that command a premium in the housing market.

Importantly also, the occupational variables show that those working in social sciences, arts and culture, and sales and service occupations tend to have shorter commuting distances and that immigrants have shorter commutes as compared to the general population. Therefore, it is Vancouver's young adult labour force, which includes a higher share of immigrants and workers in occupations characteristic of the post-Fordist economy, that contributes to the shorter commuting distances not merely the proactive planning culture, although the latter is likely a precondition.

The findings point to the important links between changes in the socio-economic composition of the labour force and changing commute characteristics in cities. It suggests that changes in built form that led to increasing densification in Vancouver are not linearly related to "more sustainable" commuting patterns. Rather, these changes facilitate processes of urban restructuring that have made the inner city increasingly attractive to young, non-family households in service sector occupations (Champion, 2001; Kern, 2010; Quastel, 2009; Skaburskis, 2006) (figure 15.3). These findings show the decreasing capacity of motility capital to counteract the power of finance capital and bring to the forefront the question, "sustainable for

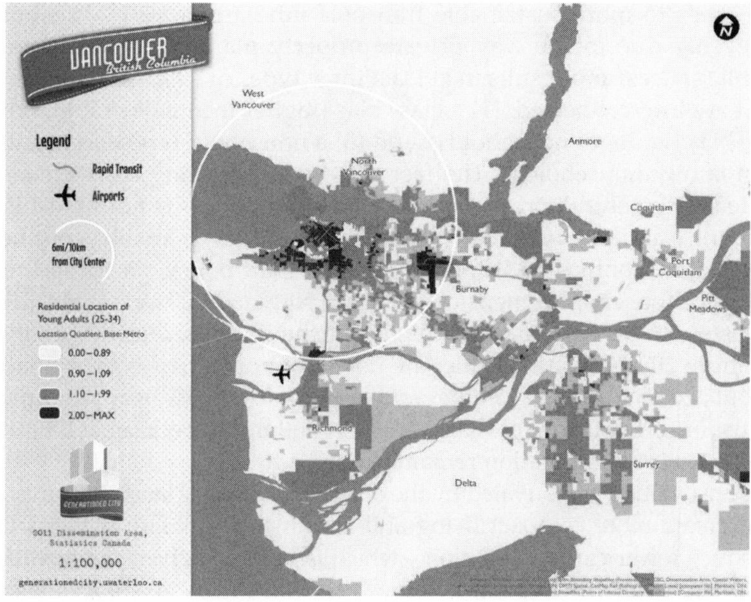

15.3 Young adult residential locations in Vancouver coincide with high-density and transit-accessible locations.
Source: generationedcity.uwaterloo.ca

whom?" (Marcuse, 1998), as shorter and less auto-oriented commutes are defined by the rising cost of housing.

Conclusions

The planning solution to environmental concerns has largely focused on promoting sustainable transport infrastructures. Where housing markets have supported the large-scale pursuit of the sustainability-as-density strategy, such as in Vancouver as compared to Montreal, demographic transitions have taken place in high-density areas. Larger households and those with children are less likely to reside in locations near sustainable infrastructure, and this finding is more pronounced in Vancouver, where urban development outcomes are dictated to a larger degree by the neoliberal approach to implementing sustainability-as-density through the private sector housing market, than in Montreal.

Access to more sustainable transport infrastructures is distributed unevenly due to the way private property markets shape access to public spaces, and results in at least three types of sustainability biases that are interconnected: (1) a class bias (higher income earners), (2) an age bias (younger households), and (3) a household type bias (smaller, non-family households). The fact that higher housing costs are associated with central locations and shorter commutes is not new, but is evidently an issue for urban sustainability strategies that has not been adequately confronted in planning and urban policy, particularly the age and household dimensions (Dale & Newman, 2009; Gurran, 2008; Quastel, 2009). It should be noted that the data used in this analysis are from 2006, and many cities are now more proactive in their policies about accommodating families and larger households near sustainable transport infrastructures (Kirk, 2017). Whether the policies will be successful in implementation remains to be seen.

Densification is motivated by the objective to reduce land consumption and create more compact living and sustainable mobility patterns that produce fewer carbon emissions, which is laudable given the threats of climate change (Ewing, Bartholomew, & Winkelman, 2008). In this context, proximate locations become scarcer, and thus the price of central locations rises, inducing a price increase throughout the metropolitan land value gradient, which reduces the amount of housing that households can consume (Skaburskis & Moos, 2008). Paying more for housing is thus perhaps an inherent quality of urban sustainability in a market economy. But evidently the cost of housing is also a limiting factor of shorter commuting distances and accessing more sustainable transport infrastructures, which are important policy goals to pursue if not at least for their environmental benefits. However, pursuing sustainability planning in a market framework without intervention in the housing market amounts to a denial of the right to the city for those who cannot afford proximate locations.

The chapter explores the social inequality aspect of infrastructures, raised by most chapters in this volume. It thus contributes to the discourse on infrastructure-induced inequity reverberating throughout the book. But while other chapters attribute social inequity caused by infrastructures to intentional policymaking biases, chapter 15 depicts the connection between infrastructures and inequity as mediated by the residential market. In its view, it is not transportation infrastructures themselves that are responsible for inequitable social outcomes, but the attraction they exert on gentrification-inducing upper- and middle-income households. The social impacts of transportation infrastructures portrayed by this chapter are thus indirect. In this light, the difference in the social repercussions of public transit in Vancouver and Montreal is not so much a function of the extent and nature of the respective public transit systems of these two urban regions as it is of the different buoyancy of their real estate markets.

NOTE

Parts of this chapter are taken from Moos (2012).

REFERENCES

Amato, L. (2004). *On foot: A history of walking*. New York: New York University Press.

Barnes, T., Hutton, T., Ley, D., & Moos, M. (2011). Vancouver: An entrepreneurial economy in a transitional city. In L. Bourne & T. Hutton (Eds.), *Canadian urban regions: Trajectories of growth and change* (pp. 291–328). Don Mills, ON: Oxford University Press.

Bromley, R., Tallon, A., & Thomas, C. (2005). City centre regeneration through residential development: Contributing to sustainability. *Urban Studies, 42*(13), 2407–29. https://doi.org/10.1080/00420980500379537

Brown, D. (2006). Back to the basics: The influence of sustainable development on urban planning with special reference to Montreal. *Canadian Journal of Urban Research, 15*(1), 99–117.

Burton, E. (2000). The compact city: Just or just compact? A preliminary analysis. *Urban Studies, 37*(11), 1969–2006. https://doi.org/10.1080/00420980050162184

Buzzelli, M. (2008). *Environmental justice in Canada – It matters where you live.* Canadian Policy Research Networks (CPRN) Research Report, December 2008, Document 50875. Ottawa, ON: CPRN. Retrieved from http://oaresource.library.carleton.ca/cprn/50875_en.pdf

Champion, A.G. (2001). A changing demographic regime and evolving polycentric urban regions: Consequences for the size, composition and distribution of city populations. *Urban Studies, 38*(4), 657–77. https://doi.org/10.1080/00420980120035277

Chatterton, P., & Hollands, R. (2002). Theorizing urban playscapes: Producing, regulating and consuming youthful nightlife city spaces. *Urban Studies, 39*(1), 95–116. https://doi.org/10.1080/00420980220099096

Choko, M., & Harris, R. (1990). The local culture of property: A comparative history of housing tenure in Montreal and Toronto. *Annals of the Association of American Geographers*, *80*(1), 73–95. https://doi.org/10.1111/j.1467-8306.1990.tb00004.x

Dale, A., & Newman, L. (2009). Sustainable development for some: Green urban development and affordability. *Local Environment*, *14*(7), 669–81. https://doi.org/10.1080/13549830903089283

Danyluk, M., & Ley, D. (2007). Modalities of the new middle class: Ideology and behaviour in the journey-to-work from gentrified neighbourhoods. *Urban Studies*, *44*(10), 2195–2210. https://doi.org/10.1080/00420980701520277

Dillahunt, T.R., & Malone, A.R. (2015). The promise of the sharing economy among disadvantaged communities. In *CHI 2015, Proceedings of the 33rd Annual ACM Conference on Human Factors in Computing Systems, 18–23 April 2015, Seoul, Republic of Korea* (pp. 2285–94). New York: Association for Computing Machinery. https://doi.org/10.1145/2702123.2702189

Dupuy, G. (2008). *Urban networks – Network urbanism*. Amsterdam: Techne Press.

Ewing, R.H., Bartholomew, K., & Winkelman, S. (2008). *Growing cooler: The evidence of urban development on climate change*. Washington, DC: Urban Land Institute.

Fainstein, S. (2010). *The just city*. Ithaca, NY: Cornell University Press.

Filion, P., & Bunting, T. (2010). Transition in the city. In T. Bunting, P. Filion, & R. Walker (Eds.), *Canadian cities in transition: New directions in the twenty-first century* (pp. 39–55). Don Mills, ON: Oxford University Press.

Filion P., & Kramer, A. (2011). Metropolitan-scale planning in neo-liberal times: Financial and political obstacles to urban form transition. *Space and Polity*, *15*(3), 197–212. https://doi.org/10.1080/13562576.2011.692567

Fischler, R., & Wolfe, J.M. (2012). Planning for sustainable development in Montréal: A "qualified success." In I. Vojnovic (Ed.), *Sustainability: A global urban context* (pp. 531–59. East Lansing, MI: Michigan State University Press.

Frisken, F. (1994). *The changing Canadian metropolis: A public policy perspective*. Berkeley, CA: Institute of Governmental Studies Press.

Germain, A., & Rose, D. (2000). *Montreal: The quest for a metropolis*. New York: John Wiley & Sons.

Gilderbloom, J., Riggs, W., & Meares, W. (2015). Does walkability matter? An examination of walkability's impact on housing values, foreclosures, and crime. *Cities*, *42*(Part A), 13–24. https://doi.org/10.1016/j.cities.2014.08.001

Grube-Cavers, A., & Patterson, Z. (2015). Urban rapid rail transit and gentrification in Canadian urban centres: A survival analysis approach. *Urban Studies*, *52*(1), 178–94. https://doi.org/10.1177/0042098014524287

Gunder, S. (2006). Sustainability: Planning's saving grace or road to perdition. *Journal of Planning Education and Research, 26*(2), 208–21. https://doi.org/10.1177/0739456X06289359

Gurran, N. (2008). Affordable housing: A dilemma for metropolitan planning? *Urban Policy and Research, 26*(1), 101–10. https://doi.org/10.1080/08111140701851985

Hanlon, B. (2010). *Once the American dream: Inner-ring suburbs of the metropolitan United States*. Philadelphia, PA: Temple University Press.

Hanlon, B., & Vicino, T. (2007). The fate of inner suburbs: Evidence from metropolitan Baltimore. *Urban Geography, 28*(3), 249–75. https://doi.org/10.2747/0272-3638.28.3.249

Harcourt, M., & Cameron, K. (2007). *City making in paradise: Nine decisions that saved Vancouver*. Vancouver, BC: Douglas & McIntyre.

Harris, D. (2011). Condominium and the city: The rise of property in Vancouver. *Law & Social Inquiry, 36*(3), 694–726. https://doi.org/10.1111/j.1747-4469.2011.01247.x

Harvey, D. (2008 [1973]). *Social justice and the city* (revised ed.). Athens, GA: The University of Georgia Press.

Hulchanski, D. (2004). How did we get here? The evolution of Canada's "exclusionary" housing system. In D. Hulchanski & M. Shapcott (Eds.), *Finding room: Policy options for a Canadian rental housing strategy* (pp. 179–94). Toronto, ON: Centre for Urban and Community Studies, University of Toronto Press.

Hulchanski, D. (2010). *The three cities within Toronto: Income polarization among Toronto's neighbourhoods, 1970–2005*. Toronto, ON: Cities Centre Press.

Hutton, T. (2004). Post-industrialism, post-modernism and the reproduction of Vancouver's central area: Retheorising the 21st-century city. *Urban Studies, 41*(10), 1953–82. https://doi.org/10.1080/0042098042000256332

Hutton, T. (2008). *The new economy of the inner city*. New York: Routledge.

Jarvis, H. (2001). Urban sustainability as a function of compromises households make deciding where and how to live: Portland and Seattle compared. *Local Environment, 6*(3), 239–56. https://doi.org/10.1080/13549830120073257

Jones, C., & Ley, D. (2016). Transit-oriented development and gentrification along Metro Vancouver's low-income SkyTrain corridor. *The Canadian Geographer, 16*(1), 9–22. https://doi.org/10.1111/cag.12256

Kaufmann, V. (2002). *Re-thinking mobility: Contemporary sociology*. Burlington, VT: Ashgate Publishing Company.

Kaufmann, V., Bergman, M., & Joye, D. (2004). Motility: Mobility as capital. *International Journal of Urban and Regional Research, 24*(4), 745–56. https://doi.org/10.1111/j.0309-1317.2004.00549.x

Kenworthy, J.R. (2006). The eco-city: Ten key transport and planning dimensions for sustainable city development. *Environment and Urbanization, 18*(1), 67–85. https://doi.org/10.1177/0956247806063947

Kern, L. (2010). Gendering re-urbanisation: Women and new-build gentrification in Toronto. *Population, Space and Place, 16*(5), 363–79. https://doi.org/10.1002/psp.581

Kesteloot, C. (2000). Brussels: Post-Fordist polarization in a Fordist spatial canvas. In P. Marcuse & R. van Kempen (Eds.), *Globalizing cities: A new spatial order?* (pp. 186–210). Oxford: Blackwell Publishers. https://doi.org/10.1002/9780470712887.ch9

Kirk, M. (2017, 6 July). The quest to vertical living family friendly. *CityLab*. Retrieved from https://www.citylab.com/life/2017/07/how-to-make-vertical-living-family-friendly/532752/

Krueger, R., & Gibbs, D. (Eds.). (2007). *The sustainable development paradox: Urban political economy in the United States and Europe*. London: Guilford Press.

Lefebvre, H. (1968). *Le droit à la ville*. Paris: Anthopos.

Lefebvre, H. (1991). *The production of space*. Oxford: Blackwell.

Leitner, H., Peck, J., & Sheppard, E. (Eds.). (2007). *Contesting neo-liberalism: Urban frontiers*. New York: The Guilford Press.

Léonard, J.-F., & Léveillé, J. (1986). *Montreal after Drapeau*. Montreal, QC: Black Rose Books.

Lucy, W., & Phillips, D. (2006). *Tomorrow's city, tomorrow's suburb*. Chicago, IL: Planner's Press.

Marcuse, P. (1998). Sustainability is not enough. *Environment and Urbanization, 10*(2), 103–11. https://doi.org/10.1177/095624789801000201

McGee, T. (2001). From village on the edge of the rainforest to Cascadia: Issues in the emergence of a livable sub-global world city. In F.C. Lo & P.J. Maractullio (Eds.), *Globalization and the sustainability of cities in the Asia Pacific region* (pp. 428–54). Tokyo: The United Nations University.

McLaren, D., & Agyeman, J. (2015). *Sharing cities: A case for truly smart and sustainable cities*. Cambridge, MA: MIT Press.

Mendez, P., Moos, M., & Osolen, R. (2014). Driving the commute: Getting to work in the auto-mobility city. In A. Walks (Ed.), *The urban political economy and ecology of automobility: Driving cities, driving inequality, driving politics* (pp. 103–28). New York: Routledge.

Mitchell, K. (2004). *Crossing the neoliberal line: Pacific Rim migration and the metropolis*. Philadelphia, PA: Temple University Press.

Molin, E., Mokhtarian, P., & Kroesen, M. (2016). Multimodal travel groups and attitudes: A latent class cluster analysis of Dutch travelers. *Transportation Research Part A, 83*, 14–29. https://doi.org/10.1016/j.tra.2015.11.001

Moos, M. (2012). *Housing and location of young adults, then and now: Consequences of urban restructuring in Vancouver and Montreal.* (Unpublished doctoral dissertation). Department of Geography, University of British Columbia, Vancouver, BC.

Moos, M. (2013). Generational dimensions of neoliberal and post-Fordist restructuring: The changing characteristics of young adults and growing income inequality in Montreal and Vancouver. *International Journal of Urban and Regional Research, 38*(6), 2078–2102. https://doi.org/10.1111/1468-2427.12088

Moos, M. (2014). "Generationed" space: Societal restructuring and young adults' changing location patterns. *The Canadian Geographer, 58*(1), 11–33. https://doi.org/10.1111/j.1541-0064.2013.12052.x

Moos, M. (2016). From gentrification to youthification? The increasing importance of young age in delineating high-density living. *Urban Studies, 53*(14), 2903–20. Retrieved from http://usj.sagepub.com/content/early/2015/09/15/0042098015603292

Myles, J., Picot, G., & Wannell, T. (1993). Does post-industrialism matter? The Canadian experience. In G. Esping-Andersen (Ed.), *Changing classes: Stratification and mobility in post-industrial societies* (pp. 171–94). London: Sage.

Neilsen. (2014). Is sharing the new buying? Reputation and trust are emerging as new currencies. Retrieved from http://www.nielsen.com/content/dam/nielsenglobal/apac/docs/reports/2014/Nielsen-Global-Share-Community-Report.pdf

Newman, P., & Kenworthy, J.R. (1999). *Sustainability and cities: Overcoming automobile dependence.* Washington, DC: Island Press.

Owyang, J., Samuel, A., & Grenville, A. (2015). *Sharing is the new buying: How to win in the collaborative economy.* Retrieved from http://www.web-strategist.com/blog/2014/03/03/report-sharing-is-the-new-buying-winning-in-the-collaborative-economy/

Quastel, N. (2009). Political ecologies of gentrification. *Urban Geography, 30*(7), 694–725. https://doi.org/10.2747/0272-3638.30.7.694

Quastel, N., Moos, M., & Lynch, N. (2012) Sustainability-as-density and the return of the social: The case of Vancouver, British Columbia. *Urban Geography, 33*(7), 1055–84. https://doi.org/10.2747/0272-3638.33.7.1055

Rayle, L. (2015). Investigating the connection between transit-oriented development and displacement: Four hypotheses. *Housing Policy Debate, 25*(3), 531–48. https://doi.org/10.1080/10511482.2014.951674

Revington, N. (2015). Gentrification, transit and land use: Moving beyond neoclassical theory. *Geography Compass, 9*(3), 152–63. https://doi.org/10.1111/gec3.12203

Sewell, J. (2003). Breaking the suburban habit: The right incentives for developers could transform suburban sprawl into more affordable, diverse and healthy neighbourhoods. *Alternatives Journal, 29*(3), 22–9.

Shaheen, S.A., Martin, E.W., Chan, N.D., Cohen, A.P., & Pogodzinski, M. (2014). *Public bikesharing in North America during a period of rapid expansion: Understanding business models, industry trend and user impacts.* San Jose, CA: Mineta Transportation Institute.

Shearmur, R. (2006). Travel from home: An economic geography of commuting distance in Montreal. *Urban Geography, 27*(4), 330–59. https://doi.org/10.2747/0272-3638.27.4.330

Skaburskis, A. (2006). New urbanism and sprawl: A Toronto case study. *Journal of Planning Education and Research, 25*(3), 233–48. https://doi.org/10.1177/0739456X05278985

Skaburskis, A., & Moos, M. (2008). The redistribution of residential property values in Montreal, Toronto and Vancouver: Examining neoclassical and Marxist views on changing investment patterns. *Environment and Planning A, 40*(4), 905–27. https://doi.org/10.1068/a39153

Solnit, R. (2001). *Wanderlust: A history of walking.* London: Penguin Group.

Thebault-Spieker, J., Terveen, L., & Hecht, B. (2015). Avoiding the south side and the suburbs: The geography of mobile crowdsourcing markets. In *CSCW '15, Proceedings of the 18th ACM Conference on Computer Supported Cooperative Work and Social Computing, 14–18 March 2015, Vancouver, BC* (pp. 265–75). New York: Association for Computing Machinery.

Toderian, B. (2014, 29 April). Mobilities in cities is about space – Proven powerfully in pictures! *Planetizen.* Retrieved from https://www.planetizen.com/node/68574

Urry, J. (2007). *Mobilities.* Malden, MA: Polity Press.

Urry, J. (2012). Social networks, mobile lives and social inequalities. *Journal of Transport Geography, 21*(1), 24–30. https://doi.org/10.1016/j.jtrangeo.2011.10.003

van Diepen, A., & Musterd, S. (2009). Lifestyles and the city: Connecting daily life to urbanity. *Journal of Housing and the Built Environment, 24*(3), 331–45. https://doi.org/10.1007/s10901-009-9150-4

Walks, A. (2010). New divisions: Social polarization and neighbourhood inequality in the Canadian city. In T. Bunting, P. Filion, & R. Walker (Eds.), *Canadian cities in transition: New directions in the twenty-first century* (pp. 170–90). Don Mills, ON: Oxford University Press.

Walks, A. (Ed.). (2015). *The urban political economy and ecology of automobility: Driving cities, driving inequality, driving politics.* New York: Routledge.

16 Conclusion: Unified and Diverse Perspectives on Suburban Infrastructures

PIERRE FILION AND NINA M. PULVER

The purpose of this concluding chapter is to reflect on the content of the book and synthesize the material introduced in its chapters. It therefore attempts to demonstrate how, collectively, these chapters advance knowledge on suburban infrastructures from a conceptual, empirical, and policy perspective. What does the volume tell us about the present state of infrastructures and their impact on the evolution of areas that surround cities, as well as about their likely future trajectory? The conclusion first disentangles themes raised by the chapters, pointing out that they discuss multiple facets of suburban infrastructures as a key contribution of the book. Unifying themes running through the different chapters are then identified, and a dual perspective to interpret suburban infrastructure issues is proposed, which focuses on blending general conceptual approaches to infrastructures with a detailed understanding of the characteristics of specific infrastructures and their respective suburban contexts. This final chapter also identifies a few gaps in the coverage of the chapters and suggests trajectories for further research that can build upon their investigations.

A Kaleidoscopic View of Suburban Infrastructures

Contributors to this volume were given the following directives: (1) they had to discuss infrastructures in relation to the concept of suburb; (2) they were required to adopt a critical, social science–inspired view of infrastructures; and (3) they were asked to draw from their own research work. Beyond these guidelines, authors were free to frame their chapters as they chose. The result is a collection of essays that is richly diverse in its coverage of conceptual frameworks, definitions of infrastructures and suburbs, and methodological approaches to investigating the evolution of suburban areas and their infrastructures.

Diversity as a defining characteristic is integral to the purpose of the book. As the idea of suburbs can be interpreted multiple ways, there is a rich variety of outlooks for infrastructures, suburbs, and the reverberations of these two objects of study on society.

With regard to the first directive, the selected essays illustrate a range of perspectives and scales for considering suburban infrastructures. Chapter 2 opened this discussion by providing a useful overview, demonstrating how different conceptual framings can guide investigations of suburban patterns and processes. Successive chapters reflected on various forms of suburbs, including cottage settlements in chapter 8, hubs that link transportation corridors between large cities in chapter 10, and non-traditional configurations such as megaprojects in chapter 13, among others.

Most of the chapters dealt with hard/physical infrastructures in suburban settings – primarily transportation and water distribution and treatment – exploring the consequences of an insufficient supply of such infrastructures. At the same time, chapter 13 drew attention to a possible oversupply of peri-urban infrastructures in China, which could result in "ghost infrastructures" whereby supply outstrips demand as infrastructure investment has become an instrument of macroeconomic policies aimed at stimulating domestic consumer demand. Chapter 4 was likewise an outlier, concentrating on the financial system – an intangible infrastructure – which underpins the development of social and physical infrastructures in suburbs. The chapter highlighted the role of this system as an "infrastructure for infrastructures," because it provides conditions for the development of hard and soft infrastructure networks. Chapter 4 also demonstrated the vital role of the mortgage and credit system in the past and ongoing development of suburbs. Differences in the focus of chapters also reflect different modes of infrastructure development and management. Some chapters concentrate on the enduring role of the public sector, others on the increasing presence of the private sector – either in the form of large corporations or small, even micro, businesses – and some allude to the role of non-governmental organizations and community organizations.

Social Consequences of Infrastructure Development

There was a wide range of societal impacts identified throughout the various chapters. In some cases, insufficient infrastructures or limited access to extant infrastructures result in the marginalization of poorer individuals. Such circumstances were described in chapters 5, 8, and 15. Chapter 5 addressed the rising cost of water resulting from the privatization of water systems in former Eastern Bloc countries. Chapter 8 pointed to the disproportional financial burden on poorer households provoked by upgrading water and sewerage systems to accommodate the suburbanization of former cottage settlements in the Stockholm archipelago. Chapter 15 documented how transit systems can be a factor of gentrification, and thus of urban social geography polarization. Chapter 4 also evoked a scenario of social inequality, but projected it in the future, when suburban households will no longer be able to meet the terms of the large loans they have contracted.

In many cases, infrastructure-induced marginalization translates into difficulty in participating in society, such as using efficient modes of transportation and accessing work, education, and social networks. In other instances, it affects the very conditions of survival. Chapters 6 and 9, which discussed the lack of coordination or absence of water and sewerage facilities in Gurgaon, India, and Hanoi, Vietnam, highlighted the potentially dire public health consequences of infrastructure failure. Inequality of infrastructure outcomes also involves unfair distribution of the financial load of infrastructures and the benefits that accrue from them. In the extreme case described in chapter 7, low-income individuals, who are often disproportionally burdened by taxes, use the highways funded by these taxes less than the rich, who drive more and use bigger cars.

A number of chapters considered the overall environmental and health consequences of suburban infrastructures, as these two types of consequences frequently accompany each other. Chapters brought to light the environmental damage resulting from privatization of suburban infrastructures (chapter 5) and a lack of coordinated approaches to rising demand for sewerage and wastewater treatment as need rises (chapter 8). Two chapters (chapters 11 and 12) stressed the essential role of green infrastructure in providing health benefits, environmental protection, and economic opportunity. Not only does green infrastructure generate a multitude of social, environmental, and economic benefits, but it also holds the promise to shape suburban settlements in ways that are more in harmony with the environment.

The wide range of impacts of infrastructures discussed in the chapters of this volume could be ranked in a way that echoes the Maslow (1943) hierarchy of needs. At the top would be the provision of conditions that are essential to life and, below this, requirements for good health. Then we would find the role of infrastructures in ensuring equitable access to participation in society, economic well-being, and respect for the environment. At the bottom of the scale of infrastructure consequences, there would be the means to procure quality of life, such as access to cultural and recreational opportunities. The repercussions of the infrastructure crises documented in chapters introducing global south cases tended to belong to the upper levels of this scale, as they pertained to necessities of life. In contrast, global north cases reported infrastructure outcomes concerning, for the most part, lower echelons of the scale that relate to social equity and accessibility to urban and suburban activities. However, this generalization is not meant to imply that global north suburban areas are immune to infrastructure failures with life-threatening repercussions, as the destructive effects of natural and human-made hazards demonstrate (Chang, 2003; Downey, 2015; Filion, Sands, & Skidmore, 2015; Pelling, 2003).

Some chapters identified infrastructure implementation issues responsible for adverse social repercussions and failure to operate as intended. Such difficulties were particularly manifest in the case of infrastructures pursuing environmental goals. Green infrastructure strategies, as described in chapters 11 and 12, confronted obstacles in reconciling urban planning objectives (intensification versus green infrastructure) and were vulnerable to development pressures, such as those threatening the Toronto greenbelt. Two chapters centred on the damaging social consequences of infrastructure programs with sustainability motivations, consequences that were intentional in one case and largely unanticipated in the other: chapter 14 described how a residential home retrofit program set norms that de facto excluded low-income households, those that would have most benefited from the program; chapter 15 showed how a more environmentally friendly lifestyle, fostered by transit- and walking-oriented development, often appeals to wealthier residents and translates into a displacement of poorer households. Accordingly, there is the risk that public transit improvements in suburban areas will trigger gentrification and the displacement of low-income households (see also Kramer, 2013).

Chapters in this compilation not only present a variety of perspectives on infrastructures, their mode of development, and their impact, but also highlight the different forms suburbs take. As such, the variety of suburban communities described is representative of many manifestations of the suburbs in different parts of the world. For example, chapter 4 concentrated on the North American suburb, devised as a macroeconomic mechanism to stimulate public and private consumption and thus prevent the drag effect on the economy of insufficient consumption. Chapter 10 introduced a new form of North American suburban development driven by the freight transportation and distribution (logistics) advantages of major highway corridors connecting important urban centres. Chapter 3 referred to the relation between UK suburbs and their respective urban areas from an urban governance perspective. The Chinese suburban model, shaped by the present wave of massive infrastructure investment, was outlined in chapter 13, while chapters 6 and 9 exposed stark social polarization in a context of chronic infrastructure insufficiency in India's and Vietnam's suburbs. We see how permutations of infrastructures and suburbs are vastly increased by the different types of suburbs identified throughout the chapters. Furthermore, as chapter 2 demonstrated, the relationship between suburbs and infrastructures is further complicated by the multiple roles infrastructures can play in a suburban context, implying that a same infrastructure will interact differently with a given type of suburb according to the role this infrastructure plays. The chapter showed how suburban response to a similar infrastructure proposal, a railway line improvement, varied according to whether the line was perceived as *in* the suburb (located in the suburb but without benefit for it) or *of* the suburb (both located in the suburb and with potential advantage for it).

Unifying Themes

From the multifaceted areas of study identified in the previous section, it would be easy to conclude that variations in the ways suburban infrastructures can be apprehended are too numerous for this phenomenon to be investigated as one consistent object of study. Suburban infrastructures would then appear as resistant to attempts to adopt a common conceptual approach to interpret their numerous manifestations. In this view, there would not be one conceptual approach capable of making sense of suburban infrastructures, but multiple approaches suited to the wide variety of infrastructures and the numerous forms of suburbs where they are found. However, unifying themes emerge from the different ways in which chapters deal with suburban infrastructures. Most significantly, in line with the second directive for author submissions, each chapter took a social science approach to analyse in a critical fashion the suburban infrastructure issues it addressed. In this sense, the volume resonates with the Flyvbjerg (2001) understanding of social science. Flyvbjerg ascribes an ethical purpose to social science, which entails criticizing society with the aim of improving societal outcomes. According to Flyvbjerg, social science is an ameliorative form of knowledge with practical implications. By adopting such an approach, the present volume fits within a growing body of work criticizing infrastructures, mainly from a social and environmental perspective, and seeking solutions to their deficiencies (see, for example, Bakker, 2010; Graham, 2010; Graham & Marvin, 2001). One does not need to spend much time scanning the literature on infrastructures to realize that this social science approach represents but a tiny minority of the research carried out in a field that is dominated by engineering. This volume represents a significant contribution to the social science literature on infrastructures because its chapters have largely concentrated on the *consequences* of infrastructures rather than detailing the *features* of these infrastructures. As indicated in the paragraphs to follow, the shared critical perspective has given rise to unifying themes that run through the different chapters: social justice, infrastructure deficits, suburbs as places of social transition, and the institutional deficiencies that give rise to variations in suburban infrastructures and their outcomes – including the frequent incapacity for suburban infrastructures to meet societal needs.

Social Justice

Virtually all chapters raised the social justice dimension of suburban infrastructures by focusing on who is advantaged by existing infrastructure networks and who is deprived because they are bypassed by these networks, required to pay a proportion of the cost that is disproportional to the services they receive, or exposed to the negative externalities of these infrastructures. The volume is thus largely about who benefits from the integrative effects of suburban infrastructures and who is adversely affected by their social fragmentation impacts. The more extreme situations narrated by the chapters paint the picture of large populations whose health is threatened by lacking or inadequate suburban infrastructures. Likewise, several chapters showed how participation in society is hampered by difficult access to infrastructures for large portions of suburban residents. These chapters point to the existence of different pathways tying suburban infrastructures to social inequality. Such connections can be intentional, as in the case of the blatantly unfair distribution of the financial burden of infrastructure maintenance depicted in chapter 7. These pathways can also be the outcome of neglect, causing the poor and politically marginalized to be systematically victimized by infrastructure shortfalls due to financial scarcity, as shown in chapters 6 and 9. And, as seen in chapter 15, pathways can stem from interactions between infrastructures and urban and suburban dynamics, resulting in the appropriation of infrastructure-induced benefits by wealthy and/or politically powerful residents. The book is therefore primarily given to a critique of the social impacts of suburban infrastructures.

Infrastructure Deficits

If the main theme raised by the chapters of this volume is the social inequality generated by suburban infrastructures, infrastructure deficits are a close second in terms of the importance they receive in the book. By "infrastructure deficit," we refer to a lack of accessibility to infrastructures on the part of a segment of the population due to an insufficient supply of infrastructures. This population is most often comprised of low-income households – hence the interconnection between the infrastructure deficit and the social inequality theme. Infrastructure accessibility impediments can be a function of the absence of infrastructures, their scarcity, or high access fees. Most chapters address this issue in a similar format. They first identify infrastructure deficits, which are then portrayed as a contributing factor to social inequality and, in certain instances, environmental degradation. Infrastructure deficits are attributed to the financial deficiency of national, regional, or municipal administrations and to policymaking biases detrimental to disadvantaged populations and geographical areas. Neoliberalism is presented as a major culprit responsible for infrastructure deficits. While alluded to in many chapters, this socioeconomic ideology was the central focus of two chapters. Chapter 5 narrated the intentional deployment of a privatization strategy to sell public water systems in former Eastern Bloc nations, and equated this privatization with uneven water infrastructure services and rising costs. Chapter 7 presented an extreme depiction of a retrenchment of the state affecting its capacity to maintain and operate infrastructures. It related the adoption in Michigan of tax cuts of such a magnitude that the state's capacity to maintain and upgrade the highway network was fundamentally crippled, thus opening the way to partial privatization of the network. While these two chapters describe deliberate state efforts to encourage neoliberal policies, elsewhere neoliberalism was imposed on communities out of necessity in order to compete economically with other jurisdictions or by outside forces such as international financial agencies.

In other situations, perceived infrastructure deficits can be the outcome of frustrations inflicted by high expectations about infrastructures. Such expectations relate to the enabling nature of infrastructures. The fundamental role of infrastructures is to make possible all sorts of human activities essential to the functioning of society, with direct life chances and quality of life implications. The public is therefore more interested in the enabling consequences of infrastructures than in the infrastructures themselves. From this viewpoint, infrastructures will never appear to be up to the task of supporting needed and desired societal activities. For example, cities are agglomerations of activities compressing space to maximize interactions between these activities. A key goal of cities can thus appear to be to minimize transportation time. In an urban technological utopia, cities would overcome the friction of space and offer instantaneity, an ideal that transportation infrastructures can never fully achieve. It is the same thing for necessities of life, such as water. The ideal for suburban residents is to have access to plentiful clean water for no or little cost, even though human-made systems for water conveyance and treatment by necessity incur construction and maintenance costs. Consequently, perceptions may exist that we are always in a situation of infrastructure deficit relative to need, and that existing infrastructures do not live up to expectations.

Suburbs in Transition

Another theme raised by many chapters is that of the suburb as a place in transition, undergoing infrastructure crises that point to the need for infrastructure innovation to adapt to changing circumstances. Many of the infrastructure issues chronicled in the volume stem from high growth rates experienced in suburban areas, as most urban expansion takes place there. Suburban expansion is at the core of the infrastructure problems raised in chapters 6, 8, 9, and 10. In some chapters, suburbs come out as test beds for new forms of infrastructures or modes of infrastructure delivery. For example, chapter 12 demonstrated how the greenbelt in the Toronto region was created to check suburban sprawl and provide urban and suburban amenity in a high-growth metropolitan region.

Institutional Deficiencies

There is also agreement among most chapters about the complexity of decision-making related to suburban infrastructures. Chapters showed that infrastructure biases and deficiencies are not solely due to dominant interests using their control over infrastructure decision-making to favour certain constituencies at the expense of others. There are other factors accounting for the infrastructure shortcomings depicted in the volume, most notably, insufficient financial means and the absence of institutional structures capable of delivering the level of infrastructures required to address needs. When not championed by a sufficiently powerful organization, suburban infrastructure requirements indeed run the risk of getting lost in a maze of institutions, each with its own objectives, operating at various scales. Chapters 3, 6, 9, and 14 concentrated on the absence in the global north and global south suburban contexts they investigated of institutional structures capable of harnessing resources towards infrastructure systems and providing required coordination. Chapter 6, which perhaps painted the most dramatic picture of infrastructure deprivation, ascribed the absence of conditions for the development of sewerage systems to a deficiency in institutional structure, a form of benign neglect. Chapter 14 was more blatant when it exposed how a home heating efficiency program was deliberately framed to exclude poorer households. Chapters 6 and 14 together illustrated how both interest-based and institutional impediments can undermine equitable suburban infrastructure development.

A Bi-Scalar Perspective

The findings of the chapters in this volume point to important differences between infrastructures, related to the many forms they take, the decision-making processes from which they emanate, and the context in which they unfold. But, at the same time, the chapters bring to light features common to all infrastructures that illustrate their enabling role. In order to understand the nature, role, and impact of important infrastructural features, be they unique to specific infrastructures or common to all infrastructures, they must be apprehended at two scales. The first scale is that of infrastructures in general and how they operate as enabling networks. The second scale is that of the specific contexts in which infrastructures operate and wherein decisions concerning them are made. Without general concepts, it is impossible to grasp the essence of infrastructures and their impact on different aspects of society – in other words, what distinguishes them from other built forms and services. But this general conceptual level yields only a broad understanding of infrastructures, which is of limited use to understand their manifestations in specific settings. Focusing on the enabling role of infrastructures helps us make sense of what this role should be in contexts as different as, for example, that of the relatively wealthy recreational/suburban Stockholm archipelago (chapter 8) and the sewerage-deprived context of Gurgaon, India (chapter 6). At the same time, such a general perspective is of little assistance in casting light on issues specific to these contrasting infrastructure circumstances. What are the governance difficulties preventing a coordinated infrastructure strategy in these respective contexts? What forms do infrastructures take in these two locales? What are the consequences of the presence, ineffectiveness, or absence of infrastructures in the Stockholm archipelago and in Gurgaon? The case studies presented throughout this book have attempted to take on these complex questions.

Further Research

In this section, we explore areas of investigation touched upon by the book chapters, but which, because of their importance and unique relationship to suburban development, deserve further reflection. Most chapters in the book have concentrated on the prevailing infrastructure landscape in sectors under investigation and the consequences of current infrastructure conditions on residents, urban development, and the environment. As mentioned earlier in the conclusion, the main emphasis of the book is on the social splintering outcomes of infrastructure strategies resulting in the loss, or the absence, of universalist principles guiding infrastructure decisions. As a result, decision-making processes undergirding infrastructure developments are not as fully investigated as they could be. An understanding of the institutional structures that oversee the funding, coordination, planning, and implementation of infrastructures and how they shape these processes is critical. Indeed, some chapters begin to tease apart the relation between institutions and infrastructures. In an attempt to examine the different facets of the infrastructure phenomenon, chapter 1 referred to the classic work of Harold Innis, describing how the need for transportation infrastructures to export staples has played a determining role in shaping Canadian political institutions. Chapter 5 examined the impact of an unparalleled institutional transformation (the shift from communism to market economics) on water infrastructures. Chapter 6 described the health consequences of the failure to set up the institutional apparatus essential to the provision of sanitary infrastructures in Gurgaon, India, and chapter 10 explored how new suburban forms taking shape in the space between urban regions require adapted municipal institutions. Chapter 3 provided perhaps the most elaborate discussion of the relation between municipal institutions and infrastructures, examining the tripartite connection between regional administrations, economic competitiveness, and infrastructures.

Further research could provide a deeper and more systematic exploration of the institutional prerequisites for infrastructures in suburbs. It could investigate how institutional arrangements influence the form infrastructures take, and how this accounts for the oft-observed social inequity induced by infrastructures. Additional reflection on institutions, inspired by Innis, could verify how the need to create infrastructures abets institutional forms capable of providing conditions necessary for the existence of these infrastructures. Such a reflection could probe how the provision of suburban infrastructures in the contemporary context influences the creation and evolution of public sector administrative structures and of new neoliberal-inspired public–private sector interfaces.

Alongside the role of institutions, a broad perspective on infrastructure decision-making specific to the unique conditions of suburbs could also include that of decision processes, coalitions of interests, and major actors. Findings from multiple infrastructure decision case studies could result in the framing of decision-making models exposing the connection between the different factors involved in these decisions. To be comprehensive, such a model would also need to incorporate more occult aspects of infrastructure-related decision-making. Such decisions are indeed vulnerable to corruption, a reflection of efforts to snatch short- and long-term rewards accruing from infrastructures. For example, these rewards can take the form of overpriced submissions in a fraudulent bidding system, where builders pocket the extra money and pass some of it to the politicians and bureaucrats attributing contracts (Commission d'enquête, 2015). They could also involve the location of infrastructures so they benefit landowners with connections to decision-makers. Corruption can have serious impacts on infrastructure outcomes. By inflating prices, it results in the provision of fewer infrastructures within a given budget. Another possible impact is the construction of poor-quality infrastructures with a reduced life span and burdened with heavy maintenance costs. There is also the introduction of biases in the form and location of infrastructures, responsible for deviations from need-based choices (Kenny, 2006).

Just like decision-making processes, financial conditions constitute an upstream factor determining the presence, absence, and form of infrastructures. In fact, these two factors are intimately interconnected, as available funding weighs heavily on decisions concerning infrastructures. In this volume, financial considerations are addressed directly by three chapters, although we can sense their influence on the case studies reported in all chapters. Chapter 4 sounded the alarm about heavy reliance on debt to finance North American suburban development, both infrastructures and private consumption, especially homes and automobiles. The warning from this chapter is disheartening: if debt financing has become essential to the perpetuation of this urban form, which accounts for a large majority of urban development, and for enduring gross domestic product growth, in time resulting prosperity can only lead to over-indebtedness and a collapse of property values and the entire financial system. Chapter 13 told a similar story. It showed how debt financing of infrastructures in China is used to stimulate the economy by propping up consumption. While also exploring the financial aspects of infrastructures, the perspective advanced by chapter 7 was different. Whereas chapters 4 and 13 considered infrastructures as part of macroeconomic stimulus strategies, chapter 7 pointed to the effects of efforts to shrink the state and therefore curtail the public sector financing of infrastructures.

These chapters demonstrated the critical impact of financial arrangements on infrastructure development and its societal repercussions. They also opened the way for further analysis of the macroeconomic dimension of infrastructures. Such an examination would consider infrastructures as macroeconomic instruments, but set within suburban contexts (see, for example, Crain & Oakley, 1995; Cwick & Wieland, 2011; Davison, 2009; Lin, 2012; Ramsay & Lloyd, 2010; Stilwell & Primrose, 2010; Wylie, 1996). This analysis would address issues inherent in the macroeconomic role of infrastructures: notably, difficulties in calibrating the provision of infrastructures according to needs when macroeconomic objectives drive their construction. Further investigation of the financing of infrastructures could explore the connection between neoliberalism, reduced government investment capacity, and additional reliance on private investment for suburban infrastructures.

What are the implications on suburban infrastructures now that we are experiencing an over-accumulation of capital seeking investment outlets? Hence the current low interest rates. Moreover, pension funds and other long-term financial vehicles look for secure, if relatively low-yield, investments such as public–private partnerships and private infrastructures (Della Croce, 2011).

Ascending costs may also be a factor causing infrastructure financing stress. We have broached in the introduction the possibility that infrastructure deficits could partly be the result of infrastructure costs climbing faster than the consumer price index (CPI). If this is the case, why is it happening? Is it, as hypothesized in the introduction, the outcome of relatively slow productivity gains in the infrastructure field compared to other sectors of the economy and higher production costs when work takes place domestically, as in the case of infrastructures, than when it is outsourced? To clarify these matters, a longitudinal study could verify how infrastructure expenses compared to the CPI over the long term. If a discrepancy is indeed detected, further analysis could identify reasons for the infrastructure cost gap with the CPI. Finally, an investigation of the financial dimension of infrastructures could also expose its effects on the social inequality theme repeated throughout the volume.

It would be instructive to merge the social science approach adopted in the volume with the more mainstream engineering-dominated perspective on infrastructures. The critical tools of social science could serve to display biases embedded in the technical literature on infrastructures. These tools would expose the social inequity present within what parades as socially neutral technical discourses by linking their content with observable or anticipated social impacts of infrastructures. Such inquiries could be targeted at technical innovations on the horizon. What will their implications be in terms of employment, the environment, investment, and the overall functioning of society? Predicting the future of infrastructures and their social impacts requires a blending of technical and social science approaches, hence the need for a theorization of the social impacts of technologies, of how different technical innovations translate into different social impacts.

The last area for further research we acknowledge is the semiology of infrastructures – the messages that emanate from infrastructures (see Dourish & Bell, 2007). The semiology of infrastructures was raised in chapter 6, which demonstrated how access to sewerage was perceived as an expression of social status. Chapter 6 thus illustrated the relation between access to different infrastructures and the perception of one's position in society. But this dimension of infrastructures could be further expanded by exploring how messages associated with infrastructures can be manipulated. Such manipulation can be used to encourage reliance on more expensive infrastructures, rather than cheaper ones, as a mark of social standing. Likewise, infrastructure decisions themselves can be driven by efforts to project certain messages, as in the case of decisions favouring infrastructures that symbolize progress and modernism (airports, high-speed trains, cutting-edge healthcare centres) over those such as basic water and sewerage systems in improvised developments that are more effective at improving the well-being of large segments of the population. Moreover, when inequality of access is built into infrastructures, they serve to reinforce the image of a divided and individualistic society, which sanctions the segmentation of all aspects of social life according to personal wealth differences.

The Future of Infrastructures

We close the book by identifying suburban infrastructure futures suggested by the content of its chapters. As these chapters concentrated on contemporary suburban infrastructure issues, we must infer possible futures from their treatment of these issues. The two chapters discussing green infrastructures are probably the most future oriented, as they discuss a relatively new approach to infrastructures, which is gaining momentum. It is not inconceivable that, in the future, suburban areas can be a major source of green infrastructure assets, in the form of carbon sinks, biodiversity protection, air and water cleansing, active transportation corridors, and recreational opportunity that enhances quality of life for all. At the same time, capitalizing on these opportunities would require preserving areas of suburban space from development and taking a more deliberate, systematic approach to suburban land planning than is currently the norm in many regions. Future-oriented solutions can also be found in some chapters that point to infrastructure innovations emanating from the crisis situations they describe. Most of these innovations are small-scale and private sector alternatives to failing or absent government infrastructure programs, but they indicate larger problems that will need to be addressed in the future.

In general, the chapters put more faith in large-scale centralized approaches to suburban infrastructures. They tend to depict alternatives to these approaches as prone to failure and sources of social inequity. Privatization and small-scale responses to needs, especially when concerning water and sewerage, are seen as contributing to social polarization and environmental degradation. Such portrayals fit the volume's critique of insufficient state intervention and privatization because of the inefficiencies they cause and their adverse environmental and social impacts. This critique raises questions about future infrastructure innovation trajectories. Does an efficient and equitable delivery of infrastructures require a return to large-scale centralized public sector strategies, or could infrastructure innovations, which result in improvements in terms of cost, efficiency, environmental protection, and social equity, take place in a privatized and/or small-scale intervention context (Simone, 2009)? What about the role of community organizations in setting up local infrastructure networks? Could it constitute an alternative to centralized state interventions?

We cannot discuss infrastructure innovation without looking at technological change, such as the phenomenal evolution of information technology over the last decades (Brown, 2014). Advances in information technology can improve infrastructure delivery by making it possible to target the needs of different segments of the public, coordinate complex networks, and automate pricing and fare collection. These advances are also conducive to the development of the sharing economy, which extends the possibility to rely on individuals and microenterprises for the provision of infrastructure-related services (see, for example, Heinrichs, 2013). We already witness, in poor global south contexts, dependence on people for tasks normally assumed by physical infrastructures, but information technology promises a wider and more efficient involvement of individuals and microenterprises in such matters. We must, however, take care not to exaggerate the extent to which information technology can influence infrastructures. There is not much in terms of information technology that can fully replace conventional hard infrastructures, such as pipes for water distribution and highways and rail for transportation, or green infrastructures. Self-driven vehicles, a significant technological innovation likely to materialize in the near future, will profoundly impact the way that suburbs function. However, regardless of their impact on suburban life, these new innovations will continue to rely on existing roads and highways.

REFERENCES

Bakker, K. (2010). *Privatizing water: Governance failure and the world's urban water crisis*. Ithaca, NY: Cornell University Press.

Brown, A. (2014). *Next generation infrastructure*. Washington, DC: Island Press.

Chang, S. (2003). Evaluating disaster mitigations: Methodology for urban infrastructure systems. *Natural Hazard Review*, 4(4), 186–96. https://doi.org/10.1061/(asce)1527-6988(2003)4:4(186)

Commission d'enquête sur l'octroi et la gestion des contrats publics dans l'industrie de la construction. (2015). *Rapport final*. Québec, QC: Author. Retrieved from https://www.ceic.gouv.qc.ca/la-commission/rapport-final.html

Crain, W.M., & Oakley, L.K. (1995). The politics of infrastructures. *The Journal of Law and Economics*, *38*(1), 1–17.

Cwik, T., & Wieland, V. (2011). Keynesian government spending multipliers and spillovers in the Euro area. *Economic Policy*, *26*(67), 493–549. https://doi.org/10.1111/j.1468-0327.2011.00268.x

Davidson, P. (2009). *The Keynes solution: The path to global economic prosperity*. New York: Palgrave Macmillan.

Della Croce, R. (2011). *Pension funds investment in infrastructure: Policy actions*. (OECD Working Papers on Finance, Insurance and Private Pensions, No. 13). Paris: OECD Publishing

Dourish, P., & Bell, G. (2007). The infrastructure of experience and the experience of infrastructure: Meaning and structure in everyday encounters with space. *Environment and Planning B*, *34*(3), 414–30. https://doi.org/10.1068/b32035t

Downey, D.C. (2015). *Cities and disasters*. London: CRC Press.

Filion, P., Sands, G., & Skidmore, M. (Eds.). (2015). *Cities at risk: Planning for and recovering from natural disasters*. London: Routledge.

Flyvbjerg, B. (2001). *Making social science matter: Why social inquiry fails and how it can succeed again*. Cambridge: Cambridge University Press.

Graham, S. (Ed.). (2010). *Disrupted cities: When infrastructure fails*. London: Routledge.

Graham, S., & Marvin, S. (2001). *Splintering urbanism: Networked infrastructures, technological mobilities and the urban condition*. London: Routledge.

Heinrichs, H. (2013). Sharing economy: A potential new pathway to sustainability. *GAIA – Ecological Perspectives for Science and Society*, 22(4), 228–31. https://doi.org/10.14512/gaia.22.4.5

Kenny, C. (2006). *Measuring and reducing the impact of corruption in infrastructure*. Washington, DC: World Bank Publications (Vol. 4099).

Kramer, A. (2013). *Divergent affordability: Transit access and housing in North American cities.* (Doctoral dissertation). School of Planning, University of Waterloo, Waterloo, ON.

Lin, Y.J. (2012). Beyond Keynesianism: Global infrastructure investments in times of crisis. *Journal of International Commerce, Economics and Policy, 3*(3), 1250015. https://doi.org/10.1142/s1793993312500159

Maslow, A.H. (1943). A theory of human motivation. *Psychological Review, 50*(4), 370–96. http://dx.doi.org/10.1037/h0054346

Pelling, M. (2003). *The vulnerability of cities: Natural disasters and social resilience.* London: Earthscan.

Ramsay, T., & Lloyd, C. (2010). Infrastructure investment for full employment: A social democratic program of funds regulation. *The Australian Journal of Political Economy, 65*, 59–87.

Simone, A.M. (2009). *City life from Jakarta to Dakar: Movements at the crossroads.* New York: Routledge.

Stilwell, F., & Primrose, D. (2010). Economic stimulus and restructuring: Infrastructure, green jobs and spatial impacts. *Urban Policy and Research, 28*(1), 5–25. https://doi.org/10.1080/08111141003610046

Wylie, P.J. (1996). Infrastructure and Canadian economic growth 1946–1991. *The Canadian Journal of Economics, 29*(1), S350–5. https://doi.org/10.2307/136015

Contributors

Jean-Paul D. Addie is an assistant professor in the Urban Studies Institute at Georgia State University. His research lies at the intersection of city-regional urbanization, the politics of infrastructure, and a critical account of the production of space. He is currently exploring how universities are adapting their institutional structures, pedagogies, and spatial strategies in response to the changing dynamics of global urbanization. Addie has worked on topics including university urbanism, transportation governance, suburbanization, city-regionalism, neoliberal policy, and sociospatial theory, and published in journals such as *IJURR*, *Transactions of the Institute of British Geographers*, *Environment and Planning A*, *Regional Studies*, and *CITY*. Prior to joining Georgia State, he was a Marie Curie Fellow in the Department of Geography at University College London, and Provost Fellow and Lecturer in UCL's Department of Science, Technology, Engineering, and Public Policy. He holds a PhD in geography from York University.

Jeanette Eckert has a PhD in geography from Michigan State University. Her research interests include spatial inequalities, dimensions of race and class in urban areas, urban sustainability, local food systems, and human-animal interactions. Her previous research has been published in journals including *Applied Geography*, and she has co-authored multiple book chapters. She currently works in the Office of Research Compliance at the University of Toledo.

Pierre Filion is a professor at the School of Planning of the University of Waterloo. His areas of research include metropolitan-scale planning, downtown areas and suburban centres, and infrastructures. He has served on the Planning and Real Estate Advisory Committee of the

National Capital Commission, the Central Zone Strategy Panel of the Ontario Government Ministry of Municipal Affairs (which led to the proposal of most Growth Plan policies), and the Scientific Advisory Committee of the International Joint Commission.

Shubhra Gururani is the chair of and an associate professor in the Department of Social Anthropology at York University. She is an affiliated member of the CITY Institute, York University Centre for Asian Research, and the Centre for Feminist Research. Her current research focuses on the changing political ecology of urban peripheries and examines the politics of land acquisition, planning, and infrastructure in Gurgaon, on the outskirts of New Delhi. She is part of the SSHRC-funded Major Collaborative Research Initiative on Global Suburbanism. Her previous research focused on the politics of conservation and history of forestry in the Indian Himalayas.

Rebecca Ince is a Research Fellow at the Open University. Her research covers health and environmental policy, combining the disciplinary and theoretical perspectives of both geography and social policy. Her projects illuminate how networks of people and organizations provide services in different and unique contexts, and how relational dynamics between actors in those networks impact service provision and equality of access. She is committed to co-producing research and has worked closely with service users, cooperatives, national and local voluntary sector organizations, and local and national government throughout her research.

Roger Keil is York Research Chair in Global Sub/Urban Studies at the Faculty of Environmental Studies, York University, in Toronto. He researches global suburbanization, urban political ecology, cities and infectious disease, and regional governance. He is the author of *Suburban Planet* (Polity, 2018), editor of *Suburban Constellations* (Jovis, 2013), and co-editor, with Xuefei Ren, of *The Globalizing Cities Reader* (Routledge, 2017), of *Suburban Governance: A Global View* (with Pierre Hamel), and *Massive Suburbanization* (with K. Murat Güney and Murat Üçoglu), both with UTP.

Zeenat Kotval-K is an assistant professor of urban and regional planning with the School of Planning, Design and Construction at Michigan State University. Her research interests lie in the environmental and health impacts of the built environment and human behaviour. Additionally, her recent research areas extend transportation and travel accessibility to food security (focusing on India specifically) and well-being through increased mobility for older adults (focusing specifically on US cities).

Xiaomeng Li is a PhD candidate in the Department of Geography, Environment and Spatial Sciences at Michigan State University. Her major research interests include urban travel behaviour, urban networks, and urban modelling and simulation. Her current research focuses on the leisure and service travel behaviour of residents in the Detroit region. The research goal is to provide a deeper understanding of urban travel in distressed urban regions and assist in the achievement of efficiency and social equity.

Lucy Lynch (Master of Environmental Studies, York University) is the project coordinator of the MCRI "Global Suburbanisms: Governance, Land and Infrastructure in the 21st Century" project based at York University. Her research focuses on urban and environmental planning, green infrastructure, greenbelt, boundaries, urban political ecology, and public space planning. She is an active member of the Parkdale Neighbourhood Land Trust in Toronto, where she helps to facilitate the Parkdale Free School, an initiative that organizes free and informal classes and workshops across the neighbourhood.

Sara Macdonald is a PhD candidate in human geography and spatial planning at Utrecht University in the Netherlands. Her research explores the factors that influence the regional governance and planning of greenbelts in the Greater Golden Horseshoe region of Southern Ontario, Canada, and the Frankfurt RheinMain region, Germany, and how they shape institutional interaction and policy implementation. This PhD research is part of the "Global Suburbanisms: Governance, Land and Infrastructure in the 21st century" project. She was the coordinator of the City Institute at York University (CITY) and the project coordinator for the "Global Suburbanisms" project from 2006 to 2017.

Simon Marvin is a professor and director of the Urban Institute at the University of Sheffield. His research interests focus on sociotechnical change in different urban and regional contexts. His recent research focuses on the role of computational logic, automation and robotics in urban restructuring, and the strategic development of controlled environments in recreating artificial natures.

Markus Moos is an associate professor in the School of Planning at the University of Waterloo. His research focuses on the economies, housing markets, and social structures of cities and suburbs, including youthification and the generational dimensions of urban change.

Janice Morphet is a visiting professor at the Bartlett School of Planning, University College London. Janice was on the Planning Committee of the London 2012 Olympic Games. She was a senior advisor on local government for central government, local authority chief executive, and professorial head of the School of Planning and Landscape. Janice has been a trustee of the RTPI and TCPA. She has published several articles and books on local government, spatial planning, infrastructure delivery planning, and Brexit. Her current research is on national infrastructure delivery, governance, and local authority housing provision.

Frederick Peters, PhD, is a Research Fellow at the City Institute at York University in Toronto. His research projects address cities and networked infrastructure restructuring, urban environmental politics and political economy, and riparian environmental history and politics. He also sails.

Nina M. Pulver, MLA, OALA, CSLA, is a PhD candidate in planning at the University of Waterloo, whose research focuses on municipal policymaking around green infrastructure management and stewardship by private citizens. She works concurrently as a licensed landscape architect, specializing in urban design and cultural heritage landscape assessment. She is a recipient of the SSHRC Joseph Armand Bombardier Canada Graduate Doctoral Scholarship (2014), the Ontario Graduate Scholarship (2013, 2014), and has served on the Canadian Association of Landscape Architects (CSLA) Climate Adaptation Task Force. She has widely presented on subjects relating to green infrastructure and climate adaptation.

Xuefei Ren is an associate professor of sociology and global urban studies at Michigan State University. Her research interests are urban sociology, political economy, governance, historical and comparative sociology, architecture, and the built environment, with specific focus on China, India, and Brazil. She is the author of *Building Globalization: Transnational Architecture Production in Urban China* (University of Chicago Press, 2011), *Urban China* (Polity Press, 2013), and co-editor of

The Globalizing Cities Reader (Routledge, 2018). She is completing a new book comparing urban governance in China and India.

Jonathan Rutherford holds a research post at the Laboratoire Techniques, Territoires et Sociétés (LATTS), Université Paris Est and École des Ponts ParisTech (France). His research interests are in the processes and politics of urban sociotechnical change through a focus on the shifting relations between infrastructure and cities. He co-edited recent special issues of *Urban Studies* and *Energy Policy* on urban energy, and the 2016 Routledge volume *Beyond the Networked City* on new emerging urban infrastructure forms, dynamics, and outcomes.

Sara Saboonian is a PhD candidate in the School of Planning of the University of Waterloo. She holds a Master of Science in Sustainable Design from the University of Minnesota and a Master of Architecture from the Tehran University of Art. Her research considers sustainable urban development with a focus on green infrastructure and storm water management. She is currently investigating how to blend green infrastructure design and planning with smart growth and address trade-offs between green infrastructure and intensification strategies within the context of Greater Toronto Area municipalities.

Sophie Schramm's research focuses on urban infrastructure systems in the global south. In 2014, she published her dissertation, *City in Flow – Hanoi's Wastewater System in Light of Social and Spatial Transformations* (in German). From 2013 to 2016, she studied the translation of infrastructures in African cities at the Chair for Spatial and Infrastructure Planning, TU Darmstadt. From 2016 to 2017, she was a junior research group leader at Kassel University for a project on dynamics of "dis/ordering" African cities. Currently, she is an assistant professor at the Chair of Spatial Planning and Human Geography at Utrecht University.

Igor Vojnovic is a professor and acting director of the Global Urban Studies Program at Michigan State University. His main area of research focuses on urban (re)development processes and the study of resulting socioeconomic, physical, environmental, and health impacts. He holds appointments in the Department of Geography, Environment and Spatial Sciences, the School of Planning, Design, and Construction, and the Global Urban Studies Program. He is also the editor-in-chief of the *Journal of Urban Affairs*.

David Wachsmuth is the Canada Research Chair in Urban Governance at McGill University, where he is also an assistant professor in the School of Urban Planning and an associate member in the Department of Geography. He directs UPGo, the Urban Politics and Governance research group at McGill, where he investigates urban governance problems that extend beyond the boundaries of the traditional city and thus challenge how scholars, policymakers, and the public alike understand cities as social systems. He has published widely in top journals in urban studies and geography. He is the Early Career Editor of the journal *Territory, Politics, Governance* and serves on the editorial boards of *Urban Geography* and *Urban Planning*.

Alan Walks is an associate professor of urban planning and geography at the University of Toronto. His research is concerned with understanding the causes and consequences of urban inequalities, including the roles played by automobility and financialization. He is editor of *The Urban Political Economy and Ecology of Automobility: Driving Cities, Driving Inequality, Driving Politics* (Rutledge, 2015), and co-editor of *The Political Ecology of the Metropolis* (ECPR, 2013).

Jonathan Woodside is a PhD candidate with the School of Planning at the University of Waterloo. His research examines models of municipal governance and urban design that respond to changes in the structure of the economy, including the development of digital platforms and the proliferation of precarious employment.

Lucía Wright-Contreras is a doctoral candidate at the Technical University of Darmstadt in Germany. She holds the Erasmus Mundus double-degree Master, International Cooperation in Urban Development – Mundus Urbano. Her experience includes collaborative research at the Nanyang Environment & Water Research Institute (NEWRI) of the Nanyang Technological University in Singapore and a current position as research consultant for UN-Habitat's initiative Global Water Operators' Partnerships Alliance (GWOPA) in Barcelona. Her aim is to contribute to the discussions on global development issues, specifically concerning water service provision in areas where vulnerable populations are challenged with access to safe drinking water.

Index

GLOBAL SUBURBANISMS

Series Editor: Roger Keil, York University

Published to date:

Suburban Governance: A Global View / Edited by Pierre Hamel and Roger Keil (2015)

What's in a Name? Talking about Urban Peripheries / Edited by Richard Harris and Charlotte Vorms (2017)

Old Europe, New Suburbanization? Governance, Land, and Infrastructure in European Suburbanization / Edited by Nicholas A. Phelps (2017)

The Suburban Land Question: A Global Survey / Edited by Richard Harris and Ute Lehrer (2018)

Critical Perspectives on Suburban Infrastructures: Contemporary International Cases / Edited by Pierre Filion and Nina M. Pulver (2019)

Massive Suburbanization: (Re)Building the Global Periphery / Edited by K. Murat Güney, Roger Keil, and Murat Üçoğlu (2019)